Green Energy and Technology

João Cruz

Ocean Wave Energy

Current Status and Future Prepectives

With 202 Figures and 13 Tables

 Springer

João Cruz
Garrard Hassan and Partners Limited
St Vincent's Works
Silverthorne Lane
Bristol
BS2 0QD
United Kingdom
joao.cruz@garradhassan.com

ISBN 978-3-540-74894-6   e-ISBN 978-3-540-74895-3

Springer Series in Green Energy and Technology ISSN 1865-3529

Library of Congress Control Number: 2007936359

Production: LE-TEX Jelonek, Schmidt & Vöckler GbR, Leipzig
Cover design: WMXDesign GmbH, Heidelberg

Printed on acid-free paper

9 8 7 6 5 4 3 2   SPIN 12551774

springer.com

Preface

Wave energy is reaching a critical stage, following over three decades of intensive research and development. The first few full-scale prototypes have been tested at sea and the first pre-commercial orders were placed. The first offshore wave farm is to be installed in the near future and it is likely that similar schemes will shortly follow. Such projects will in the medium term provide a comparable output to the conventional wind farms, allowing an alternative approach when trying to overcome the technological challenge of finding alternative renewable energy sources. It will also fulfil one of the oldest desires of civilization: to harness the power of ocean waves.

This book compiles a number of contributions prepared with the aim of providing the reader with an updated and global view on ocean wave energy conversion. Given the topics covered and the link between of all them, it can be considered one of the first textbooks (or handbooks) related to this field. The authors are recognised individuals within the wave energy community with different backgrounds, and their contributions try to give an overall perspective of the state of the art of different technologies. The book does not intend to point to a specific technology; the market will be responsible for that. The main motivation is to provide, to a wide engineering audience, a first contact with the current status of wave energy conversion technologies, hopefully inspiring the next generation of engineers and scientists. The assembly of the experiences from a panel who has knowledge about the design and operation of such machines enhances the practical nature of the textbook, filling a gap in the available literature.

It would be unfair not to acknowledge Springer-Verlag, namely Dr. Christoph Baumann, for the continuous support and enthusiasm throughout this project. It would also be extremely unfair not to point out that all merit and credit should be given to the authors, without whom the book would not exist.

João Cruz

Editor

Contents

1 Introduction

João Cruz

Garrad Hassan and Partners Ltd
Bristol
England, UK

When thinking about renewable energies, wind, solar and hydro energy typically come to mind. To tackle climate change and all the challenges imposed by the need to find alternative and reliable energy sources, there is one major resource that has remained untapped until now: wave energy. Its potential has been recognised for long, and mostly associated with a destructive nature. No solution has yet been found to harness it. Or has it? This book will show the reader not only the principles of wave energy conversion but also (several) technological solutions to solve the problem. Some are now in a pre-commercial stage.

Wave energy is a concentrated form of solar energy: the sun produces temperature differences across the globe, causing winds that blow over the ocean surface. These cause ripples, which grow into swells. Such waves can then travel thousands of miles with virtually no loss of energy. The power density is much higher than for wind or solar power. These deep-water waves should not be confused with the waves that are seen breaking on the beach. When a wave reaches shallow water (roughly when the water depth is less than half a wavelength), it slows down, its wavelength decreases and it grows in height, which leads to breaking. The major losses of energy are through breaking and through friction with the seabed, so only a fraction of the resource reaches the shore.

A wave carries both kinetic and gravitational potential energy. The total energy of a wave depends roughly on two factors: its height (H) and its period (T). The power carried by the wave is proportional to H^2 and to T, and is usually given in Watt per metre of incident wave front. For example, the coastline of Western Europe is 'blessed' with an average wave climate of about $50\,kW$ of power for each metre width of wave front. The overall resource (around $2\,TW$) is of the same order of magnitude as the world's electricity consumption. A conservative estimate is that it is possible extract 10–25 % of this, suggesting that wave power could make a significant contribution to the energy mix. On a typical day, about $1\,TWh$ of wave energy enters the coastal waters of the British Isles. This corresponds to approximately the average daily electricity consumption in the UK, and is about the same amount of energy as that of the Indian Ocean tsunami of the 26th December 2004 (see Chapter 4). These numbers put into perspective the sort of demand that human beings apply on natural resources, and the urgent need to find sustainable solutions.

Although the first patent dates back to 1799, wave energy research was intensified during the 1970s, particularly in the United Kingdom. It would be unfair to neglect the pioneer work conducted in the 1960s in Japan, when the Japanese navy built a marker buoy which used waves to power its lamp. The turning point that spurred research in several countries was the publication in 1974 of an article in the widely-read scientific journal *Nature* by Prof. Stephen Salter, from the University of Edinburgh. This came as a direct reply to the oil crisis of the 1970s, and its conclusions meant that the attraction to wave energy was immediate. The concept, a cam-shaped floating body known as the Salter duck, is still renowned as one of the most efficient at absorbing waves.

Why didn't wave power take off after this? Around the early 1980s, the UK government made the bold decision to focus all funding on large generating systems rated at 2 *GW*, the capacity of a large coal-fired or nuclear station. Many scientists and researchers believed that this was not the way forward, and that it would be better to think in terms of arrays of smaller units, each rated at a few *MW*. Experience suggests they were right. Some supporters of alternative energy claim that the government's policy was designed to stop wave energy research and to justify the route to nuclear power. The lack of funding virtually halted the significant progress that was taking place (see Chapter 2).

Starting in the mid-1990s, there have been significant achievements in the development of offshore wave power systems, with several full-scale prototypes being tested and connected to national grids (see Chapter 6). As with wind energy fifteen years ago, there are still several competing approaches, and it is unclear which one (if any) will make the final leap towards commercial applications. Pre-commercial schemes are being supported by national governments, namely in the UK and Portugal, with several test centres built or planned (see Chapter 7). This political will is pivotal to ensure the development of the industry.

1.1 Wave Energy Literature

When initiating studies in this field and looking for references, there is a clear lack of suitable textbooks which simultaneously approach wave theory, modelling techniques and results from technology developers. Good starting points are several conference proceedings that have been published over the past few decades. The ones from the biennial European Wave and Tidal Energy Conference (EWTEC) are the most relevant example, dating back to an early symposium held in Gothenburg in 1979. The proceedings are a proof of the continuous interest that wave energy has produced in both the scientific and the industrial communities over the years, a legacy for future generations and valuable lessons for those who are willing to develop new concepts. Other relevant publications include the proceedings of the 1985 IUTAM Symposium in Lisbon edited by DV Evans and AF Falcão, which focus many different subjects of the hydrodynamics of wave energy converters that are remarkably up-to-date (from survivability to optimisation). One final reference to one of the annex reports to the 1993 Generic Technical Evaluation Study of Wave Energy Converters, sponsored by the European Union (at the time Commission of the European Communities), entitled 'Device fundamentals Hydrodynamics' (coordinated by University College Cork). The contributions relative to basic hydrodynamic aspects, optimum control and laboratory testing are particularly significant to those who are beginners in this field. The latter contribution is updated in this book (section 5.3). Finally there are also many scientific journal publications, spread throughout different journals and focusing many different subjects. Although extremely valuable, if the objective is to gather the technical basis of a variety of options regarding wave energy conversion, with examples of different technologies, while simultaneously obtain the theoretical background, these are clearly not the best option to start. Textbooks that overview all these points are in fact fairly rare.

Books which focus linear wave theory are in greater number, and provide the basic theory which underlines wave energy research. Some of these titles have specific chapters dedicated to the subject. Standard examples are Lé Mehauté (1976), Newman (1977), Mei (1989; revised and extended edition in 2005), and Falnes (2002), among many others. In the first two, the underlying theoretical principles are thoroughly reviewed. Le Méhauté (1976) presents a complete survey of wave theories and general hydrodynamic aspects. Waves and wave effects are also discussed in Newman (1977), with particular emphasis to the definitions of damping and added mass, exciting force and moment, and also the response (or motion) of floating bodies. Mei (1989) dedicates a sub-chapter (7.9 in the 1989 edition; 8.9 in the 2005 edition) to the absorption of wave energy by floating bodies. The basic principles of the energy conversion chain are described, and examples of concepts that can be classified as terminators (beam-sea absorbers), attenuators (head-sea absorbers) and omnidirectional absorbers are given (see Chapter 3). Also in Mei (1989) the case of a special two-dimensional terminator (Salter's duck) is presented in detail, particularly with regard to the equations of motion and to the capture width, the length of wave crest that contains the absorbed power. Such contribution follows directly from the work of Mynett et al. (1979), who presented this first comprehensive hydrodynamic numerical modelling exercise related to a wave energy converter.

A thorough review on the theoretical principles of wave energy conversion is given in Evans (1976; 1981). However the most complete compilation of mathematical work related to the absorption of waves by oscillating bodies can be found in Falnes (2002), where the basics of wave-body interactions are presented alongside the principles of optimum control for maximisation of converted energy. A special chapter is dedicated to oscillating water columns, emphasising the amount of work carried out for this specific technology at an early stage.

There have been attempts to provide textbooks which are more suitable to a wider engineering audience. In Ross (1995) an interesting yet generic account of events in wave energy research up to the early 1980s is given, from the journalist's point of view. It is not a technical textbook and it does not intend to be; still it is relevant to those interested in the field, as it describes the issues which lead to a halt in wave energy research, mostly in the United Kingdom. Earlier, McCormick (1981) had presented a detailed description of some initial concepts. The content and clear layout of this textbook are still valuable, but inevitably as time passes it becomes more and more outdated, as other technologies emerge; Shaw (1982) provided a similar overview. More recently Brooke (2003) compiled and edited the work from the members of the Engineering Committee on Oceanic Resources (ECOR – working group on wave energy conversion), with emphasis to resource assessment and providing a short introduction to selected technologies and power take-off mechanisms. Economics and environmental impacts are also focused, and the book is concluded with contributions focusing different geographical areas, which aim to characterise the activities which have been conducted worldwide.

Thus there is a need to address in detail the main energy conversion possibilities and exemplify them with the concepts that have endured through all stages of development, reaching full-scale. Additionally, an account of the operational experience gathered by the technology developers is of great value to engineers and scientists who wish to work in the area or increase their knowledge of the subject. To a certain extent, such experiences should also be shared between all involved in the (young) wave energy industry, so that mistakes can be avoided and lessons learned. If an updated resource assessment contribution is added along with an account of passed events and a review of wave theory, all the basic components of a textbook become defined. This project was envisaged following these guidelines, and its main objective is clear: to provide a solid first reference, which can be used either as a starting point for novices to the field or by any interested reader with an engineering background.

1.2 Chapter Layout

Following the Introduction (Chapter 1), in which the main objectives of the book are explained, an historical review is conducted in Chapter 2. Written by one of the most prominent minds in the wave energy world, Chapter 2 provides insight to the innovative design methodology that lead to the development of several wave energy converters, namely the Edinburgh (or Salter) duck. This account of more than 30 years of activity also focuses on the politics which lead to the halt of funding in the 1980s and a 'Looking Forward' section, in which the challenges which the new wave industry faces are addressed.

Chapter 3 is devoted to the major theoretical aspects that have underlined the research and development of concepts over the past decades. The hydrodynamic principles which rule the optimisation procedure when studying a new concept are described, along with tentative classifications of types of devices, terminology, control methodologies, etc. The importance of designing a device to match the wave climate is also emphasised.

Chapter 4 describes the wave energy resource, its origin and the factors which most influence it. Detail is given to the instruments required to measure and estimate the wave climate, namely buoys, satellite altimeters and wave models, and to the mathematical methods behind the estimation procedures. Some case studies, which refer to specific evaluations and quantify the worldwide resource, are presented. The methodologies available in this chapter are relevant when planning wave farm projects. The quantification of the expected annual energy output for a given location is subject to a number of factors (e.g.: local bathymetry, seasonal variability, etc.), that cannot be neglected, as the accurate estimation of the wave climate is one of the most critical aspects to ensure the success of a project.

Chapter 5 describes the modelling options when studying a wave energy converter (WEC), from numerical to experimental approaches. Comparisons between such models allow valuable conclusions and often drive the development of specific configurations. The design, construction and operation of wavemakers and wave tanks are also addressed, along with guidelines for the testing of scale models. The chapter is concluded with a case study related to one of the concepts that is later described in Chapter 7, reviewing all the stages of development and emphasising the need of both numerical and experimental modelling.

Chapter 6 explains the several technical possibilities to harness the power from ocean waves. The power take-off (or power conversion) alternatives presented roughly cover all the technical proposals currently being tested. Alternative applications, like seawater desalination, are also addressed. The review is directly linked with the technologies that are presented in Chapter 7.

Chapter 7 describes some of the concepts that have reached the full-scale stage. Other examples could be given, but the analysis was limited to the four main competing technologies: OWC (Oscillating Water Column), Archimedes Wave Swing (AWS), Pelamis and Wave Dragon. Each illustrates one particular power conversion mechanism that was addressed in Chapter 6, namely air turbines, direct drive linear generators and hydraulics. For the Wave Dragon case, a sub-section regarding low-head turbines and overtopping theory is presented. Section 7.5 gives an account of the operational experience gathered by the several technology developers. To conclude, section 7.6 provides a brief update on test centres, pilot zones, and also on the most relevant EU funded projects. A case study based on one of the technologies concludes the chapter.

Finally, Chapter 8 drafts some tentative guidelines with regard to Environmental Impact Assessments, based on the experience gathered with onshore and offshore wind farms. The need to establish a common legislation throughout the EU member states, which would in turn become the standard when deploying wave energy farms in other locations, is addressed. Socio-economic aspects are also focused. Predictions regarding the future costs of wave energy, a relevant aspect for the development of the wave energy industry, are not addressed as they are already targeted in other references (e.g.: Callaghan and Bould, 2006). Although with some limitations given the early stage of the industry, the overall conclusion of the majority of the predictions is that it is fair to expect a cost reduction trend with regard to installed capacity similar to what wind energy has been experiencing over the past 20–30 years.

References

Brooke J (2003) Wave Energy Conversion. Elsevier Science

Callaghan J, Bould R (2006) Future Marine Energy. The Carbon Trust

Evans DV (1976) A theory for wave-power absorption by oscillating bodies. J Fluid Mech 77(1):1–25

Evans DV (1981) Power from Water Wave. Annu Rev Fluid Mech 13:157–187

Falnes J (2002) Ocean Waves and Oscillating Systems. Cambridge University Press

Le Méhauté B (1976) An Introduction to Hydrodynamics & Water Waves. Springer Verlag

McCormick M (1981) Ocean Wave Energy Conversion. John Wiley & Sons

Mei CC (1989) The Applied Dynamics of Ocean Surface Waves. Adv Ser Ocean Eng 1. World Scientific (revised edition in 2005)

Mynett AE, Serman DD, Mei CC (1979) Characteristics of Salter's cam for extracting energy from ocean waves. Appl Ocean Res 1. pp. 13–20

Newman JN (1977) Marine Hydrodynamics. MIT Press

Ross D (1995) Power from the Waves. Oxford University Press

Shaw R (1982) Wave Energy – A Design Challenge, Ellis Harwood Ltd, John Wiley & Sons

Evans DV, Falcão AF (1986) Proceedings IUTAM Symposium on Hydrodynamics of Ocean Wave-Energy Utilization. Lisbon, Portugal, 8–11 July 1985. Springer Verlag

2 Looking Back

Stephen Salter

School of Engineering and Electronics
University of Edinburgh
Edinburgh
Scotland, UK

'... *if you can hear the truth you've spoken twisted by knaves to make a trap for fools ...*'

In the autumn of 1973 the western economies were given the rare chance of a ride in a time machine and saw what the world would be like when there was no longer cheap oil. Most people thought it looked rather uncomfortable but a few very powerful people made a great deal of money by exaggerating the crisis. Others, who had previously been regarded as eccentric, increased their efforts to develop what were then called alternative, and are now called renewable, energy sources. Still others set out to destroy what they saw to be a threat.

Waves were only one of many possible sources and there are many possible ways in which waves can be harnessed. There are floats, flaps, ramps, funnels, cylinders, air-bags and liquid pistons. Devices can be at the surface, the sea bed or anywhere between. They can face backwards, forwards, sideways or obliquely and move in heave, surge, sway, pitch and roll. They can use oil, air, water, steam, gearing or electro-magnetics for generation. They make a range of different demands on attachments to the sea bed and connections of power cables. They have a range of methods to survive extreme conditions but perhaps not quite enough.

Their inventors, myself included, invariably claim at first that they are simple and, after experience with the dreadful friction of reality, invariably discover that this is not totally true when they come to test in the correct wave spectra with a Gaussian distribution of wave amplitudes. An easy way to detect beginners is to see if they draw waves the same size on both sides of their device.

Appeals to simplicity are widespread and have a strong appeal to non-engineers and particularly to political decision-makers and investors. But it is hard to find any field of technology which does not get steadily more complicated as it gets faster, lighter, cheaper, more powerful and more efficient. The complications are all introduced for good reasons and, if the necessary hardware is properly re-searched, will produce good results. Who would abandon railways for wheel bar-rows because of the smaller number of wheels? Only a simpleton.

Although almost everyone knows which of the devices proposed so far will ul-timately prove the best it is not certain that no improvement could be invented. This chapter describes some of the work done on several devices at Edinburgh University in the hope that future generations of wave inventors can save time and avoid mistakes.

2.1 Wave Energy at the University of Edinburgh

Many inventors of wave power devices, going back to Girard père et fils in 1799, start with heaving floats. Apart from a brief flirtation with oscillating water col-umns (see Chapters 3 and 7), so did I. But I had the advantage of a workshop in which I could make any mechanical or electronic instrument that I was able to de-sign and there was a narrow tank that I could borrow. As so often in physics and engineering, a full understanding of all the energy flows leads to a full understand-ing of the problem and points to suitable solutions.

It was necessary to make something against which a float could do work that could be accurately measured and compared with the energy transfers from in-coming, transmitted and reflected waves. While the Girards proposed the use of a ship of the line, I thought it would initially be cheaper to begin with a length of 100 *mm* by 25 *mm* varnished balsa wood, just fitting inside the 300 *mm* width of a small wave tank. Rotating bearings are much nicer than translating ones. But if they are at the end of a long arm they give a good approximation to a translating constraint. If you grind a 70-degree cone on the end of a length of tool steel and use it to punch the end of a light alloy or brass rod you get a beautiful socket into which you can place a 60-degree conical-point screw with friction acting at a very short radius. Grease will slow, if not stop corrosion long enough for plenty of tests. The first heaving buoy model is sketched in Fig. 2.1a.

For the power measurement I used two very strong bar magnets in a magnetic circuit which excited two coils wound like an oversize galvanometer movement and linked together in a parallelogram using the same spike bearings pulled into cones in the end of a strut by elastic bands. The parallelogram could be coupled to the float with another strut and elastic band. These acted like a universal joint with very low friction and no backlash.

Moving the float generated a nice velocity signal in one of the coils. This could be amplified and fed back to the second coil with polarity chosen so as to oppose the movement. Changing the gain of the amplifier would change the damping co-efficient. A high gain made it feel as if it was in very thick honey. If the amplifier feedback connections are such that it delivers an output current proportional to the input voltage, then temperature changes in the galvanometer coils do not change the calibration.

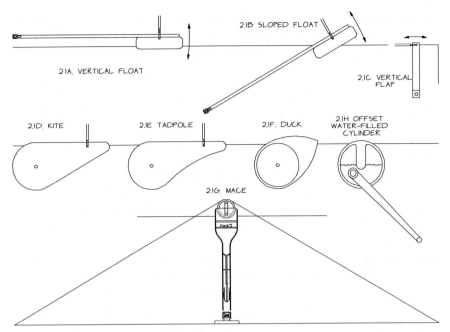

Fig. 2.1. In the beginning was a vertical heaving float...

From calculus we know that the position of an object is the time integral of its velocity history plus some constant. If the signal from the velocity coil is put into an operational amplifier circuit connected as an integrator we get an accurate position. If the parallelogram is moved backwards and forward between the jaws of a Vernier gauge, the integrator output signal will be a square wave. The field-effect transistor operational amplifiers of 1973 had low enough offset currents to allow this position signal to be read on a digital voltmeter. The force was calibrated by making the pushrod drive the pan of a weighing machine.

Measuring the waves could be done with a light float made from expanded polystyrene foam mounted on a swinging arm. A pair of micro-ammeters, coaxial with the linkage bearings, with their needles glued to the float arms, gave a very clean velocity signal from even the smallest waves. Integrating float velocity gave an even cleaner wave-amplitude signal. The float averaged wave measurements across the width of the tank and so was insensitive to cross waves. It could measure waves down to 0.01 *mm* which we could not even see, far less than the meniscus hysteresis of resistive-wire gauges which we later had to use for very steep waves.

To calculate the power you just multiply the instantaneous force signal by the instantaneous velocity signal, which will give you an offset sine wave at twice the wave frequency. You then take a long-term average with a low-pass filter.

This equipment allowed the measurement of model efficiency. The first result for the vertical heaving balsa wood float in Fig. 2.1a was disappointing – only 15 %whatever adjustments were made to the damping coefficient. Some of the energy was reflected but most went straight past the model. However the depth of the hinge was very easy to adjust. If it was pushed down so that the movement was along a slope as in Fig. 2.1b, the performance shot up to 50 %, much higher than most people would have predicted.

A vertical flap hinged below the water as in Fig. 2.1c could also be coupled to the dynamometer. This showed an efficiency of about 40 % with 25 % being transmitted on to the beach and 25 % sent back to the wavemaker. It looked as though the horizontal motion of a wave, which almost all new wave inventors ignore, was better than the vertical one. Despite rich vocabularies of nautical terms we have no word in any language for this movement of a wave.

The borrowed narrow tank had a commercial hinged-flap wavemaker with amplitude set by a crank radius and frequency set by a mechanical variable-speed mechanism. One problem was that there was no way to make mixed seas. But a more serious one was that the drive to the flap was rigidly fixed by the crank eccentricity so that the flap reflected waves just like a rigid vertical cliff. Test tank beaches are not perfect and the first designs of any wave device are likely to reflect a substantial fraction of the incoming waves. It was even worse because the amplitude of a wave created by a hinged flap for a given angular movement depends on the square of the depth of the hinge and this would be increased during the crest of any reflection and reduced during the trough, together with some Doppler shifting. Even if we could not make irregular waves with the spectrum of our choice the tank reflections would make one with a spectrum of their own. Trying to make a regular wave could lead to amplitude variations of three to one.

The vertical flap showed that it was wrong to allow the model to transmit waves behind. Was it possible to make a model with a front but no back? Figure 2.1d shows an attempt, code-named Kite. This showed an efficiency of 70 % and very low onward wave transmission. Figure 2.1e, code-named Tadpole, was meant to allow the circular motion of water particles to continue but had the same result. But waves are very good at sending energy to the next volume of water with almost no loss: the idea of allowing the water motion to continue in the way it would do in the absence of a model was powerful. Could the circular backs of 2.1d and 2.1e be combined with a shape which let the decaying orbital motion of water take place just as it would in the open sea?

I asked a computer-minded PhD student, Peter Buneman, to help while I struggled with a slide rule and drawing board. We converged on the same shape shown in 2.1f, code-named Duck. Its efficiency was measured at 90 %, which even we did not believe despite many calibrations cross-checked by Jim Leishman from the National Engineering Laboratory, Gordon Goodwin from the Department of Energy and Brian Count from the CEGB, then the big English electricity monopoly. Later, photographs by Jamie Taylor in Fig. 2.2 allowed visual proof that the calibrations were correct. It is a one-second exposure of a duck model on a fixed mounting in a narrow tank. The two wires are connections to part of an electromagnetic dynamometer, which is absorbing power. Waves are approaching from the right. Drops of a neutrally buoyant tracer-fluid consisting of a mixture of carbon tetrachloride and xylene with titanium oxide pigment have been injected to show the decaying orbits of wave motion.

Fig. 2.2. Jamie Taylor's photograph taken in 1976 which convinced people who really knew about waves that high efficiency could be achieved

The amplitude of the incoming waves can be measured from the thickness of the bright band. Nodes and anti-nodes due to the small amount of reflection are evident. However the thickness of the bright band to the left of the model is largely due to the meniscus, as is confirmed by the very small orbits of tracer fluid in this region.

As the energy in a wave is proportional to the square of wave amplitude we can use the photograph to do energy accounting. If nodes and anti-nodes show that the reflected wave is one-fifth of the amplitude of the input it would have one twenty-fifth, or 4%, of its energy. This means that 96% has gone into the movement of the test model. The dynamometer showed that just over 90% of the power in the full width of the tank had been absorbed by the power take-off, leaving 6% loss through viscous skin friction and vortex shedding. We joked that the rate of improvement might slow because of some impenetrable barrier around 100%.

One should be careful about such jokes. Johannes Falnes and Kjell Budal in Trondheim had found that point absorbers in wide tanks or the open sea could absorb more energy than was contained in their own geometrical width, just as the signal from a radio aerial does not depend on the wire diameter (Budal and Falnes, 1975). The terms 'capture width' and 'capture width ratio' replaced efficiency for devices in wide tanks. The Falnes Budal findings were simultaneously and independently confirmed by David Evans at Bristol and by Nick Newman and Chiang Mei at MIT.

Because absorbing energy from waves was the whole objective and making waves was very similar to absorbing them, it seemed an obvious step to build a wavemaker with the same control of force and velocity as an absorbing model.

The motors available then had too much brush friction to allow the use of current as a control so a force-sensing strain gauge was built into a drive arm. A tacho-generator measured the velocity. The displacer was the same shape as a duck but with a hollow cylindrical interior to avoid the large vertical buoyancy force. The shape was rather expensive to make in the large numbers planned for a wide tank and later versions used flaps with a textile rolling-seal gusset to maintain a 'front with no back'. Either design allowed the generation of very accurate waves even with 100 % reflecting models and gave repeatability to one or two parts per thousand.

Force-sensing does not suffer the phase lag, 90° at about 8 Hz, of the meniscus of a wire wave-gauge so, with a stiff drive path, high loop gains can be achieved. By using force and velocity to control energy and giving that energy to the water at the right frequency, we allow the water to choose the shape it likes to transmit that energy even if what are called 'evanescent modes' have the wrong waveform close to the wavemaker. The chief design problem is getting rid of any friction that could corrupt the force measurement. Many more absorbing wavemakers have been sold by a spin-off company, Edinburgh Designs, run by Matthew Rea.

The next task was to widen the band of high efficiency and move it to longer wave periods, equivalent to having a smaller device. This was done by Jamie Taylor who used systematic variations of the hub depth, ballast position and power take-off torque for various duck shapes. We built a sliding mounting with a clamp and adjustable stop, which allowed one person to remove and reinstall a model to the exact position using only one hand in three seconds. This is likely to be harder, slower and more expensive at larger scales. The models had tubes running through them into which stainless steel rods of various lengths could be inserted to adjust ballast. They had Aeroflex moving-magnet torque-motors at each end. One gave a velocity signal which could be processed by analogue operational amplifier networks built by David Jeffrey. These could implement variable damping, torque-limiting, positive or negative spring and inertia, indeed any power take-off algorithm we could specify.

Analogue multipliers needed for power calculations can perform a useful job with large input signals. The usual transfer function is 0.1 (A x B). With 10 volts on both inputs giving 10 V output, an error of 0.1 V is only 1 % and would be tolerable. But if A and B are only 1 V the product is 0.1 V and the error is 100 %. The solution is to arrange a system of pre-amplifiers and post-attenuators on a double-bank rotary switch before and after the multiplier and manually adjust gain and attenuation so that the two input signals do not quite clip.

To measure waves we used a pair of heaving-floats on mountings which could be clamped to each other at distances of one quarter or three quarters of a wavelength. The pair could slide along ground stainless steel rails aligned parallel to the calm water surface. This rail alignment had been done with a capacitance proximity sensor and fine adjustment screws with everything finally locked by a metal-filled epoxy putty. The sensor was just sensitive enough for us to pretend that the rails followed the curvature of the earth rather than being quite straight. By sliding the pair of gauges to the position which maximised the difference of their outputs we could put one gauge on a node and the other on an anti-node. Half the sum gave the amplitude of the incoming wave and half the difference gave the amplitude of any reflection.

If we set very high damping the model would be locked almost stationary and would reflect nearly all the incoming energy like a cliff with an anti-node at its front surface. If we set the damping to zero it would move violently but still reflect with a node at the front. It was easy to find the best match because David's electronics, described in Jeffrey et al. (1976), could calculate the instantaneous efficiency and Jamie would know immediately if his choice of damping, hub-depth or ballast position was good or bad. He would have accurate measurements after two tank transmission times.

Playing with different damping settings showed that wave devices were like loads on transmission lines which should be matched to the line impedance. A mismatch by a factor of two either way was tolerable but more than this would progressively lose much more output from reflections.

By integrating the velocity signal with a very low drift operational amplifier we could get a good position signal and we could combine either polarity of this with the damping feedback signal to get positive or negative spring. Although this needed a small investment of energy back to the model it was repaid with large interest, widening the efficiency band and moving it to longer waves. Rapid changes with rapid results make for rapid progress. Jamie Taylor pushed the performance band from a peak at a wavelength of four duck diameters to fifteen diameters with creditable performance at twenty-five.

David Jeffrey built two more electronic systems which turned out to be immensely useful and should be copied by others, perhaps using computer graphics. We had nearly sixty signal sources from wave gauges and model which could be sent to thirty signal destinations, such as meters, signal processors and oscilloscope displays. Getting any connections confused could negate an entire experiment and waste days of work. David built a pin-board matrix with signal sources along the top and destinations along the left vertical. Any source could be connected to any destination by the insertion of a pin at the corresponding intersection of row and column. A new experiment could be planned, set up and checked in about a minute with first results a minute later.

The second system was a display of two oscilloscopes. One had a long-persistence phosphor while the second had a storage tube which used electrostatic technology to retain a trace for about an hour. The conventional oscilloscope time-base was replaced by one which was locked to the wavemaker drive frequency. The sweep time was exactly the full wave period but also the start of the trace was always at an upward zero crossing of a wave, the crest always at 25% of the screen width and the trough always at 75%. We could also plot any variable against any other.

When the long-persistence tube showed that the tank conditions were steady, the press of a button would write the next trace to the storage tube. The conditions could be changed for the next test and the next trace written. Provided we could finish a series within the tube storage time we could build up families of curves and take a Polaroid photograph such as the ones in Fig. 2.3.

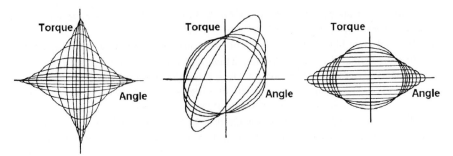

Fig. 2.3. Families of Lissajous plots of duck torque against angle for variable damping, variable amounts of negative spring giving reactive loading and a selection of torque limits. These are from actual oscilloscope photographs of tank models

This shows torque to angle diagrams for variations in damping, torque limit and reactive loading with negative spring. The area inside the loop measures useful work. These are analogous to pressure-volume indicator diagrams for steam engines.

Fig. 2.4. The all-analogue tank-control bench with direct-reading efficiency calculation, pin board, transfer-function analyser and wave-locked pair of oscilloscopes

Another very useful commercial instrument was a transfer-function analyser which combined a very accurate, crystal-locked low-frequency signal generator with two digital voltmeters giving the in-phase and quadrature magnitudes of signals at that frequency or at harmonics of it.

The control desk allowed two people to sit in comfort within reach of every control knob and with eyes at wave level. It is shown in Fig. 2.4. Some people think that this photograph was contrived but this was the actual working setup. Despite enormous advances in digital computing power since 1976 and wonderful data collection and analysis software, I have never since worked with such a fast and convenient tank control system as one using entirely analogue electronics. Glen Keller even built analogue circuitry which allowed to control design features of a gyro power take-off with correct torques fed back to the model in the water.

Until then all data analysis had been performed with the Hewlett Packard HP 65 hand-calculator which had a magnetic strip reader that could store programmes with as many as 64 steps. In order to work with multiple spectra we went to the dreadful expense of £7000 of getting a Tektronix 4051 computer which had an enormous memory of 16 k, a graphics display and even a cassette tape reader for programmes and data. This cost the annual salaries of three research associates but allowed measurements of every possible wave and model signal in realistic wave spectra.

If the large forces from waves are to do useful work there must be some reaction path to oppose them. By now we knew enough about wave forces to realise that providing this with a rigid tower for the largest waves in deep water would be very expensive and we wanted a way in which the structures would never be stressed to any level above that which would arise at their economic power limit. We wanted something that would experience large forces and high relative velocities in small waves but not in large ones.

The only solution for deep water seemed to be a spine long enough to span many wave crests to get stability but with joints that could flex before the bending moments could cause any damage. We needed to know how such an elastic and yielding system would behave. We built the nearest approximation to replicate a spine in a narrow tank. It was a mounting called a pitch-heave-surge rig, shown in Fig. 2.5, which allowed the support stiffness, damping and inertia to be set to any desired value but also to yield at forces above a chosen value. It could also be used to drive a model in calm water to measure the relationship between force and velocity so as to give hydrodynamic coefficients of damping and added mass.

The rig proved to be ideal for testing the Bristol cylinder invented by David Evans (Evans et al., 1979). Whereas we had worked for days to discover the best ballast position and power take-off settings of a new model shape, he was able to calculate directly what the values should be. We already had a 100 mm diameter neutrally buoyant cylinder which we had used for force measurements. We set the stiffness and damping to his values and the model achieved almost 100% efficiency immediately. The Bristol cylinder does this by combining movements in both horizontal and vertical directions so that a long wave, which might be expected to propagate below the cylinder, is cancelled by the wave generated by the cylinder movements. David Evans suggested that this would also be true for our duck system and so it was. The long wave performance could be greatly improved by reducing the mounting stiffness. Fortunately the correct stiffness values were lower than those which could be supplied by post-tensioned concrete at full-scale.

Jamie Taylor explored the effects of mounting stiffness and produced a map with two regions of high efficiency separated by a valley of very low efficiency at a particular heave stiffness. We called this Death Valley. The angular movements of the duck and its movement relative to the water surface could be reduced to almost zero in quite large waves. This could be very convenient for gaining access.

Computers are like bacteria. Once you have one it breeds others at exponentially increasing rates. The Tektronix was joined by a Commodore Pet which could generate seas in which the phases of each component could be combined with cunning malevolence to produce extreme wave events such as those as shown in Fig. 2.5. It could also trigger flash photographs at any time with microsecond precision. The force records against time in Fig. 2.6 and as heave against surge forces in Fig. 2.7 are the result of freak waves hitting the model placed at a series of positions relative to the nominal break point. It was a surprise to discover that there was a strong downward and seaward tendency, that the most dramatic production of white water could occur with quite low forces and that the peak force occurred during the second trough following the instant of wave breaking. We clocked up half a million years worth of hundred-year waves. Any developer who does not follow this path does not deserve insurance but will certainly need it badly.

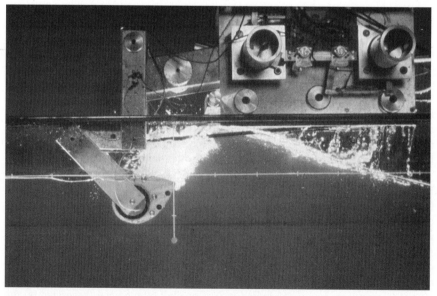

Fig. 2.5. The hundred year wave with maximum possible steepness achieved by selection of the phases of a mixed sea hitting a duck on a locked pitch-heave-surge rig

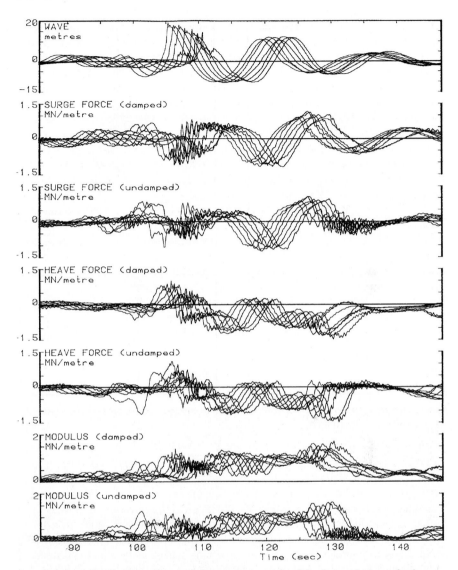

Fig. 2.6. The superposition of a set of time series records of the forces during a freak wave on a duck on a rigid mounting. The records are taken with the model axis at each of the vertical tick points along the water line. Note the downward forces and the larger total force at 130 seconds – long after the nominal break. 'Damped' means the normal operation of the duck power take-off which had rather little effect in such large waves. Half the testing was done with none

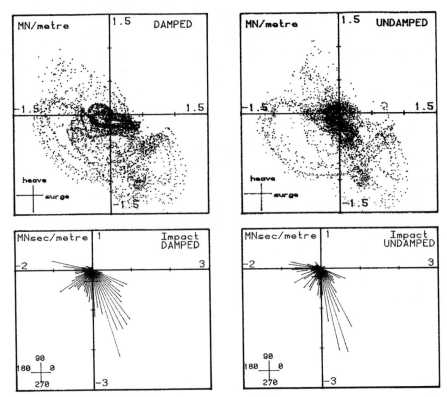

Fig. 2.7. Lissajous plots of the vertical heave forces plotted against surge forces. Note the low upward and forward force

There were always anxieties about whether results from small models at around 1:100 scale could apply to full-scale but the leading experts assured us that we were just clear of the scale where surface tension becomes a significant restoring force and insects can walk on water. We did hire a 1:10 scale tank for a week. The results were within 2% of our narrow tank ones but while you could lift a 1:100 scale model with one hand and make it in a day, dropping a 1:10 scale model could easily kill somebody. The 1:10 scale tank took twenty minutes instead of forty seconds to settle. Everything was far slower and more expensive but, for shapes like those of most wave devices, no more accurate.

The results of the work with the pitch-heave-surge rig were convincing enough to justify building a wide tank to test long-spine models. This had to be designed backwards from £100,000, the maximum amount of money which could be authorised by the programme manager, Clive Grove-Palmer, without going to a superior committee which had a member who was certain to oppose it. We got the go signal on 1 June 1977.

Meeting the cost was made possible only by the purchase of 120 scrapped printed-motors which had been stripped out of ancient IBM disk drives. Some of the armatures had been overheated and had come unglued. Opening the case to re-glue them broke the magnetic path and destroyed the magnetisation of the alnico disks which energised the gap. Each of these magnets was wrapped by one and a half turns of wire leading to terminals outside the case. We calculated that it needed a current pulse of $7000\,A$ for one millisecond to reset the magnets but that a pulse of $10\,ms$ would melt the wire. The resetting of the magnets was done by Glen Keller but the method he used has been removed from this contribution for fear of prosecution by the Health and Safety Executive. He recovered over 100 good motors, most of which are working still, thirty years later.

While the tank building was being put up we built the wavemakers and drive electronics with the help of students and school leavers, many of whom are now successful engineers. Filling with water was complete on 1 January 1978. The electronics for 89 wavemakers and the drive software were debugged in two weeks and a rival wave power team began testing on 1 February 1978. A second tank with 60 identical wavemakers was built near Southampton and soon both were working 24 hour shifts. An even bigger one with flatteringly similar but very much bigger wavemakers was built in Trondheim. It took us 18 months to get money to build duck models to test in our wide tank because politics had reared its ugly head.

2.1.1 Politics

To understand any research programmes you must understand money flow. Money for wave energy came from several sources. Firstly there were the pockets of private inventors with enough confidence in themselves and their devices to spend their own money rather than other people's. Secondly there were firms who are risking some of their shareholders money. Thirdly there were the Foundations such as Nuffield and Wolfson. Fourthly there was the Science Research Council, now renamed the Engineering and Physical Sciences Research Council. Largest of all were the Government Departments, initially the Department of Industry, later the Department of Energy and now the Department of Trade and Industry. Usually devices with promise were taken over by the Department of Energy when it appeared that more substantial amounts of money were needed. While private sources of money enjoy flexibility, the Department of Energy was and is locked into the timing of the Treasury Financial Year which begins in April.

The Department of Energy had a better chance of defending its decisions if they had been supported by advice from outside and so there existed a committee known as ACORD. This stands for the Advisory Council on Research and Development (in the fuel and power industries). It gave advice on fission, fusion, oil, gas, coal, tidal, geothermal, wind, hydro and wave power, and, one also hopes, conservation. As it met quarterly and as its membership was selected from the busiest, most senior experts, one can readily calculate how much time could have been devoted to any one topic.

ACORD did not suggest programmes or experiments. It passed opinions on those submitted to it by the Wave Energy Steering Committee, or WESC. This met at monthly intervals and dealt solely with waves. But its members were all fully employed in other fields and so received advice from six separate advisory groups, specialising on technical aspects, and two groups of consultants: Rendel Palmer and Tritton, on marine and civil aspects, and Kennedy and Donkin on electrical problems. These consultants worked closely with one another and became abbreviated to RPT and K and D, or "the consultants'. They could assign people to work full-time on assessing proposed designs and visited the device teams regularly. Their mandate was to provide professional criticism, to spot flaws in the arguments and mistakes in the calculations of the starry-eyed enthusiasts in the laboratories. We had to fight hard for every milliwatt against people paid to act the part of pessimistic misers who gave us no benefit of any doubts. This meant that the consultants' opinion would always reflect a maximum price and a minimum resource size. The distinction between proven and probable reserves of oil is relevant. I do not know whether our comrades in the other renewable energy fields have ever been subjected to such hard-nosed scrutiny, but it would have been very good for improving design.

The day-to-day administration of the programme was carried out by the Energy Technology Support Unit (ETSU) at Harwell, part of the UK Atomic Energy Authority which controlled research into all the renewable sources. The Programme Manager had a number of Project Officers who actually visited the laboratories. They helped device teams shape research proposals, they monitored progress and they approved claims for expenditure.

Six Technical Advisory Groups (abbreviated TAGS) dealt with the assessment of new devices, the acquisition of wave data, the measurement and calculations of fluid loading, the problems of mooring, the problems of generation and transmission and finally the subject of environmental impact, which seemed to be the very least of our difficulties. There were somewhere between six and nine rungs in the ladder between the men in the laboratory and the men with the money and power.

The financial year

The cycle of events began with the Treasury Financial Year in April. There had to be time for the Department of Energy officials to consider the ACORD advice and for ACORD to approve its own minutes. This meant that the advice must be given at an ACORD meeting in February. The proposals put forward to ACORD had to be discussed by one meeting of WESC and modified for approval by a second. This meant that WESC must have all the information it needed in early December. The most important piece of information required was the report by the Consultants. If they worked flat out they could finalise reports on a number of devices in about a month, but this meant that they must bring down the chopper on the work of the device teams by the beginning of November. Everything they saw was a flash photograph of the position in October. There is no chance of a device team saying "There. It is finished. Nothing can improve it. We have spoken." The drawings and graphs carried long streaks as the paper was wrenched from beneath their pencils at 23.59 on October 31st.

After April the Department of Energy would tell the Programme Manager how much he would have to spend. This would be unlikely to be the same as the amount he wanted and so he would have to talk with device teams, Project Officers, Consultants and TAGS and arrive at a new revised programme. If he worked with the tireless devotion for which programme managers are selected he might have this done by the end of May, ready for discussions by WESC in June and for modifications and re-approval in July. The sums of money involved exceeded the amount which could be authorised without signature by officials of the Department of Energy, who are of course on holiday in August. But when they returned in September it took no time at all to authorise and issue the formal contracts from the Harwell contracts branch. It was just possible to get one out by mid October, leaving two weeks for the ordering of equipment and the recruitment if not the training of staff before the consultants' axe descended. A single hiccup in any part of the procedure could make the official working time go negative and often did. When the contracts arrived they could be amazingly complicated. In one the work programme was split into four time periods and four different work topics giving sixteen different pots of money and no certainty that it could be transferred between them.

The delays in issuing contracts were matched only by the delays in paying for the work done. Harwell had a rule that if there was any irregularity in an invoice sent by a contractor, all subsequent invoices would be blocked until the matter had been cleared up. I can quite see that this would be a good way to encourage contractors to avoid irregularities. However there was no obligation on Harwell to tell the contractor the nature of the irregularity. All we knew was that the cheque was not in the post. In 1979 the Atomic Energy Authority set up an account to pay for feeding members of its committees including the Wave Energy Steering Committee but did not trouble to tell me anything about it or the numbers to use. I went on paying for their lunches from my research grant as before. In 1980, when UK annual inflation rate was 18 % and a senior University researcher was paid £12,000 a year, the backlog in payments reached nearly half a million pounds, all because of a lunch bill for £25 had the wrong account number.

Eventually, in desperation, I told Harwell payments branch that I would have to get help from the University Rector. In England the word 'rector' means a slightly senior vicar or parish priest and no doubt the Harwell payments branch imagined that I would be seeking tea and sympathy. But in Scottish universities the Rector is elected by the students to defend their interests. He also defends those of research staff. While students from some unmentionable universities have tried to elect a pig as their rector, Edinburgh students have much higher standards and former holders of the post include Gladstone, Lloyd George, Baldwin and Churchill and Gordon Brown. In 1982 our Rector was David Steel, then leader of the Liberal party when the Liberal SDP Alliance was on the rise. When this became known by the Atomic Energy Authority, the problems of getting paid vanished over a weekend like the morning dew.

Jumping the gun

The traditional, and sound, engineering approach for many projects has been to measure or calculate all the loads on a structure before finishing the design and then to make a series of design modifications in the light of cost calculations before arriving at the final optimised result. But politicians and investors want to know the bottom line before making any initial investment and are in a position to enforce their wishes for continuous assessment and early Figures for the cost of electricity. This is very much like people wanting to know the winning horse before placing a bet.

Work on the full-scale design was carried out long before we knew enough about bending moments and mooring forces. We had help from the big civil engineering company Laing, who taught us lots about the advantages of post-tensioned concrete in sea water. The first power take-off was based on getting a torque reaction from a pair of gyros spinning in opposite directions. If they were allowed to precess freely they would lock a frame against which a ring-cam pump could do useful work. Two advantages were that the gyros could also be used as flywheels to store energy for tens of minutes and that everything was hermetically sealed in a super clean vacuum. The disadvantage was that the full duck torque had to go as a radial load through high speed gyro bearings. Robert Clerk designed some amazingly efficient hydrostatic ones with active impedances and fine clearances despite large deflections.

The choice of a gyro reference frame called for new types of hydrostatic bearing (Salter, 1982). Digital control had profound effects of the design of high pressure oil pumps and motors (Salter, 1984; 2005). We tried to design for the level of technology which would be available at the time that the energy crisis really hit, rather than for things that would be obsolete by then. Many of the ideas, such as the use of microchips to change mechanical design, seemed wild at the time and were questioned by people responsible for power generation issues. All were outside the field of the civil and heavy electrical engineers who were employed to assess our work. Accordingly the task of assessing ducks was transferred from Rendell Palmer and Tritton to an outside consultant Gordon Senior. He subjected us to a sharper scrutiny than the civil engineers, who had missed a serious mistake we had made with the 1979 reference design. He checked calculations, quotations and data from tank experiments. His questions and comments were a great help in improving the design.

The consultants had to consider many sorts of data. There were the heights, spectral shapes and angular distributions of the raw wave input. There was the hydrodynamic performance of the devices. There was the conversion efficiency of the mechanism used for generating electricity, collecting it and transmitting it ashore. There was the reliability of the overall system. There was the capital cost of building yards and of the devices and transmission cables. There was the rate of interest charged for the loans. There were charges for installing the devices and charges for maintenance. Finally there was the ultimate life. Some of these data are well known. Some can be measured by experiment. Some have to be guessed. Some are unalterable. Some can be changed by better understanding or more intelligent design. Many can be misinterpreted through accident, malevolence or enthusiasm. Some remain unknown. If input data are false, no amount of subsequent processing can improve the conclusion. But it has always been necessary to decide policy with imperfect assumptions. With skill and luck some of the mistakes cancel others. The history of official cost predictions up to 1982 is shown in Fig. 2.8.

The accounts of the Central Electricity Board for 1979–80 showed that generation from oil was 6.63 pence, coal was 3.35 pence and nuclear 2.2 pence per kilowatt hour. We now know that this latter Figure could not have included the correct amount (£90 billion) for waste disposal and decommissioning but, at the time, this cost of nuclear was accepted as gospel truth. Even so the gap between waves and conventional sources was closing and the trend of cost reductions made us confident of further ones.

Our confidence was misplaced. It was on the basis of this information that the ACORD committee recommended the closure of the wave programme at their meeting of 19 March 1982, a meeting from which Clive Grove-Palmer was excluded. He resigned as programme manager. The Consultants 1981 report had been circulated in draft but withdrawn from publication. The report that they released in June 1983 showed that ACORD had been very wise to recommend closure.

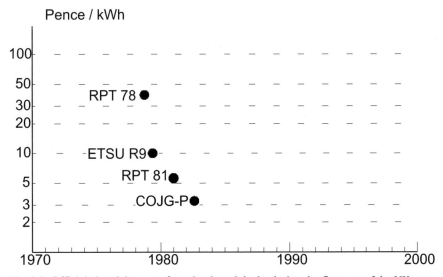

Fig. 2.8. Official electricity costs for spine-based ducks during the first part of the UK wave energy programme from Rendel Palmer and Tritton, the Energy Technology Support Unit and the programme manager, Clive Grove-Palmer, whose Figure was based on final development

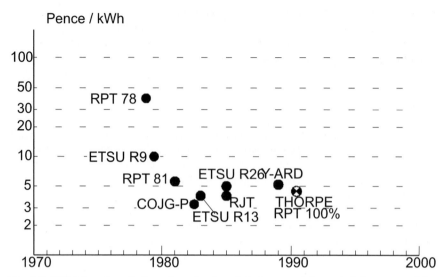

Fig. 2.9. Official cost predictions for spine-based ducks except for an infinite one resulting from the cable failures. The 2007 payments in Portugal are about 16 pence

Many arguments followed such as whether the cost per tonne of steel work for anchors of a wave device really was three to five times higher than that of a Colchester Magnum Lathe. There was also the problem that Harwell's cost estimating consultants insisted that the cost of a stack of very big steel washers with a weight of 100 tonnes had to be based on navigation gyros. The cost predictions from this series of re-estimates are drawn in Fig. 2.9 again in the money values of the date concerned.

But the real killer for deep water devices like the duck was the values used for reliability and, in particular, the failure rates of marine cables. In an early consultants report (Clark, 1980) they suggested a cable survival rate of 333 kilometre years of operation per fault. But in the final report (Clark, 1983) this was reduced to just 10 kilometre years. This was much worse than the data from the North of Scotland Hydro Electric board who operated 80 undersea cables some of which had never failed since installation in the 'thirties. It was far worse than the then Figure of 625 kilometre years per fault of the large Norwegian marine cable network which was easily available.

By an ironic stroke of fate in the summer of 1982, at the very time that the Consultants were adjusting their numbers, a cable was laid from the mainland across the Pentland Firth to Orkney. Its length was 43 kilometres and so by June 2007 it had achieved more than 1000 kilometre years in similar waves and much worse currents.

Gordon Senior reported to a House of Lord Select Committee that somebody in the Rendel Palmer and Triton office had reversed what he had written about Duck technology (Senior, 1988). The reversals even included the insertion of the word NOT in the middle of one of his sentences. Strenuous but unsuccessful actions had been taken to prevent him discovering the changes. The correspondence can be downloaded from http://www.see.ed.ac.uk/~ies/.

2.1.2 Life after politics

During the delay in getting money for a proper wide tank model we tested bits of plastic drain pipe and learned that spanning wave crests was indeed a good way to get a stable reference. We also found that long, free-floating, low freeboard spines would move gracefully out to sea when waves began to break over them instead of ending up on the beach and that they liked to lie beam-on rather than head-on to waves.

The model we did eventually get to build had electronic control of stiffness and limiting bending moment at the joints and realistic power take-off for each duck. Figure 2.10 shows David Jeffrey with the set of spine joints and Fig. 2.11 the model in the tank. But by this time we were told by Harwell that we were not to do any duck tests and to merely confine ourselves to 'generic spine research'.

We found that bending moments were highest about half a crest length in from each end of a very long spine, rather than in the middle. There were also some interesting results with some oblique sea states inducing very large bending mo-

Fig. 2.10. David Jeffrey with the complete spine model. Beam elasticity could be varied electronically from the control bench. Two illicit generic absorbers can be seen, lower left

ments at the down-wave end. The reason was revealed when Jamie Taylor plotted results of bending moments in the same matrix format as a 'pox-plot' diagram which showed the distribution of period and angle in the 46 sea states selected for testing all UK wave devices (Taylor, 1984).

In Fig. 2.12 upper, each sea has been represented by 75 points carrying equal energy. One coordinate of a point represents the period of the energy from zero to 20 s. The other represents the angle from which it comes as 'hindcast' by the Institute of Oceanographic Sciences from knowledge of weather systems at the time. In Fig. 2.12 lower we can see the surge and heave bending moments along the spine for a 46 joint model with joint stiffness set to 1000 Nm/rad, model scale. The surge bending moment is always larger than the heave. The high down-wave bending moments are evident in sea-states 220, 360, 366 and 371 with their obvious cause in the corresponding pox plots. The build-up occurs when the propagation velocity of the flexure wave along the spine coincides with the velocity of the wave crest which causes it. As the flexure wave is a function of joint stiffness, which is under our control, it is not a cause for concern and can indeed be turned to advantage as in Pelamis. The largest credible wave at the most sensitive joint produced a deflection angle of only 4 degrees giving the full-scale design a factor of safety of three. Measurements of the fatigue bending angles showed that we had a factor of 400 relative to cable bending tests carried out by Pirelli.

Despite the official Atomic Energy Authority ban on testing ducks, careless management and an unfortunate breakdown of internal communications meant that some ducks were in fact tested with proper moorings and a realistic power take-off. The performance was in line with what had been predicted from the narrow tank models at the time that the long spine models had been designed. It was a surprise that a fault in one duck could not be detected in the total power generation of a group of them because the neighbours teamed up to help. The group also produced an efficiency of 25 % based on the spine length when the waves ran directly parallel to it.

Work had continued in the narrow tank. We found that sharp corners shed far more energy in vortex shedding than we expected. We found that the benefits of negative spring could be achieved without reverse power flow. Henry Young developed an iterative learning program that today would be called a genetic algorithm (Young, 1982). It started with Jamie's best settings and then ran the same pseudo-random sea repeatedly with slight random changes to the power take-off and mount stiffness, keeping the good changes and abandoning the bad ones. Overnight Henry's model could 'learn' to increase performance by as much as 20 %. It was clear that improvements to control strategy would never cease and that the power take-off hardware would have to be compatible with unforeseen future improvements. Figure 2.13 shows his results.

Fig. 2.11. The long spine model on the wide tank. The dynamics of each joint had electronic control of stiffness, damping and yielding bending moment with measurement of bending moment and joint angle

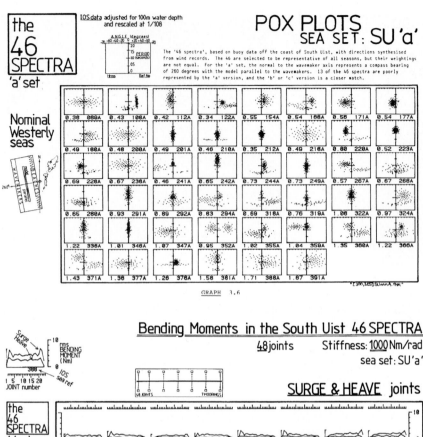

GRAPH 3.6

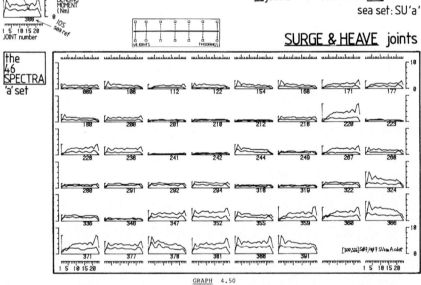

GRAPH 4.50

Fig. 2.12. Top: shows a pox-plot of the 46 sea states. Lower: resulting bending moments

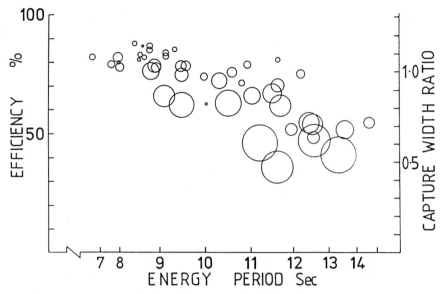

Fig. 2.13. Henry Young's narrow tank results for duck efficiency in the 46 sea states specified for the UK programme. Circle diameter is proportional to output power with the effects
of torque and power limits showing in the largest seas. The period axis has been stretched
to reflect the energy content of the Atlantic South Uist wave climate with clipping of any
sea states above $100\,kW/m$. Harwell reported to the UK Department of Energy that 'efficiency of wave plant was typically 40 %.'

However it was by then clear that the Edinburgh strict adherence to the UK target of designing a 2000 *MW* power station as a first step was a serious political
mistake and certain to make ducks appear far too risky to investors. To get adequate stability from a crest-spanning spine needed an initial installation of at least
ten units with a power rating of 60 *MW*. In contrast the wind industry had started
with units of a few kilowatt and most other countries were building wave devices
of a few hundred kilowatt. While Johannes Falnes had frequently urged very small
devices in large numbers spaced well apart, I lived in a country with much less per
capita sea front than Norway and wanted to use every millimetre to best advantage. While this might eventually be the right way, in 1983 it was as wrong as giving Bleriot the specifications for a Boeing 747. We wanted to build smaller systems of solo ducks, or even just parts of solo ducks to build confidence.

David Skyner moved the pitch-heave-surge rig to the wide tank and achieved
capture widths of 1.8 for most of the useful Atlantic spectrum with a scale model
of a 10 *m* diameter unit (Skyner, 1987). However, for the first time we were facing
forces with nothing like spine bending to limit them and the tension-leg moorings
of the solo duck showed nasty snatching if ever they went slack and then retight-
ened. We badly needed a small system to build confidence in components even if
it was nothing like a duck.

The next attempt was the Mace, Fig. 2.1 g, a vertical, inverted pendulum meant for testing ring-cam power take-off and driven through tapes wound round the cam leading down to sea bed attachments. It had a very wide but rather low efficiency band at much longer wave periods than any heaving buoy but extraordinary survival features and no need for end stops.

If buoys moving vertically were too stiff and flaps moving horizontally not stiff enough, it was interesting to ask if movement along a slope direction would be a happy compromise. David Pizer used his own numerical prediction software to show that this indeed was so for a wide range of device shapes (Pizer, 1994). My phobia about translations and end-stops in wave devices was reduced by the stroke-limiting feature of the Swedish IPS buoy.

Chia-Po Lin built a test rig to find out how a sloped version would behave (Salter and Lin, 1995; Lin, 1999). He used a half cylinder to avoid rear transmission of waves as shown in Fig. 2.14. He supported it on a straight slide with water-fed hydrostatic bearings and was easily able to adjust the slope of the slide. He drove it in calm water to measure the hydrodynamic coefficients and used these to draw theoretical efficiency curves for a selection of slope angles as shown in the middle graph and then confirmed them with true power generation.

The 45-degree prediction shows a capture width ratio above unity for a two-to-one range of period, so wide that we would not really need to vary the slope to suit changes in wave spectrum. The results are in agreement with experimental ones as shown the lower graph. It is clear that that movement along a slope increases efficiency and widens the efficiency band in both directions but especially towards longer periods. It is not easy to make sloped slides at full-scale and our attempts to make a free-floating one stabilised by an inertia plate have not so far been as good. But it is clear that water displacement in the slope direction, as shown in the 1973 models which led to the duck, make for good wave devices.

The change from testing in regular waves to more realistic irregular ones with a Gaussian distribution of wave amplitudes is an unpleasant experience for wave inventors. The power signal is the square of the Gaussian distribution with frightening peaks of energy which are determined to go somewhere but are totally incompatible with electrical grids. It was clear that storing energy for about $100\,s$ would make the output much more acceptable. The only way to do this and still retain intelligent power take-off seemed to be with high pressure oil hydraulics. But the designs then on the market were too low in power rating, too low in efficiency especially at part load and were bad at combining energy flows from multiple uncorrelated sources. We did a rigorous energy analysis of every loss mechanism and ended up with a design using digital control of displacement with electro-magnetically controlled poppet-valves on each chamber (Salter and Rampen, 1993). It did for hydraulics what the thyristor and switching-mode control have done for electronics. It allowed us to move away from swash-plate and port-faces in an axial configuration to a radial one with eight or even more separate machines on a common shaft, some motoring, some pumping and some idling all under the control of a microcomputer costing a few euros. Figure 2.15 shows the design. There has also been work on ring cam pumps for absorbing the very high torques at low speeds needed for wind turbines and tidal stream generators, (Salter, 1984; 1988).

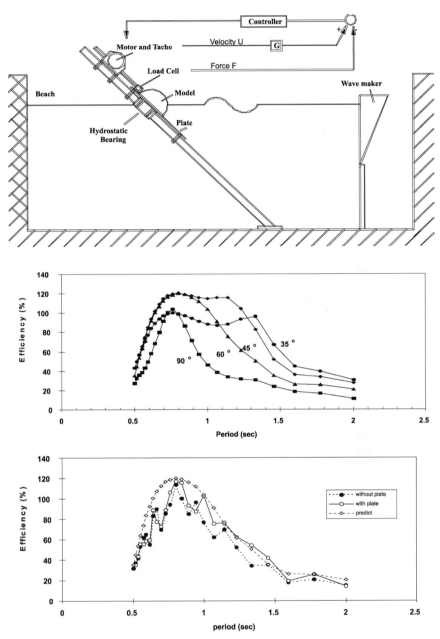

Fig. 2.14. Chia-Po Lin's results for a half cylinder moving on a fixed slope. Top, model set-up. Centre, efficiency from hydrodynamic coefficients. Bottom, measured efficiency at 60-degrees

There is now a growing need for flexible control and high part-load efficiency in vehicle transmissions and suspensions and it turned out to be possible to fund development by work on machines for the motor industry. A good regenerative braking system could reduce urban fuel consumption and pollution by 30 %.This development is being carried out by Artemis Intelligent Power, who hope to increase sizes to suit wind and tidal-stream generation as well as waves.

While the new digital hydraulics was being developed it seemed interesting to investigate a device which would inherently give the ideal torque proportional to velocity but deliver an output which in some places could be even more valuable than energy. The result was a desalination device using vapour compression rather than the normal reverse osmosis (Salter, 2005; Cruz and Salter, 2006). Figure 2.1h shows the arrangement. A plain cylinder which can rotate about an offset axis has a vertical partition and is half filled with water. The inertia of the water makes it tend to stay fixed. Movements of the cylinder about the offset axis will vary the volumes either side of the partition so that they become enormous double-acting pumps but with no machined parts. The pump chambers are full of steam and the movements suck and then compress it from and then to opposite sides of a large heat transfer surface contained in the partition. The result is two or three thousand cubic metres of pharmaceutically pure water a day in a tropical wave climate. A cross-section view showing internal details is given in Fig. 2.16. Figure 2.17 shows a 1/30 scale model under test pumping air through a blanket instead of steam in an out of a heat exchanger.

Whales and elephants do not need fur coats or wet suits because objects of that size and shape lose heat very slowly. With a metre of foam concrete for thermal insulation, the desalination system will cool at only $4°C$ per month. With a heat exchanger to transfer heat from the outgoing to the incoming flows, and all the heat from internal fluid movements tending to raise the temperature, we have to be careful not to overheat.

The main application will be in places with severe water problems which usually do not have such large waves as the Scottish Atlantic climate. However this is the first Edinburgh device to have a stiff mooring and so we are extremely concerned about peak loads. One possible approach is to allow the hull or legs to fill with water if severe weather is expected and lower the system below wave action. Many sites have sandy sea beds and it may be possible to use a combination of water jetting to sink a tripod anchor into the sand and then suction to consolidate it. This would mean that deployment and recovery could be done from light inflatable work boats with water pumps and air compressors but no heavy lifting gear.

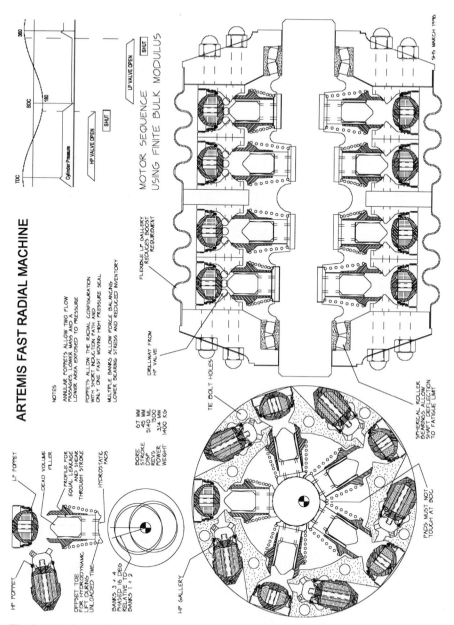

Fig. 2.15. A fast multi-bank, radial piston machine with digital poppet-valve control of pumping, motoring and idling

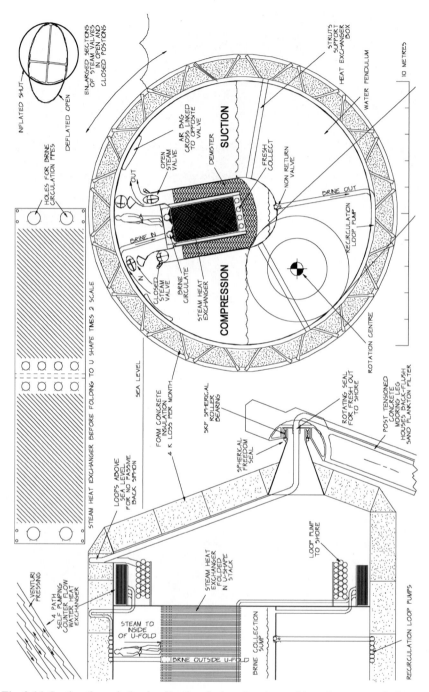

Fig. 2.16. Section through the desalination duck… People would not be present during operation!

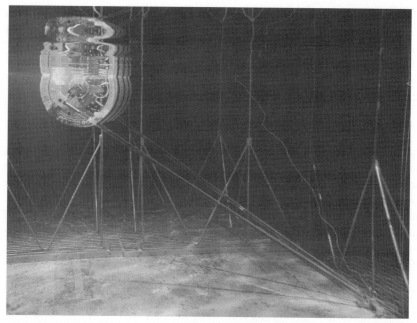

Fig. 2.17. A fish eye view of a 1/30 scale offset cylinder. The mooring will be a V-pair of post-tensioned concrete tubes with adjustable buoyancy going to a single-point attachment at the sea bed via a universal joint. The structures behind the Vee-legs are work platforms for use in the tank

2.2 Looking Forward

The challenge eventually must be to reduce costs of wave energy by a factor of about two. We may not be able to wait for the costs of fossil fuel generation to rise because that rise may push our construction costs up in direct proportion. However the initial task is to ensure that early wave power devices survive and produce the output predicted by their designers, even if the first wave electricity is as expensive as the first from coal or wind.

Survival depends on the full understanding of the statistics of the loads induced by waves and the strength of our parts. Every wave is a random experiment. We must understand the overlap of the upper asymptotic skirt of the load histogram and the lower skirt of the part endurance histogram. Separating them with large safety factors (which are really factors of ignorance and waste) is too expensive. If possible we must try to clip the load skirt to the economic limit and narrow the standard deviation of the strength histogram.

Large systems can fail because of very small components. While we can do lots in computers and indoor laboratories we must test large numbers of quite basic components such as bearings, seals, grommets, fasteners, surface coatings, cables and connectors in parallel in the chemistry and biology of the sea to know which ones will actually work. Building entire generating devices to test cheap parts, one by one, is a very expensive and very slow way to relearn the painful lessons of marine engineering. We should therefore have a test raft on which parts and sub-assemblies can be subjected to accelerated life tests.

We must try to maximise the ratio of the swept volume of any displacer to its own volume and the idle volume of the supporting structure. Low freeboards allow waves to break over a structure and so reduce, or even reverse, the mooring forces. The freeboard should therefore be chosen to suit the economic power limit. Concave shapes, like the corner at the foot of a breakwater or the focus of a shaped explosive charge, can amplify peak stresses in breaking waves so everything must be convex. Sharp edges, which we see in a great many designs of wave energy devices, waste lots of energy by shedding vortices so the convexities should have a large radius like a sucked toffee.

We must realise that it will never be possible to apply a restraining force for the largest waves and that a loss of the grid connection, even one due to some event on land, will mean that sometimes we cannot apply any restraining force at all. Devices with a low inherent radiation damping, such as smooth cornered buoys, can move an order of magnitude more than the wave amplitude and can build up very large amounts of kinetic energy which no end-stop can absorb. If we cannot use rotary mechanisms we must provide some other means for load shedding.

When we know how to make devices survive we can start to make them more productive. It has been a long term dream to design and test wave power devices in a computer with seamless links between the original drawing and the final results and with new ideas tried as quickly as Jamie Taylor could change models in a narrow tank. It is still a dream but we may be getting closer.

I predict that this will show that we must overcome the instinctive preference for movements in the vertical direction. Just because vertical motions are obvious to eyes and cameras and because we have instruments to measure them and a vocabulary to name them, does not make vertical the best mode. The horizontal forces and velocities can be just as useful. Movement in a slope direction or a combination of both as in the Bristol cylinder can give more than twice the power for much less than twice the cost. It is wrong to pay to resist large horizontal forces and then not get any power from them but nearly every beginner does exactly that.

Phase is the key to efficient transfer of energy. We must understand the inertia, damping and spring of our displacing mechanisms. We want a large swept volume for waves to move into but without the spring and inertia that is the usual accompaniment, except for Even Mehlum's Tapchan and the over-toppers. Work that is put into accelerating masses or deflecting springs will have to be returned. Only those forces that are in phase with a velocity are useful. We must maximise damping, reduce inertia to the minimum and have only enough spring to resonate with the undesirable inertia at the most useful part of the spectrum.

We must find ways to choose and control, instant by instant, the amplitude, phase and upper limit of the force going to the power-conversion mechanism. Asymmetry can shed about half the added inertia and slope reduces spring by a controllable amount. It is quite wrong to think about 'tuning' wave power devices. In radio terms tuning is the way not to get signals from unwanted transmitters. Good wave devices would be very 'low Q' resonators with a high ratio of damping resistance to reactive impedance. Half the spectra in Fig. 2.12 show that energy is coming from more than one source. Unlike the designers of radio receivers we want to receive simultaneous signals from as many transmitters on different frequencies as possible. We need to understand why the performance in irregular waves falls off with increasing amplitudes at a rate larger that would be predicted by non-linearity or torque and power limits of regular wave tests.

We should all use accurate common transparent costing methods, such as those developed by Tom Thorpe, based on material weight and safe working stresses. He even produced graphs of cost prediction plotted against the rate of interest so that people could see the effect of the 15% return required of wave energy investments and compare it with the 2.6% achieved by the CEGB or the 0.5% required by Japanese banks.

We must find ways to install and remove devices more quickly and much more cheaply than the towing methods inherited from the offshore oil industry. This may require the design of special vessels with high thrust, agile manoeuvring, instant connection and disconnection but short range.

Some wave devices may be vulnerable to currents and many marine current devices may be vulnerable to waves. Waves and current interact with one another in complicated and often dangerous ways. We must build tanks and develop software to understand the effects of these interactions.

Every new technology makes many painful mistakes. Many boilers burst, ships sank and planes crashed before we got them reliable. The mistakes only become less painful if people learn from them. They will learn only if full details of every mistake are circulated throughout the industry. This is certainly not happening now. The requirements for raising private investment require the concealment of expensive disasters in the hope that commercial rivals will repeat the mistakes.

We must find ways to get the right amount of money to front-line engineers as and when they need it. Over elaborate rules for tenders and contract management will not stop crooks embezzling public money but they certainly are too complicated for honest engineers to follow unless they also have a PhD in contract law. Perhaps we should try flexible agreements for people who have shown that they have earned trust with cruel and unusual punishments if they betray it.

We must have a management structure which can reach sensible decisions in a few days not the year or more required by the Dof E, ACORD, ETSU, RPT, TAG maze. Such committee trees are designed to make the post-disaster audit trail so complicated that no individual can be identified to take the blame when things go wrong as they so often do when decisions take so long. The community needs to believe that political leaders and officials genuinely want the technology to succeed rather than appearing to want it because they feel that this will win votes. When instead, they write letters to The Times boasting of how they stopped the programme or are given promotions to senior positions in the Nuclear industry, we all feel betrayed.

Acknowledgements

Wave research at Edinburgh University was enormously helped by a succession of vacation students and school leavers who learned engineering skills envied by local industry in an amazingly short time. Many reported that they learned as much as from a degree and many are now senior industrialists.

I cannot name several people working for the Atomic Energy Authority who risked their jobs to send us documents about dirty tricks which were very clearly not intended for us. One included a note to say that just because I was not paranoid that did not prove they were not out to get me!

The resurrection of the UK wave programme would not have occurred but for the evidence given by Gordon Senior to the House of Lords. Gordon Senior died in 2007 but will be remembered by his many friends in the wave community he saved.

References

Bott A (1975) Power Plus Proteins from the Sea. J RSA, pp 486–503
Budal K, Falnes J (1975) A resonant point absorber of ocean-wave power. Nature 257:478–479
Clark P (1980) Consultants working paper 22 WESC(80). Future Energy Solutions AEA Technology
Clark P (1983) Final Report on Wave Energy (2A/B/4) – A.3.2(5). Future Energy Solutions AEA Technology
Cruz J, Salter S (2006) Numerical and Experimental Modelling of a Modified Version of the Edinburgh Duck Wave Energy Device. Proc IMechE Part M. J Eng Marit Environ 220(3):129–147
Dixon A, Greated C, Salter SH (1979) Wave Forces on Partially Submerged Cylinders. J Waterway Port Coastal Ocean Div Am Soc Civil Eng 105:421–438
Evans D (1976) A Theory for Wave-power Absorption by Oscillating Bodies. J Fluid Mech 77:1–25
Evans D, Jeffrey D, Salter S, Taylor J (1979) Submerged Cylinder Wave Energy Device: Theory and Experiment. Appl Ocean Res 1:3–12
Jeffrey DC (1976) Edinburgh wave power project second year interim report, pp 24.1–24.39 (Download from http://www.see.ed.ac.uk/ ~ ies/)
Lin C-P (1999) Experimental studies of the hydrodynamic characteristics of a sloped wave energy device. PhD Thesis, The University of Edinburgh
Pizer D (1994) Numerical Modelling of Wave Energy Absorbers. University of Edinburgh wave project report 1994, from http://www.see.ed.ac.uk/ ~ ies. (Note: Compare his Fig. 39 with Fig. 42)
Skyner D (1987) Solo Duck Linear Analysis. Edinburgh wave power project, 1987. Download from http://www.see.ed.ac.uk/ ~ ies/

Salter S (1982) The Use of Gyros as a Reference Frame in Wave Energy Converters. In: Berge H (ed) Proceedings of 2nd International Symposium on Wave Energy, pp 99–115

Salter S, Rea M (1985) Hydraulics for Wind. In: Palz W (ed) Proceedings of European Wind Energy Conference, pp 534–541

Salter S, Rampen W (1993) The Wedding Cake Multi-eccentric Radial Piston Hydraulic Machine with Direct Computer Control of Displacement. BHR Group 10th International Conference on Fluid Power

Salter S, Lin C (1995) The sloped IPS Wave Energy Converter. Proc Second Eur Wave Power Conf, pp 337–344

Salter S (1998) Proposal for a Large, Vertical-Axis Tidal Stream Generator with Ring-Cam Hydraulics. Proc Third European Wave Energy Conference

Salter S (2002) Why is Water so Good at Suppressing the Effects of Explosions? UK Explosives Mitigation Workshop

Salter S (2005) Digital Hydraulics for renewable energy. World Renewable Energy Conference

Senior G (1988) Evidence to Lords Select Committee on European Communities. 16th report on Alternative Energy Sources, pp 204. 16 HL paper 88. HMSO, London, June 1988. Can be downloaded from http://www.see.ed.ac.uk/ ~ shs

Taylor J (1978) Edinburgh University wave project report 1978. Download from http://www.see.ed.ac.uk/ ~ ies/

Taylor J (1984) Edinburgh University wave power project 1984. Download from http://www.see.ed.ac.uk/ ~ ies/

Young H (1982) Parrots and Monkeys. Edinburgh wave energy project report 1982. Download from http://www.see.ed.ac.uk/ ~ ies

Weblinks

Aeroflex torque motor http://www.aeroflex.com/products/motioncontrol/torque-intro.cfm

Interproject Service website http://www.ips-ab.com/

Office of Naval Research. http://www.onr.navy.mil/Focus/ocean/motion/currents1.htm

Edinburgh Designs web home page. http://www.edesign.co.uk/

3 The Theory Behind the Conversion of Ocean Wave Energy: a Review

Gareth Thomas

Dept. of Applied Mathematics
University College Cork
Cork
Ireland

3.1 Introduction

The development of any new device is usually accompanied by extravagant claims that it has the potential to solve all of (or more modestly, a significant percentage of) the world's (or a nation's) energy problems. The inventor's strategy is usually to insist that the basic device is a magnificent absorber of energy, inexpensive to construct, resilient enough to survive the most violent storms and, if placed in rows or arrays around the coastline, the sum of the total output would provide the claimed power output. In the unlikely circumstance that all potential structural engineering problems have been solved, permissions granted and that grid connection really is a mere formality, then there are three simple modelling considerations that obstruct immediate success. These are in addition to host other practical obstacles that quickly arise.

The fundamental optimal performance of any device, based upon optimal performance criteria, usually counters any widely excessive claims being validated. Arrays of devices are subject to rather surprising constraints and will almost always not behave in the way they are intended to do. The third is concerned with wave climate and device design; site resource is an important consideration in device design and this is not rarely included in a preliminary implementation plan. All three are crucial aspects of device awareness and provide major pitfalls for newcomers to the field! In the present context, the question to be addressed is how these restrictions came to be known. An attempt is made within this review to provide an explanation as to these results were obtained, how they may be employed in a beneficial sense and the importance of designing a device to match the wave climate.

3.1.1 Historical and Parallel Perspectives

The pioneering work on the mathematical modelling of *Wave Energy Converters* (WECs) assumed that the waves were of small amplitude, relative to both the wavelength and the water depth, and of permanent regular form (the basic hypotheses of linear wave theory). These two assumptions permitted the extensive body of work that already existed in the fields of ship hydrodynamics and offshore structures, which are combined here under the umbrella heading of *Marine Hydrodynamics*, to be utilised immediately. Skilled practitioners became enthusiastic about the proposed technology and applied their expertise to wave energy utilisation, although the transfer of the existing theory to wave energy was itself a major task. This enabled important global results, such as the maximum power that a WEC could extract in one or more modes of motion, to be established at a very early stage. Such global modelling results have played an important part in assessing the behaviour and hydrodynamic viability of devices. In a similar manner, numerical methods developed in marine hydrodynamics have proved beneficial in wave energy applications.

The continuing importance of marine hydrodynamics to both commercial and military interests has ensured that as both theory and numerical modelling have developed, they have done so with obvious benefit to the wave energy community.

There is another aspect too. Marine hydrodynamics has progressed with both theory and experiment playing important roles. Substantial and comprehensive programmes have been implemented to develop experimental facilities with the intention of validating theoretical or numerical predictions. This has been to the obvious benefit of wave energy; it allows physical scale models to be built both with confidence and with an understanding of the strengths and limitations of the experimental testing programme.

If there has been a disadvantage associated with the maritime link, it is that the power take-off and associated technologies have not been developed at the same pace as the hydrodynamic modelling. Whilst this deficiency is now being addressed, there remains a slightly skewed approach with perhaps an over-emphasis upon the hydrodynamic input rather than the controlled output.

3.1.2 Scope of the Review

There are a surprisingly wide range of devices and it is not possible to consider each individually within a short review, nor is it possible or desirable to concentrate solely upon a single device or a family of devices. A good summary of device operation and status is given by Brooke (2003) and much more detailed information can be obtained from the accompanying chapters of this volume. The intent is to focus upon generic modelling from a hydrodynamic perspective with the aim of presenting the conversion process in as uniform a manner as possible over all families of devices. Applications will be given wherever possible but the focus is upon the broad concepts of operation and design.

A detailed knowledge of wave mechanics or hydrodynamics is not assumed but some knowledge of oscillating mechanical systems is considered desirable. The approach adopted is to provide a general background to the subject initially and then progress to the application of modelling wave energy devices. One of the principal needs of the newcomer is to acquire a working knowledge of the requisite maritime and hydrodynamic disciplines and this is addressed by identifying the appropriate textbooks in the distinct specialities. The fundamental modelling strategies and results are referenced to the key papers in the discipline and the few existing review articles are referenced whenever possible.

In terms of structure, a broad description of the terminology and concepts is given first and followed by a description of the resource to enable the conversion challenge and environment to be identified. A restriction to floating devices is then enforced, enabling the broad concepts of hydrodynamic modelling, optimal power absorption, control and design to be described for generic categories of devices. The review is completed by a final section on the modelling of the *Oscillating Water Column* (OWC) device, given separately to indicate its historical dominance presently in application and instalment but also in recognition of its distinctive modelling requirements. Analogies in approaches and results between the various sections are drawn whenever possible.

One limiting restriction is imposed throughout and this concerns permissible power take-off systems: only those devices that operate by utilising the oscillatory nature of resource directly are considered. The principle class of devices excluded is comprised of those devices, floating or fixed, that employ an overtopping principle and collecting chamber. Such a restriction is reasonable when an emphasis is placed upon time-dependent wave fields and motions.

3.2 Terminology and Concepts

3.2.1 Conversion Terminology

The standard terminology illustrates the strengths and weaknesses of WEC modelling. Many of the original concepts and terminology introduced in the 1970s remain in common use and this itself a tribute to the very high quality of the early work although others indicate perhaps a common lack of understanding at that time.

For an isolated body in three dimensions, the fundamental quantity employed to evaluate device performance is the *Capture Width*. At a given frequency this is defined to be the ratio of the total mean power absorbed by the body to the mean power per unit crest wave width of the incident wave train, where *mean* refers to the average value per wave period for regular waves or per energy period for irregular waves. Some early papers employ the descriptor *Absorption Length* or *Absorption Width* in place of capture width. Capture width has the dimension of length and sometimes the non-dimensional measure denoting the ratio of the capture width to the width of the device is a useful quantity to assess device performance. There is no standard term to indicate this quantity and *Non-dimensional Capture Width*, *Relative Capture Width* and *Non-dimensional Absorption Length* have all been used.

The two-dimensional historical analogue of capture width is *Efficiency* and this has played an important role in laboratory comparisons of theory and experiment in narrow wave tanks. Additionally, it is usually easier to develop two-dimensional numerical models rather than three-dimensional ones and this has also played a role in the adopting the given notation. It may seem that efficiency and relative capture width are similar non-dimensional measures but this is not the case. Efficiency is the ratio of output power to input power for a two-dimensional system, with a unit width of the device able to extract power from only a unit width of the incident wavefront; thus it has a maximum value of unity, more often quoted as a maximum percentage value of 100%. The relative capture width may possess a value of greater than one, as three-dimensional effects permit the device to absorb power from the total wavefront incident upon the device and not restricted to a wavefront possessing just the same width as the device. Both capture width and efficiency were originally derived for regular waves and have been extended readily to irregular waves.

Within the broader remit of wave energy extraction, the definition of capture width remains unambiguous but this is not true for efficiency. It is possible to provide a number of definitions of efficiency, based upon the consideration of various measures of the total system or particular subsystems. To avoid confusion, the traditional use of efficiency from a hydrodynamic perspective will be labelled the *Hydrodynamic Efficiency* and is the quantity considered in this review. This permits the single word *Efficiency* to follow the more general definition of the ratio of power output to power absorbed, related to the practical implementation and no longer restricted to two dimensions.

The *Maximum Capture Width* of a device of specified geometry, at a given frequency, is obtained by optimising the capture width with respect to the parameters of the power take-off mechanism. This corresponds to the mean absorbed power taking its maximum value and this will generally be strongly frequency dependent. In linear theory the power take-off mechanisms are often modelled by black-box models in which the system is represented by an applied linear damping term.

If a device is tuned to operate optimally at a specified frequency, then this will determine the power take-off parameters and the *Bandwidth* curve is found by plotting the capture width against frequency for the fixed power take-off parameters. This will coincide with the maximum capture width at the tuning frequency but not generally at other frequencies, when the values on the bandwidth curve will usually be below the maximum capture width. The character of the bandwidth curve is an important indicator of the device performance: a broad bandwidth suggests that the device will work well over a wide range of conditions, whereas a narrow bandwidth suggests that its performance capabilities will be good close to the tuning frequency but poor elsewhere.

3.2.2 Classification of Devices

Early work was targeted primarily at floating devices and typically classified a given device as being a *Point Absorber*, a *Terminator* or an *Attenuator* and the descriptors are still used to a certain extent. They are intended to describe the principle of operation and provide information on the geometry of the device and are shown schematically in Fig. 3.1. *Point Absorbers* are devices, usually axisymmetric about a vertical axis, which are small in the sense that the horizontal physical dimensions of the device are small relative to the wavelength of the incident waves. The concept of such a device is very appealing from a modelling viewpoint because the scattered wave field can be neglected and forces on the body are only due to the incident waves. Point absorbers are capable of absorbing the energy from a wavefront many times the key horizontal dimension of the absorber and so possess a large potential capture width. The theory predicts that such a performance can only be achieved if the device undergoes oscillations whose magnitude may be many times that of the incident wave amplitude. This behaviour is not permissible in practice and has led to the development of theories, not solely restricted to point absorbers, to predict the maximum capture width when the amplitude of the device oscillation is constrained in magnitude but permitted to maintain the frequency of oscillation.

Attenuators and *Terminators* are WECs which have finite dimensions relative to the incident wave field and moreover have one dominant horizontal dimension. A simple way to envisage the concept in plan is to consider a rectangle or ellipse that has a much greater length than breadth. *Attenuators* are aligned with the incident wave direction with their beam much smaller than their length and *Terminators* are positioned with the dominant direction perpendicular to the incident waves, with beam much greater than length. It is usual for attenuators to be compliant or articulated structures and often the initial design concept was that the waves would attenuate along the device as power was extracted; this concept is generally incorrect and the motion of the attenuator may be almost symmetric about the mid-point of the device, so that the fore and aft portions of the device work equally hard. Terminators can be rigid or compliant. There is little hydrody-

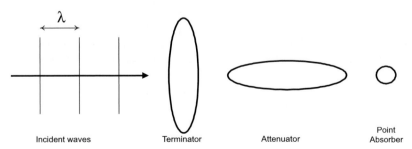

Fig. 3.1. Schematic showing scale and orientation of a *Terminator*, *Attenuator* and *Point Absorber*

namic difference in the behaviour of a compliant terminator and attenuator; it is essentially that the incident wave directions differ by a right angle. This illustrates an important point: the mode of operation is linked closely to the incident wave field and this will vary. Without controlling mechanisms, an elongated device could be forced to act as an attenuator or a terminator at the same site and different sea states. The alignment process is linked to the mooring configuration and this identifies one of the key requirements of the moorings for a WEC. It may not be sufficient for the mooring(s) to ensure that a device is maintained at a selected site, the mooring may also be required to ensure device alignment as well. In this context, it is salient to make that observation that an elongated device in attenuator mode will usually experience considerably lower mooring forces than the same device on terminator mode.

3.2.3 Alternative Device Classification

The classification system above is targeted at floating devices but this is not sufficient if all device categories are to be encompassed. For example, the first successful OWCs were introduced by Masuda, as described by Brooke (2003), and these are genuine residents of the point absorber family. More recent OWCs have been shore-mounted; any shore-mounted device can be described as a terminator but there are obvious differences between devices onshore and offshore. Thus although the usual method of classifying wave energy devices is based upon the mode of operation primarily, it is often informative to add one or more qualifiers to describe the device, its proposed method of working and intended site. Thus, for example, the descriptor *Oscillating Water Column* (OWC) describes how the device operates but does not provide information concerning the location where the device would best be employed. This shortcoming can be remedied by the inclusion of an additional qualifier, such as *Onshore, Nearshore* or *Offshore,* to specify the location.

For the present purpose it is expedient to use a slightly different classification system to the physically precise one outlined previously; the new system was developed in the EU-funded OWEC-1 project and reported by Randløv (1996). This classification is based upon the present status of a device, the development timescale and economic investment cost; the mode of operation is not used as a defining quantity. If these new considerations are utilised, then any device can be classified as being a *First Generation System*, a *Second Generation System* or a *Third Generation System*. The three categories are not mutually exclusive and share common features; this difficulty is acknowledged but further details will not be discussed.

Onshore or nearshore OWC devices are considered to be *First Generation Systems* and such devices are installed presently or under development in the UK, Portugal, India and Japan. The dominance of OWCs partly reflects an ability to build such devices with conventional technology and power take-off equipment, although this remark is tempered by the fact that considerable development work has been required in both technology and power take-off. In some sense, OWCs are the most difficult to model of the three categories, as confirmed in Section 3.7, and considerable modelling effort is still required.

Second Generation Systems, represented by float pumps, are designed to operate at a wide variety of offshore and nearshore sites where high levels of energy are available. Installation is usually considered possible in water depths of between thirty and a hundred meters. Float pumps may be slack-moored or tight-moored but all possess a favourable ratio between absorbed energy and volume. They clearly do not represent, nor are intended to represent, all future categories of offshore devices, but these devices are relatively small both in physical size and power output; as such they are ideal for a relatively short and inexpensive development period. These devices belong to the point absorber category as the horizontal physical dimensions of the device are much smaller than the wavelength of the waves from which the device is designed to extract energy. It is worth noting that it is not always possible to make an additional classification based upon power take-off characteristics. Innovative hydraulic machines promise power take-off systems with a means of energy storage and increased power output by control of device motion for some devices, whereas others are essentially offshore floating OWCs with a pneumatic power take-off mechanism.

The defining property of *Third Generation Systems* is that they are large-scale offshore devices, both in terms of physical size and power output. Such advanced systems could well be seen as the final stage of device development following the successful implementation of float pump systems. It is important to recognise that a large power output is attainable potentially from either a single device of large physical dimensions or a large array of devices, which individually are of much smaller size. Large single devices correspond to the terminators and attenuators of the device classification standard adopted in Section 3.2.2 and an array of smaller float-type devices requires the array theories described in Section 3.4.

Any proposed classification system cannot be entirely satisfactory and the task is riddled with difficulties. The Pelamis device described in Chapter 7 can be considered as an attenuator with regard to the initial classification system and as a third generation device in the context of the alternative classification. Thus both first and third generation devices have been installed and deployed successfully but no second generation device has yet been awarded similar status.

3.3 Preliminary Considerations

3.3.1 The Conversion Requirement

The instantaneous resource can be measured at or close to a particular site by an appropriate measuring device, such as a wave recorder buoy, and is usually recorded as a discrete time series. The challenge is to convert the energy contained within the wave motion described by this time series into useful electrical energy.

An assessment of a favourable site is made usually by monitoring the wave climate over a considerable period to determine the variety and power contained within the various sea-states. It is instructive to review some basic properties of the resource from a hydrodynamic perspective and without impinging upon the detailed assessment of the resource presented in Chapter 4.

This challenge can be represented by considering two power spectra, each associated with the sites deemed favourable for energy extraction. Both spectra are shown in Fig. 3.2 and may be considered as being typical of the sites and containing at least a moderate level of power. The first is the "select spectrum" of Crabb (1980) and provides a model representative spectrum, based upon site measurements, from the South Uist site off the west coast of Scotland; the water depth is $42\,m$ and the mean resource is estimated as $47.8\,kW/m$. The second spectrum is a measured spectrum from the island of Pico in the Azores, which was obtained as part of the development programme for the OWC built upon Pico and was supplied by Pontes and Oliveira (1992). The site is onshore with $8\,m$ water depth, an estimated mean resource of $26.5\,kW/m$ and the spectrum is considered to be reasonably energetic for the given site. These spectra possess similar peak values at almost the same frequency but the South Uist spectrum is broader and contains almost $80\,\%$ more power than the Pico one. A device designed to operate at either site must be capable of operating efficiently within the frequency range and power level.

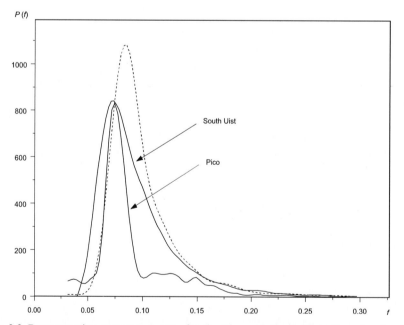

Fig. 3.2. Representative resource spectra for the Pico and South Uist sites, showing the power density $P(f)$ (in $kW/m/Hz$) against frequency f (in Hz). The solid lines show the site resource and the broken line corresponds to the offshore resource at Pico

Another Pico curve is also shown in Fig. 3.2 as a dotted line and this is the deepwater resource prior to reaching the shoreline, measured in 104 *m* of water and with a power level of 40.3 *kW/m*. A comparison of the deepwater and onshore resources shows that approximately one third of the power has been lost in the transition region between the two measuring sites, with the loss occurring in the shorter wavelengths. Another interesting feature is that the maximum value in the two spectra occurs at different frequencies, demonstrating the importance of establishing power levels at the installation site.

The power spectra in Fig. 3.2 provide good assessments of the resource and are obtained from surface elevation records that may or may not contain any measure of wave directionality. They provide a summary of individual sea-states but not of the water surface movements. A typical surface elevation associated with a particular spectrum can be obtained by extracting wave amplitude values at prescribed frequencies and then summing the contributions from the component frequencies, assigning a random phase difference to each frequency component. Such a surface elevation is shown for the Pico spectrum in Fig. 3.3. This represents the instantaneous resource and identifies the real challenge: to convert the time-varying incident power to useful electrical power or into some useful repository of power storage.

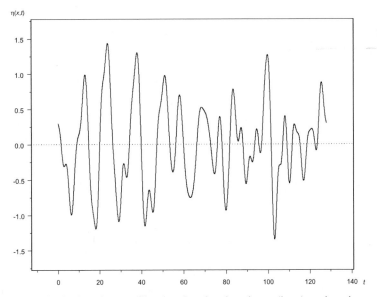

Fig. 3.3. A synthesised surface profile, showing the elevation η (in *m*) against time *t* (in *s*), corresponding to the Pico spectrum shown in Fig. 3.2

3.3.2 Modelling the Resource

To understand the requirement identified above, begin with a very simplified representation of the resource and employ elementary wave mechanics. Information on general water wave mechanics can be obtained from Dean and Dalrymple (1991) and more specialist applications to ocean waves from Goda (2000) and Tucker and Pitt (2001).

Choose co-ordinates (x, y, z) with x, y horizontal and z vertical, measured positive in an upward direction. The surface elevation in water of mean depth h is measured by a wave height recorder at a fixed point in space with horizontal co-ordinates (x_0, y_0) and the signal $z = \eta(x_0, y_0, t)$ may be continuous or discrete. As the measuring point is both fixed and general, the (x_0, y_0) part of the description can be neglected, provided it is recognised that a different signal will be found for each reference point. Thus the elevation may be considered just as $\eta(t)$ and is simply a continuous function of time that is known at any time t. Quantities requiring physical measurement typically employ f, the frequency in Hz, whereas it is often easier in modelling studies to employ ω, measured in rad/s. The two are related by $\omega = 2\pi f$ and both are used herein, as appropriate.

In a formal sense the signal at the fixed point can be regarded as being composed of a continuous spectrum of frequencies, each characterised by an amplitude density measure $a(\omega)$ and a phase function $\gamma(\omega)$, so that the surface can be characterised by

$$\eta(t) = \int_0^\infty a(\omega) \cos(\omega t - \gamma(\omega)) \, d\omega . \tag{3.1}$$

At any time $\eta(t)$ represents the integral under a curve in the frequency domain and so may also be represented approximately as the linear sum of an infinite number of frequency components,

$$\eta(t) = \sum_{m=0}^{\infty} a_m \cos(\omega_m t - \gamma_m) . \tag{3.2}$$

A comparison of Eqs. (3.1) and (3.2) suggests that the discrete amplitude a_m and continuous measure $a(\omega)$ are related by $a_m = a(\omega_m) d\omega$. Figure 3.3 was obtained from the Pico spectrum in Fig. 3.2 by employing the representation Eq. (3.2) with a finite rather than an infinite sum of components and with the phases chosen randomly.

If all of the amplitudes a_m in Eq. (3.2) are small in some sense, so that individual amplitude components are not bound to others within the series, then Eq. (3.2) can be regarded as a series of individual sinusoidal waves. The influence of each may be considered separately and then summed appropriately to obtain the influence of the complete spectrum. More specifically, each component will lie within the linear regime provided that $a_m k_m \ll 1$ and $a_m / h \ll 1$, where k_m is the magnitude of the wavenumber vector and related to the wavelength by $k_m = 2\pi / \lambda_m$. If the problem cannot be treated in this manner then it is considered to be nonlinear.

The analysis above takes no account of directionality in the wave field and it is very rare for this to be unimportant. It may be of particular importance to a chosen site, as the bed topography may provide a focusing mechanism to enhance the resource. This can be included for linear wave components by interpreting Eq. (3.2) as

$$\eta(x,y,t) = \sum_{m=0}^{\infty} a_m \cos\left(k_m x \cos \beta_m + k_m y \sin \beta_m - \omega_m t + \psi_m\right) \qquad (3.3)$$

evaluated at $x = x_0, y = y_0$ and $\gamma_m = -\left(k_m x_0 \cos \beta_m + k_m y_0 \sin \beta_m + \psi_m\right)$, with β_m being the angle between the direction of wave propagation and the x–axis. Thus Eq. (3.3) can be considered as being composed as a number of long-crested wave components propagating in arbitrary directions.

A further consequence of Eq. (3.3) is that it is sufficient to consider the influence of a single frequency component alone at an arbitrary angle of incidence within the linear wave regime, i.e., to assume that the surface elevation is given by

$$\eta(x,y,t) = a\cos\left(kx\cos\beta + ky\sin\beta - \omega t\right), \qquad (3.4)$$

with the phase ψ_m taken to be zero. This does not mean that the time series approach is unimportant, and this is certainly not the case, but Eq. (3.4) confirms that many of the important properties of the device can be obtained by an analysis in the frequency domain.

For the regular incident wave represented by Eq. (3.4), the mean power (averaged over the wave period) per unit crest width is $\mathcal{P}_W = \dfrac{1}{2}\rho g a^2 c_g$, where c_g is the group velocity. If P is the mean power absorbed by the device, then the capture width $\mathcal{L}(\omega, \beta)$, identified in Section 3.2.1 as an important property, is defined by

$$\mathcal{L}(\omega,\beta) = \frac{P}{\mathcal{P}_W}. \qquad (3.5)$$

There will be no dependence upon β if the device is axisymmetric about a vertical axis.

The concept of a capture width is an appealing one, since it shows that the device captures an amount $L\mathcal{P}_W$ from the wavefront. This will not correspond to power taken from just a strip of width L, although the power extraction would be expected to occur primarily from the frontage area nearest to the device. However, it is not a non-dimensional measure of the optimal absorption characteristics of a device. The capture width possesses the dimension of length and this is because it employs the reference measure of mean power per unit width of wave crest. If D is a typical device dimension representing perhaps frontage to the incident waves, then an important measure for the device is L/D. It is clear that this should be as large as possible and, from an intuitive perspective, values less than unity would be regarded as poor.

3.3.3 Survivability

For many WECs, the application of models provided by the small amplitude linear wave theories, within the context above, will be accurate for the vast majority of their operating times. However, such models will not suffice for force prediction or device behaviour when the WECs are exposed to very large wave-induced forces in extreme storm seas and the question of whether or not the device will survive such forces must be addressed. Wind loading may also be considerable for those devices that possess a substantial exposure above the water surface.

The potentially disastrous effects of storm seas provide part of the engineering and design challenge. It is fortunate that although it may not be possible to predict the wave-induced loading accurately, there is a solid base of work available from the offshore and maritime engineering industry that can be utilised as a starting point for WECs. However, it must be stressed that WECs introduce particular difficulties that have not been encountered previously with offshore structures. For example, the front wall of a shore-mounted OWC may appear to be rather like a breakwater and may be sloping or vertical, depending upon the designer's choice. If the breakwater survives then it serves its purpose and protects any features behind it. This is not the case with wave energy: the device must survive and must do so in such a way that the hydrodynamic behaviour of the OWC remains favourable for generative motion to take place within the OWC chamber as conditions demand. A number of innovative approaches have been proposed to enable survival of WECs, including submergence and intelligent control. Such considerations are beyond this contribution, except perhaps by way of introducing design constraints, and an interesting assessment of possibilities is given by Chaplin and Folley (1998).

It must be recognised that there is no requirement in storm seas to provide an efficient conversion chain from resource to grid: there is increased resource in such seas to require at most a moderate efficiency. However, survival is paramount and this defines the twin requirements of a WEC: very efficient conversion in small to moderate seas moving through to survival in storm seas. The importance of survivability is acknowledged to be of paramount importance but is not considered further in the conversion process.

3.4 The Hydrodynamics of Offshore Devices

It is important to recognise at the outset that the modelling of wave energy devices, although rooted in ship and offshore hydrodynamics, must necessarily possess a different viewpoint to that which exists in the parent fields. In offshore hydrodynamics, the purpose of a mathematical model is usually to determine the wave loading on a fixed or floating structure of agreed design; in ship hydrodynamics it is the response of a specific ship design to certain sea conditions that is of interest. Neither approach is applicable in wave energy studies: the key to WEC modelling is performance. All other seemingly important quantities associated with device modelling, such as wave loading, are of secondary consideration even though they often appear as a result of, or are employed in, mathematical models of WECs. If a device cannot perform sufficiently well to absorb an acceptable level of power in small or moderate seas then other considerations are not investigated.

In the first instance, attention is restricted to those devices that are intended to operate in the offshore environment and particularly to floating structures. The principal class of devices excluded is the OWCs, of which the onshore category is more important presently. There are two reasons for separate consideration of these devices: an assessment of a fixed structure is not compatible with an analysis that places a strong emphasis upon a body movement and the power take-off mechanism of an OWC, involving aerodynamic - hydrodynamic coupling, also requires a specific treatment. A floating OWC will require a combination of both approaches to be employed; OWCs are considered separately in Section 3.7.

A general rigid body motion is composed of three translational modes of motion (surge, sway, heave) and three rotational modes of motion (roll, pitch and yaw) in the directions of and about the (x,y,z) coordinate axes. For convenience, consider the single tight-moored buoy shown in Fig. 3.4 with attention restricted to the heave or surge motions. This semi-submerged sphere is axisymmetric and can be taken to represent a generic device for the purposes of the present discussion. It can be described, in the terminology of Section 3.2, as a point absorber when it has a sufficiently small diameter, or as an attenuator or terminator if otherwise; it is classified as a second generation device although it was one of the first to be considered.

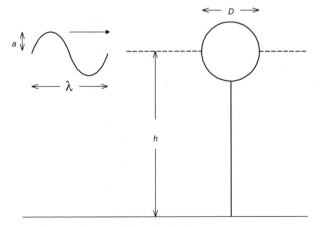

Fig. 3.4. Schematic of an axisymmetric heaving device

The modelling process is best described by considering a single translational mode in a long-crested incident wave field and the initial presentation follows that of Jefferys (1980). If m is the mass of the body and $X(t)$ represents its time-varying displacement, then the equation of motion of the body is given by

$$m\ddot{X} = F_T(t) + F_{ext}(X, \dot{X}, t),$$

(3.6)

where the total force $F_T(t)$ has components $F_f(t)$ and $F_{ext}(X, \dot{X}, t)$ denoting the wave/fluid induced forces and externally applied forces respectively acting in the direction of $X(t)$. The external forces can represent power take-off mechanisms and/or mooring constraints and the power take-off mechanism may be incorporated into the mooring system. Clearly the force will be dependent upon the type of mooring and the task that it is expected to perform. At this point, the formulation is exact.

3.4.1 Hydrodynamic Approximations

A regular incident wave field in water of depth h, as in Eq. (3.4), can be characterised by amplitude measure a, wavenumber k (or wavelength $\lambda = 2\pi/k$) and radian frequency ω. These will be related by a dispersion relation of the form $G(a, k, \omega, h) = 0$ and the character of the wave field can be described by the non-dimensional parameters ak and kh, associated with the wave-slope and water depth-to-wavelength ratio respectively. In water of intermediate depth or deep water, classified by $kh = O(1)$ or $kh \gg 1$, the nonlinearity is determined by the value of the wave slope ak. In shallow water, $kh \ll 1$, and the nonlinearity depends upon both ak and kh.

The structure can be characterised by an appropriate number of length scales dependent upon the body geometry. For the present general discussion, it is sufficient to permit just one length scale D and the diameter of the buoy in Fig. 3.4 is a good representative measure. The forces and pressures that represent the wave-structure interaction depend upon the appropriate non-dimensional ratios of the parameters that describe the individual components, so intuitively the dependency would be expected to be reflected by quantities such as kD, a/D, h/D and possibly some representation of viscosity. Combinations of these parameters are also important, with the Reynolds number R_e and Keulegan-Carpenter K_c being most prominent in the present discussion. The general definitions of these quantities can be adapted to the confines of linear wave theory, with K_c interpreted as the distance travelled by a particle relative to the length scale of the object, and taken for illustrative purposes as

$$R_e = \frac{aD\omega}{v}, \qquad K_c = \frac{2\pi}{\tanh kh} \frac{a}{D}.$$

(3.7)

Diffraction effects dominate when the flow remains laminar in the vicinity of the structure, viscous effects are negligible and no vortex shedding occurs. A consequence is that if the incident waves are irrotational, then the flow within the vicinity of the body will also be irrotational and so a formulation in terms of a velocity potential may be employed. This is very enabling from both an analytical and numerical perspective.

The parameter kD (or $D/\lambda = kD /2\pi$) is often considered to be the measure of the importance of diffraction and Sarpkaya and Isaacson (1981) consider that diffraction should be included whenever $kD > 1.3$, equivalent to $D/\lambda > 0.2$. Based upon the allowable range of parameters before wave breaking, these authors consider that the resulting Keulegan-Carpenter number is bounded by $K_c < 0.07\,kD$ and will be at most 2.2 and usually less then unity. The requirement that $D/\lambda > 0.2$ for the diffraction regime to remain valid is important from a modelling perspective and easily tested. A full discussion is given by Sarpkaya and Isaacson (1981) and useful practical information can also be obtained from Faltinsen (1990) and Goda (2000). Within the present context, the number of frequencies in the incident wavetrain, as described by Eqs. (3.2) or (3.3), must be considered. It is assumed that the diffraction regime holds for each component in the spectrum and, for a body of given dimension, this imposes restrictions subsequently upon the frequency components that can be considered to lie within the diffraction regime.

It is usually assumed in preliminary models of WECs that the forces remain within the diffraction regime and that the importance of other known forces can be considered at a later stage. With attention confined to linear water wave theory in the diffraction regime, the fluid induced forces $F_f(t)$ can be approximated by the combination

$$F_f(t) = F_S(t) + F_R(t) + F_H(t), \tag{3.8}$$

where $F_S(t)$, $F_R(t)$ and $F_H(t)$ are the exciting, radiated and hydrostatic forces respectively. In this representation the exciting and radiated forces are associated with the response of the body to the incident wave motion and the hydrostatic component is independent of the waves; a complete discussion of this decomposition is given by Newman (1977). Each component will be considered separately.

With the incident wave given by Eq. (3.4), it often convenient in linear theory to employ a complex representation, so the surface elevation can be written as $\eta(x,y,t) = \text{Re}\{a \exp i(kx\cos\beta + ky\sin\beta - \omega t)\}$. By analogy it is often assumed that the motion of the body in this single mode can be written as

$$X(t) = Re\{\xi e^{-i\omega t}\}, \tag{3.9}$$

where ξ is some unknown complex constant and which can only be determined via the solution of Eqs. (3.6) and (3.8). This complex form is a useful representation of $X(t)$. The magnitude of ξ corresponds to the magnitude of the oscillation and the phase is also important and may differ from that contained within the incident wave. The condition imposed by diffraction theory is that the quantity $|\xi|/a$ is at most of $O(1)$ and is consistent with the representation of the three terms that appear in Eq. (3.8). However, while the form Eq. (3.9) is clearly attractive, it cannot be imposed without justification and the external force $F_{ext}(X,\dot{X},t)$ must be amenable to such a representation. The accompanying velocity field is

$$\dot{X}(t)=U(t)=Re\{\mathcal{U}\,e^{-i\omega t}\}, \quad \mathcal{U}=-i\omega\xi \tag{3.10}$$

and this is included here since it will be utilised at future points.

The view has been expressed previously that linear theory will be valid for most devices in most operating circumstances and that storm conditions will provide the principal exception. This is certainly the case for floating devices but a caveat needs to be imposed when nearshore or onshore WECs are considered. The nearshore wave climate often lies within the shallow water regime and the demands placed by the condition that the *Stokes* (or *Ursell*) parameter $(ak)/(kh)^3$ is $\ll 1$ for the applicability of linear theory is severely restrictive. This does not mean that nearshore WECSs cannot be considered using linear theory, it is more that linear theory may not provide a sufficiently accurate working environment for a substantial period of operational time.

The *Exciting* (or *Scattering*) *Force* $F_S(t)$ is the force that the body would experience if it were held fixed in its mean position and, in keeping with Eq. (3.9), is usually written as

$$F_S(t)=Re\{\mathcal{X}\,e^{-i\omega t}\}, \tag{3.11}$$

for some complex constant $\mathcal{X}$. Some texts employ $\mathcal{X}_s$ instead of $\mathcal{X}$ but there is no inconsistency or confusion in using the simpler form here. This quantity can be considered as being composed of two parts,

$$\mathcal{X}=\mathcal{X}_{inc}+\mathcal{X}_{diff}, \tag{3.12}$$

corresponding to contributions from the incident and diffracted waves respectively. The quantity $\mathcal{X}_{inc}$ is usually straightforward to obtain and corresponds to integrating the known pressure due to the incident waves over the wetted body surface. In contrast, determination of the second term $\mathcal{X}_{diff}$ is often a difficult task and this quantity can only be calculated when the pressure field over the whole wetted surface has been determined. Analytical or semi-analytical solutions are rare and restricted to very simple geometries; solutions must be sought numerically in most cases of practical interest and these utilise industry standard codes. The numerical methods employed have been developed in the fields of naval hydrodynamics and offshore engineering and are described in Chapter 5, providing another good example of the benefit of retaining a link with a cognate technology. A good summary of the available mathematical techniques for problems of this type is given by Linton and McIver (2001). If $|\mathcal{X}_{diff}|\ll|\mathcal{X}_{inc}|$ then the diffracted component can be neglected and the scattered force is represented by the contribution from the incident waves alone. This is known as the *Froude-Krylov Approximation* and is clearly useful when circumstances permit.

The *Radiation Force* $F_R(t)$ corresponds to the force experienced by the body due to its own oscillatory movement in the absence of an incident wave field and is proportional to the amplitude $|\xi|$ of the displacement in the linear theory. The standard practice is to regard the force as being composed of two components: one in phase with the body acceleration and the other in phase with the body velocity, i.e.

$$F_R(t) = -\left\{A(\omega)\ddot{X} + B(\omega)\dot{X}\right\}, \tag{3.13}$$

where $A(\omega)$ and $B(\omega)$ are known as the added mass and damping coefficients respectively. In keeping with Eq. (3.9) and Eq. (3.11), write Eq. (3.13) as

$$F_R(t) = Re\left\{\mathbb{F}_R e^{-i\omega t}\right\}, \quad \mathbb{F}_R = \left[\omega^2 A(\omega) + i\omega B(\omega)\right]\xi. \tag{3.14}$$

Falnes (2002), and in earlier papers, has made extensive use of analogies with applications in other oscillatory systems. Following this approach, the radiation force is then written

$$\mathbb{F}_R = -\overline{Z}(\omega)\mathbb{U} \qquad Z(\omega) = B(\omega) + i\omega A(\omega), \tag{3.15}$$

where the complex quantity $Z(\omega)$ is termed the *Impedance* or the *Radiation Impedance*, the overbar denotes the complex conjugate and $\mathbb{U}$ has been given in Eq. (3.10). Whether $Z(\omega)$ or $\overline{Z}(\omega)$ appears in Eq. (3.15) depends upon the form of the time dependency employed in Eqs. (3.9) and (3.11); the conjugate is required if the time dependency is taken to be $e^{-i\omega t}$ but not if $e^{i\omega t}$ is used. Although the latter may appear more appropriate in standard oscillator applications, the former is employed here to retain the same time dependency as that of the incoming wave field, given immediately prior to Eq. (3.9).

As with the diffracted force $\mathbb{X}_{diff}$ in Eq. (3.12), the determination of $A(\omega)$ and $B(\omega)$ is a task that must be accomplished numerically in most cases of practical interest, with the available mathematical approaches being summarised by Linton and McIver (2001). Fortunately, as in the numerical determination of $\mathbb{X}_{diff}$, $A(\omega)$ and $B(\omega)$ can be obtained employing the same numerical algorithm so that some efficiency is possible in the solution of this difficult numerical problem.

The *Hydrostatic Force* $F_H(t)$ is the buoyancy force on the device and given by

$$F_H(t) = -CX(t) = -Re\left\{C\xi e^{-i\omega t}\right\}, \tag{3.16}$$

where C is the buoyancy coefficient. If the body is in a position of equilibrium in the absence of waves, then this is only non-zero for the heave (vertical) mode of motion and for the roll and yaw rotational modes of a floating body.

3.4.2 The Equation of Motion

Return to the exact equation of motion Eq. (3.6) and employ the approximations consistent with the assumptions that the incident waves are regular and linear, that the forces remain within the diffraction regime and the magnitude of the body motion is comparable to the incident wave amplitude. Combining Eqs. (3.6) and (3.8), for the wave-fluid interaction forces then gives the equation of motion $m\ddot{X} = F_S(t) + F_R(t) + F_H(t) + F_{ext}(X,\dot{X},t)$. Introducing the appropriate forms from Eqs. (3.11) – (3.16) for the force representations gives

$$(m + A)\ddot{X} + B\dot{X} + CX = Re\{\mathbb{X} e^{-i\omega t}\} + F_{ext}(X,\dot{X},t),\qquad(3.17)$$

which, via Eq. (3.9), can also be written as

$$Re\{[-\omega^2 (m + A) - i\omega B + C]\xi e^{-i\omega t}\} = Re\{\mathbb{X} e^{-i\omega t}\} + F_{ext}(X,\dot{X},t).\quad(3.18)$$

At first glance Eqs. (3.17) and (3.18) may appear to describe a standard harmonic oscillator but this is not the case, since the equation can only be consistent if F_{ext} has the same time dependency as those terms based upon the incident wavetrain. Jeffreys (1980) states that it is not even a differential equation in the standard sense! The form of $F_{ext}(X,\dot{X},t)$ is crucial to further progress and this has already been identified as an important factor in enabling Eq. (3.9) and upon which Eq. (3.18) is based. A simple linear damper model is often chosen to represent the power take-off mechanism but this may not always be a realistic approximation. Mooring forces should also be included whenever present and a taut mooring can provide a significant influence upon the device motions.

An extension to the impedance approach of Eqs. (3.14) and (3.15) is also possible. With $Z_T(\omega)$ defined by

$$Z_T(\omega) = Z(\omega) + i(\omega m - C/\omega^2) = B(\omega) + i\omega(A(\omega) + m - C/\omega^2),\quad(3.19)$$

Eq. (3.18) becomes

$$Re\{\overline{Z_T}(\omega)U e^{-i\omega t}\} = Re\{\mathbb{X} e^{-i\omega t}\} + F_{ext}(X,\dot{X},t).\qquad(3.20)$$

the usefulness of this form is dependent upon the structure of F_{ext}, upon its time dependence in particular and will be considered in Section 3.6.

The derivation above has assumed, for simplicity, that the body undertakes a single translational mode of motion. As stated previously, a general rigid body motion will be composed of three translational modes and three rotational modes. These can be accommodated in the present framework by extending $X(t)$, presently representing a single translational mode, to the column vector $X = \{X_j, j = 1,2,...,6\}$ and where the first three components of X represent the translational modes and the second three are the rotational modes. In a similar manner, quantities appearing as forces are also six-component column vectors and are associated with both forces and moments in the same manner as X. The corresponding extension to Eq. (3.17) is

$$\sum_{j=1}^{6}\left[\left(m_{kj}+A_{kj}\right)\ddot{X}_{j}+B_{kj}\dot{X}_{j}+C_{kj}X_{j}\right]=Re\left\{\left(\mathbb{X}\right)_{k}e^{-i\omega t}\right\}+\left(F_{ext}(X,\dot{X},t)\right)_{k}, \quad (3.21)$$

where $k=1,2,\dots,6$. A complete description for sinusoidal motion is provided by Newman (1977). The added mass and damping matrices are symmetric and again usually require numerical determination, via the codes described earlier.

While the form Eq. (3.6) is exact, all of the ensuing discussion assumes that the waves are monochromatic and linear, that the wave loading is within the diffraction regime and the external force is of an amenable form. Suppose that the latter two of the three conditions remain valid but the waves are no longer of a single frequency. In principle it is possible to adopt the approach of Eqs. (3.2) and (3.3) and consider all frequency components independently, with the final result requiring a careful combination of all components and their phases. It is more usual to employ the time-dependent form of the equation of motion and the changes moving from regular to irregular wave fields are examined.

Equation (3.18) was obtained by employing the decomposition Eq. (3.8) and then approximating each of the component terms for a regular wave motion. When an irregular wave field is present, only the hydrostatic component $F_H(t)$ in Eq. (3.16) remains unchanged. The exciting force $F_S(t)$ cannot be modelled in a harmonic manner and must be included in a general form and the radiation force $F_R(t)$ is modelled via the derivation given by Cummins (1962). For a single mode of motion the equation is

$$\left(m+A_{\infty}\right)\ddot{X}+\int_{0}^{t}K(t-\tau)\dot{X}(\tau)d\tau+CX=F_{s}(t)+F_{ext}(X,\dot{X},t), \quad (3.22)$$

where A_{∞} is a constant related to the added mass and $K(t-\tau)$ is an impulse response function related to the radiation damping. A full review of floating body hydrodynamics within the diffraction regime is given by Wehausen (1971) and the applicability to WEC modelling is described in Jefferys (1980).

3.5 Optimal Hydrodynamic Performance

Much of the original work on the extraction of power from waves drew heavily upon the expertise of practitioners in marine hydrodynamics and particularly upon those associated with mathematical modelling in the offshore environment. Newcomers to the field were able to draw upon a considerable resource of material that could be adapted readily to their own uses. It is not possible to attribute all of the many contributions made during this classical period of hydrodynamic device performance and more complete accounts are given in a major paper and review by Evans (1980, 1981a) and in the text of Falnes (2002), these being two of the most prominent contributors. The notation adopted here tends to follow that of Evans.

The instantaneous power associated with a general force $F(t)$, acting on a body in a single translational mode of motion described by the displacement $X(t)$, is the instantaneous rate of work given by

$$P(t) = F(t)\dot{X}(t) = F(t)U(t) . \tag{3.23}$$

The mean power absorbed per wave cycle P_M, or whatever time interval is specified, is

$$\mathcal{P}_M = \langle F(t)\dot{X}(t) \rangle = \langle F(t)U(t) \rangle , \tag{3.24}$$

where $\langle \ \rangle$ denotes the time average over the specified period and $U(t) = \dot{X}(t)$ has been introduced from Eq. (3.10), as it is sometimes more convenient to work with velocity components directly rather than with derivatives of the displacement.

Now assume the waves are monochromatic, linear and that the forces are within the diffraction regime. The mean power generated by the fluid (hydrodynamic) forces, using Eqs. (3.8) and (3.24), is

$$\mathcal{P}_{Mf} = \langle F_f(t)U(t) \rangle = \langle [F_S(t) + F_R(t) + F_H(t)]U(t) \rangle ,$$

with the individual forces given in Eqs. (3.11) – (3.16) and the velocity by Eq. (3.10). It is clear that the averaging will produce non-zero contributions only from those force components in phase with the velocity $U(t)$ and so neither the added mass nor the hydrostatic force will contribute.

This concept is readily extended to include all possible modes of motion in a manner similar to which the equation of motion in a single mode Eq. (3.17) was extended to Eq. (3.21) for all modes. It can be shown that the mean hydrodynamic power generated by a body in general motion, i.e. all six modes of motion, is

$$\mathcal{P}_{Mf} = \frac{1}{2}Re\{\mathbf{X}^*\mathbf{U}\} - \frac{1}{2}\mathbf{U}^*\mathbf{B}\mathbf{U} , \tag{3.25}$$

where the superscript * denotes the complex conjugate transpose when applied to square matrices or column vectors. The expression can also be written as

$$\mathcal{P}_{Mf} = \frac{1}{8}\mathbf{X}^*\mathbf{B}^{-1}\mathbf{X} - \frac{1}{2}\left(\mathbf{U} - \frac{1}{2}\mathbf{B}^{-1}\mathbf{X}\right)^*\mathbf{B}\left(\mathbf{U} - \frac{1}{2}\mathbf{B}^{-1}\mathbf{X}\right) \tag{3.26}$$

under the assumption that $\mathbf{B}^{-1}$ exists and which holds for most circumstances of interest. Both terms in this equation are always positive; the first is fixed for a particular geometry whereas the second is not, since $\mathbf{U}$ can be controlled.

Thus the maximum value of $\mathcal{P}_{Mf}$ is given by the first term and occurs when the second takes its minimum value, which is zero, to give

$$\mathcal{P}_{opt} = \frac{1}{8}\mathbf{X}^*\mathbf{B}^{-1}\mathbf{X} \tag{3.27}$$

and this occurs when

$$U = -i\omega\xi = \frac{1}{2}B^{-1}X \quad \Rightarrow \quad \xi = \frac{i}{2\omega}B^{-1}X . \tag{3.28}$$

Thus P_{opt} depends upon both the body geometry (via B) and the interaction between the incident waves and the body (via X). In addition Eq. (3.28) identifies clearly the requisite body motion in both amplitude and phase for maximum power absorption. As B is real by construction, the required phase of the velocity is that of the exciting force. The notation P_{opt} is used to denote the maximum rather than P_{max} as the quantity represents the optimal value, i.e. the best that can be achieved and not diminished by restrictions introduced by the power take-off system or other physical constraints.

The capture width $L(\omega,\beta)$ was introduced previously as a suitable measure of power absorption and given in Eq. (3.5). Newman (1976) has shown the rather surprising result that it is possible to relate the damping matrix B to the exciting force vector X, so that Eqs. (3.5)and (3.27) can be combined to give

$$\mathcal{L}_{opt}(\omega,\beta) = \frac{\lambda}{2\pi}X^*W^{-1}X \quad (W)_{mn} = \frac{1}{2\pi}\int_0^{2\pi} X_m(\theta)\overline{X_n(\theta)}d\theta, \tag{3.29}$$

where the overbar denotes the complex conjugate. Thus the optimal capture width and the body motion required to achieve maximum capture depend upon the interaction between the incident waves and the body when held in a fixed position. This is a generic and rather remarkable result!

The form of Eq. (3.29) becomes much simpler when the device is restricted to a single mode of motion and thus Eqs. (3.27) and (3.28) become

$$P_{opt} = \frac{|X(\beta)|^2}{8B} \quad \text{when } U = \frac{X}{2B} . \tag{3.30}$$

Newman (1976) has also shown that the exciting force X in this case can be related to the angular dependence $A(\beta)$ of the radiated wave field. Thus Eq. (3.29) may be replaced by either of the following alternative versions,

$$\mathcal{L}_{opt}(\omega,\beta) = \lambda\frac{|X(\beta)|^2}{\int_0^{2\pi}|X(\theta)|^2 d\theta} = \lambda\frac{|A(\beta)|^2}{\int_0^{2\pi}|A(\theta)|^2 d\theta} \tag{3.31}$$

and the choice of application depends upon which of $X(\theta)$ and $A(\theta)$ is easier to determine.

3.5.1 A Single Axisymmetric Device

The exciting force X and angular variation A will be independent of the incident direction θ. For an axisymmetric device operating in heave Eq. (3.31) reduces to the simple form

$$\mathcal{L}_{opt}(\omega, \beta) = \frac{\lambda}{2\pi} \qquad (3.32)$$

for a heaving buoy. This benchmark result was derived independently, and in slightly differing ways, by a number of contributors including Budal and Falnes (1975), Evans (1976), Mei (1976) and Newman (1976).

Writing the body motion as $X(t) = Re\{a\mathbb{D}e^{-i\omega t}\}$ enables $\mathbb{D}$ to be determined via Eqs. (3.9) and (3.28) as

$$\mathbb{D} = \frac{i}{2a\omega}\frac{X}{B} = \frac{4iP_w}{ak\omega\overline{X}} \qquad (3.33)$$

and $|\mathbb{D}|$ is known is the *Displacement Amplitude* or *Amplitude Ratio*, representing the ratio of the body displacement to the wave amplitude at optimal power take-off. The linear theory and diffraction regime permit this to be at most of O(1) but this measure may not provide substantial information on the physical displacement and the acceptability of the predicted measure must be determined from practical considerations. If the amplitude displacement induces excessively large motions in the power take-off mechanism, it may violate a physical constraint upon the possible motion and this is sometimes referred to as the end-stop problem. A ratio based upon the incident wave amplitude, such as $|\mathbb{D}|$, is a useful quantity from a mathematical modelling perspective but the physical value determining the end-stop will depend upon the device and not the non-dimensional amplitude.

The importance of Eqs. (3.32) and (3.33) is clear: the maximum power that an axisymmetric buoy can extract in heave is given by $\lambda/2\pi$ irrespective of the scale of the device. However, as a general rule, the magnitude of X will increase with body diameter and a larger body will possess a smaller displacement amplitude. Thus decreasingly smaller buoys will need to perform increasingly large oscillations and this may violate physical and modelling constraints. An assessment can be made of the applicability of the diffraction theory in such cases, using the criterion that diffraction effects become important whenever the non-dimensional parameter $kD > 1.3$ ($D/\lambda > 0.2$) and where D can be taken to be the diameter of the device. As D becomes smaller the point absorber approximation of determining X from X_{inc} alone, as discussed in Eq. (3.12), becomes more valid. However, the accuracy of the model may be reduced in such cases, since the decrease in diffractive effects is accompanied by an increase in the importance of viscosity.

There is also a further consideration. If the non-dimensional optimal capture width $\hat{\mathcal{L}}_{opt}$ is defined to be the ratio of the capture width to the width of the device, then from Eq. (3.33),

$$\hat{\mathcal{L}}_{opt}\left(\omega,\beta\right) = \frac{\lambda}{2\pi D} = \frac{\lambda}{4\pi R},\qquad(3.34)$$

where D and R are the diameter and radius of the device, respectively. This may be regarded as a measure of the structural efficiency, since it increases with decreasing radius and shows, with Eq. (3.32), that a balance must be sought between acceptable cost and realistic displacement amplitudes.

For a horizontal motion, Newman (1962) has shown that $A(\theta)$ is proportional to $\cos\theta$ and so Eq. (3.31) provides the analogous result of

$$\mathcal{L}_{opt}\left(\omega,\beta\right) = \frac{\lambda}{\pi}\cos^2\beta.\qquad(3.35)$$

This is maximum when $\beta=0$ or π, giving an optimal capture width of λ/π when the body motion is in alignment with the waves. Thus the maximum power absorption for a horizontal mode is twice that of a heave mode. The heave and surge point absorber results in Eqs. (3.32) and (35) are important results but their significance does not just lie with point absorbers, they provide a benchmark for the absorption properties of any device being compared with a point absorber.

3.5.2 Constrained Motion

A drawback of the optimal result in Eq. (3.28) is that it may only be achievable if large displacement amplitudes, such as that given by Eq. (3.32) for a single mode, are permissible. This will not usually be the case and the end-stop problem will impose a physical constraint in addition to considerations associated with ensuring the validity of modelling within the diffraction regime. For a body undertaking a single mode of motion, Evans (1981b) imposed a constraint equivalent to one upon the non-dimensional displacement amplitude of the form

$$\left|\mathbb{D}\right| \le \varepsilon.\qquad(3.36)$$

The corresponding maximum power absorbed $\mathcal{P}_{opt}^c$, obtained by constrained optimisation, can be written as $\mathcal{P}_{opt}^c(\beta) = \frac{\left|X(\beta)\right|^2}{8B}\left\{1-(1-\delta)^2 H(1-\delta)\right\}$, where $\delta = \varepsilon/\left|\mathbb{D}\right|$ is the ratio of the maximum permissible amplitude imposed by the constraint and the displacement at optimal performance. The function $H(x)$ is the Heaviside step function, defined to take the values of zero and one when the argument is negative or positive respectively. Thus the constraint becomes active whenever $\delta<1$ and there is an accompanying reduction in the maximum power.

As the optimal power P_{opt} is given in Eq. (3.30), define the non-dimensional constrained power absorption ratio by

$$\hat{\mathcal{P}}^c_{opt} = \frac{\mathcal{P}^c_{opt}(\beta)}{\mathcal{P}_{opt}(\beta)} = 1-(1-\delta)^2 H(1-\delta) \qquad (3.37)$$

and this shows the influence of the constraint explicitly. The relationship holds universally for a single mode of constrained motion and is shown in Fig. 3.5.

If $\hat{\mathcal{P}}^c_{opt}$ or the constrained capture width $\mathcal{P}^c_{opt}(\beta)$ is sought for a particular device, it can be obtained from Eq. (3.37) but evaluation may be a far from trivial task. The constraint parameter δ depends upon the exciting force and the damping, as does the optimal power P_{opt}; these two quantities need to be calculated for particular devices and evaluation may be analytic or numerical, more often the latter. Evans presents the constrained curves for the point absorber in Fig. 3.4 and also for other representative devices of generic interest.

The original constraint derived by Evans is more general than the one presented above for a single body in a single mode of motion and is a global constraint derived for a number of bodies oscillating independently, each capable of absorbing energy from the incident wave field. It is imposed via the sum of the squares of the individual velocities. Pizer (1993) has extended this approach to a global weighted constraint for a single body moving in more than one independent mode, which permits an interesting assessment to be made of the potential power absorption in multi-mode power take-off. However, it must be noted that the analytical

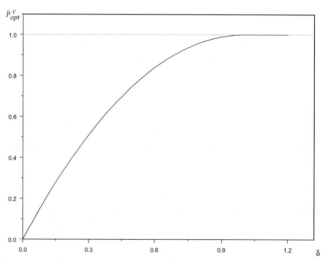

Fig. 3.5. Variation of the non-dimensional capture width $\hat{\mathcal{P}}^c_{opt}$ with constraint parameter δ

approach cannot be employed to enforce individual constraints, either on a particular body or mode of motion, and these need to be imposed numerically.

3.5.3 Arrays of Devices

Many devices are anticipated as operating in an array and the behaviour of the array members may be controlled for best individual or best array performance. Indeed when a new device is proposed, initial publicity usually contains an "artist's impression" of an array (or "farm") of the devices operating in long-crested seas. The array typically has many rows and while this may appear impressive initially, an analysis of array performance will most likely tell another story.

The fundamental modelling on arrays of wave energy devices was presented independently by Evans (1979) and Falnes (1980). Most work has assumed that the devices will be axisymmetric and this limitation will be enforced here. For a system of N bodies constrained to operate in just the heave mode of motion, the maximum power absorption is given by a form of Eq. (3.26), $\mathcal{P}_{opt} = \frac{1}{8} \boldsymbol{X}^* \boldsymbol{B}^{-1} \boldsymbol{X}$.

The complex exciting force $\boldsymbol{X}$ and damping $\boldsymbol{B}$ now describe the interaction between the N members of the array and are an N-dimensional column vector and a square $N \times N$ matrix respectively. The corresponding capture width can be written as

$$\mathcal{L}_{opt}(\omega, \beta) = \frac{\lambda}{2\pi} N q(\omega, \beta). \tag{3.38}$$

It is known from Eq. (3.32) that $\lambda/2\pi$ is the capture width of a single device in heave and hence q represents the ratio of the power absorbed by the array relative to that which would be absorbed by the N array members acting independently in heave and in isolation. Thus q is dependent upon the parameters that define the formation of the array and the geometry of its members, the frequency of the incident wave field and the orientation of the array relative to the incoming wave field. The value $q = 1$ indicates no net influence of the array formation. *Constructive Interference* occurs when the total power output of the array exceeds that of N individual devices and *Destructive Interference* occurs when there is a net loss of generated power; these correspond to $q > 1$ and $q < 1$ respectively. For an array of bodies of given dimension the aim is to obtain $q \geq 1$ by an appropriate choice of power take-off and spacing parameters.

The concept has clear analogies with the theory of radio antennae and the prediction of the interactions, via the scattered wave fields, requires the determination of $\boldsymbol{X}$ and $\boldsymbol{B}$. This is a difficult task unless some simplifying approximations are made. One of the initial investigations was by Thomas and Evans (1981), who studied single and double row arrays, with each row composed of five equally-spaced members; other unreported studies by the same authors included two, three and ten members in the array. Two body geometries were employed: the semi-submerged sphere of Fig. 3.4 and a thin ship, chosen as it possessed strong geometrical bias. Figure 3.6 below is taken from Thomas and Evans (1981) and shows the variation of q-factor with the non-dimensional measure $kd = 2\pi d/\lambda$, d being the spacing between members, for a row of five semi-submerged spheres for head, beam and $\pi/4$ incident seas. It is presented here for illustrative rather than definitive purposes, since array analysis contains many parameter combinations and conclusions cannot be drawn readily from a single figure.

One of the key features of this figure is the confirmation that regions of both constructive and destructive interference exist and that such regions cannot be avoided for the parametrical settings of the study. The performance is generally better in beam seas, though not always, consistent with the intuitive concept of greater frontage to the waves providing a greater opportunity for absorption. Although not shown here, the device members did not work equally hard, although there was symmetry about the central member in all cases corresponding to no sense of attenuation. Global and individual constraints were imposed numerically, as an extension of Eqs. (3.37) and (3.38); it was found that limiting the displacements to two or three times the incident wave amplitude was not severely restric-

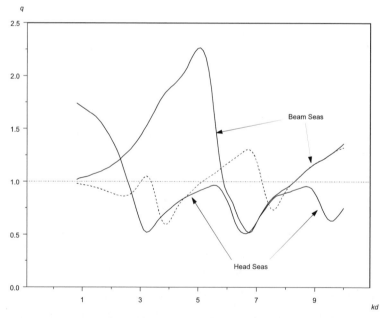

Fig. 3.6. Variation of the q-factor with the non-dimensional spacing parameter kd, for a uniformly-spaced linear array of five semi-submerged spheres and incident waves corresponding to beam and head seas (solid lines) and at $\pi/4$ (broken line). The dashed line $q = 1$ is equivalent to no interaction between array members

tive. The constructive interference decreased in magnitude but the regions of destructive interference were not significantly affected.

Later studies by McIver (1994), Mavrakos and McIver (1997) and Justino and Clément (2003) have also concentrated upon one or more linear arrays of small devices of the point absorber type, taut-moored and operating in heave. Variants on the earlier work include the influence of unequal spacing, constrained motions, full diffraction modelling and simultaneous consideration of all three translational modes of motion. However, the general features of array performance described by Thomas and Evans (1981) remain unchanged: regions of constructive and destructive interference occur generally and gains will be accompanied by losses. This remains a topic for much further work and perhaps justly so, as the alternative classification strategy of Section 3.2.3 places arrays of small devices in the third generation category.

3.5.4 Elongated Bodies

The general theory presented in the preceding sections, with the exception of the above discussion on arrays, is targeted at a single body capable of extracting power from the six independent modes of motion. However, the generic attenuator shown in Fig. 3.1 was conceived originally as being capable of extracting power as the wave moved along its length; it is also associated with low mooring forces, in contrast to those for a device of terminator type. The power take-off mechanism was not specified in the general case, although this will affect the chosen mode. Possibilities include articulated sections and a constant internal volume employing a pneumatic principle. Such devices are intended to work only in head seas and this restriction is enforced here by considering the case $\beta = 0$ only.

The seminal work is by Newman (1979), who considered a compliant body undertaking vertical displacements along its length. As a generalisation of the solid body modes of Eq. (3.9), the compliant motion is represented as a combination of possible independent modes $z = Re\{\zeta(x)e^{-i\omega t}\}, \zeta(x) = a\sum_j v_j \mathbb{Z}_j(x)$, with $\mathbb{Z}_j(x)$

denoting the j-th body mode, which is to be specified, and v_j denotes the unknown complex amplitude associated with the mode. By consideration of individual modes it was shown that the maximum capture width associated with the general j-th mode is given by

$$\mathcal{L}_{opt,j}(\omega) = \lambda \frac{\left|H_j(\beta)\right|^2}{\int_0^{2\pi}\left|H_j(\theta)\right|^2 d\theta} \tag{3.39}$$

where $H_j(\theta)$ denotes the Kochin function for the j-th mode and which is related to the radiated wave field function $A(\theta)$ in Eq. (3.31). Both Eqs. (3.32) and (3.35) can be deduced from Eq. (3.39) using similar arguments as before.

Slender body theory is valid in head seas when the beam and draft of the body are much less than its length. With this restriction placed upon the body geometry, Newman considers polynomial, trigonometric and piecewise linear modes and reaches some far-reaching conclusions, of which the two most prominent concern a hinged device and the optimal body length. For a hinged device, it is deemed sufficient from a practical viewpoint to have a single hinge and this should be placed away from the ends. If the most energetic waves are in the wavelength range of 100–200 m, then the optimal length of a compliant device appears to be of the order of the target incident wavelength.

Evans and Thomas (1981) considered a different attenuating device, the Lancaster Flexible Bag. This was described by Chaplin and French (1980) and operates by permitting a variable width along its length and subject to a global volume constraint. In addition, the device possesses small beam and draft relative to its length and is constructed to behave symmetrically about its vertical length-wise centre-plane. In contrast to Eq. (3.39), which described vertical motions, a width function $w(x) = Re\{\zeta(x)e^{-i\omega t}\}$ may be employed to describe the body motion. By considering $w(x)$ to be composed of a number of piecewise constant boxes and employing the Froude-Krylov approximation, a model was constructed very similar to the array structures of the previous section and included the capability of applying both global and (unpublished) individual constraints. Greater accuracy was achieved by increasing the number of boxes and a reasonable estimate of the maximum capture length would be half the body length. Local constraints can reduce this value and there is no attenuation along the bag, so that the rear portion must work as hard as the front portion.

The other geometry of interest in this category is the Pelamis, which is classified as an attenuator and belongs to the third generation category; the device team appears to prefer the descriptor *Line Absorber*. In its present form its dimensions are 140 m in length with a diameter of 3.5 m, so the slender body approximation is valid from a modelling perspective. The device body possesses four segments, with the three interlinking flexible joints containing the power take-off system. Those advantages attributed to elongated devices, such as low mooring and good survivability, have been confirmed by an extensive modelling and testing programme. However, there is one major difference between the Pelamis and the generic structures discussed above. This concerns the mode of motion: the Pelamis permits horizontal hinging about its joints, so that there need not be a vertical centreplane of symmetry along the length of the device when it is operation. It is suggested that the capture width of the device, extracting power in more than one mode of motion, is in the vicinity of $\lambda/2$ for typical sea conditions. Recent work on the device is contained in Pizer et al. (2005) and earlier work may be tracked from the references therein.

3.6 Control and Design

The results presented in the previous section are optimal results from a hydrodynamic analysis and there is no guarantee or expectation that such levels of power absorption can be achieved in practice. This is not to decry their usefulness and they provide both upper bounds and a filtering process that can be used to good effect. What is missing is this approach is a link between the final output and the input to the process. To establish this link it is useful to concentrate upon the processes of how to design a device and control the output from a hydrodynamic modelling perspective.

The earliest control strategies were devised to ensure that point absorbers operating in regular seas could maximise the converted power. This process required the phase and amplitude of the oscillation to be chosen to ensure optimal performance. However, the phase and the amplitude can be varied independently and so the two distinct aspects became known as *Phase Control* and *Amplitude Control*. More recently the terminology has been simplified to *Control*, as the field of application has widened and includes sub-optimal strategies. *Optimal Control* is sometimes used to refer to those cases where the aim is to convert the optimal amount of power. The maximum power absorption based upon hydrodynamic considerations alone corresponds to optimal control, as it determines both the phase and amplitude of the desired motion. This is seen clearly for a single body in multi-mode operation from Eqs. (3.27) and (3.28) and for a single mode from Eq. (3.30), where the optimal result is obtained by forcing both the amplitude and phase to take particular values.

It is clear from the equations of motion in Eqs. (3.17) and (3.22) for regular or irregular wave motion that any device can be considered as a mechanical system with two sets of parameters. For convenience these parameters are termed *Geometrical Parameters* and *Control Parameters*, though there is no accepted usage of these terms. *Geometrical Parameters* define the structure of the device and cannot be changed once the device has been built; these are represented here by the vector G. *Control Parameters* provide the power take-off mechanism and are usually variable parameters, capable of being tuned to match the wave environment; these are represented by a vector J. The number of components in G and J will not usually be the same and will be dependent upon the structure of the device and complexity of the power take-off system. The purpose of *Design* is to determine the geometrical parameters G and some measure of the control parameters J. The purpose of *Control* is to make the device run efficiently, in some sense, perhaps employing a better measure of J once it has been built.

Although the division with G and J is convenient to employ, it is not definitive and there is an intermingling of parameters at the design stage. Attention is restricted initially to a device of stipulated geometry acting in a regular wave field, then design is considered from a hydrodynamic perspective and finally, more general aspects of control are discussed.

3.6.1 Tuning and Bandwidth

The shortcoming of the optimal control approach is that it does no attempt to include a model of the power take-off mechanism; it acknowledges that a power take-off mechanism is necessary but assumes that it is a black box that can always be adjusted to ensure maximum power absorption. Thus it makes no reference to the equation that governs the device motion or to the contribution made by the external force. It has been the practice thus far to present the mean power, and other related quantities, as functions of the angle of incidence parameter β; this dependency is understood in the present discussion but not explicitly stated.

By analogy with the expression for the radiation force $F_R(t)$ in Eqs. (3.14) and (3.15), write the external force $F_{ext}(t)$ as

$$F_{ext}(t) = Re\left\{-\overline{Z}_E(\omega)U\ e^{-i\omega t}\right\} \qquad Z_E(\omega) = B_E(\omega) + i\omega A_E(\omega), \qquad (3.40)$$

where $Z_E(\omega)$ may be described as the mechanical impedance and U is the usual complex velocity, i.e. $U = -i\omega\xi$; this single mode formulation is readily extended to a multi-mode motion. Consider the equation of motion in the form Eq. (3.20) and substitute for $F_{ext}(t)$ from Eq. (3.40), to give the complex velocity and displacement as

$$U = \frac{X}{\overline{Z}_T(\omega) + \overline{Z}_E(\omega)} = -i\omega\xi. \qquad (3.41)$$

The instantaneous rate of working of the external force is $F_{ext}(t)U(t)$ and the instantaneous output power is given by $P_{ext}(t) = -F_{ext}(t)U(t)$, with the minus sign providing the link between the rate of working and the power absorbed. Employing Eqs. (3.40) and (3.41), the mean output power becomes

$$\mathscr{P}_{ext} = -\langle F_{ext}(t)U(t)\rangle = \frac{1}{4}\left(Z_E + \overline{Z}_E\right)|U|^2 = \frac{|X|^2\left(Z_E + \overline{Z}_E\right)}{4\left|Z_T + Z_E\right|^2}, \qquad (3.42)$$

with Z_T regarded as fixed, the maximum value of this quantity occurs when

$$Z_E = \overline{Z}_T \qquad (3.43)$$

to give P_{max} as

$$\mathscr{P}_{max} = \max\{\mathscr{P}_{ext}\} = \frac{|X|^2}{8B} \qquad \text{when } U = \frac{X}{2B}. \qquad (3.44)$$

This is agreement with the optimal result Eq. (3.30) and the approach can be extended to a single body in more than one mode of motion and provides confidence in the optimal results derived earlier. The control defined by Eq. (3.40), with optimal value Eq. (3.43), is usually called *Complex Conjugate Control*. The structure of the control can be seen by forming the ratio of the two terms in Eqs. (3.42) and (3.44) and then employing (3.41) to give

$$\frac{\mathcal{P}_{ext}}{\mathcal{P}_{max}} = 2B\frac{\left(Z_E + \overline{Z}_E\right)}{\left|Z_T + \overline{Z}_E\right|^2} = \left\{1 - \frac{\left|Z_T - \overline{Z}_E\right|^2}{\left|Z_T + Z_E\right|^2}\right\}, \tag{3.45}$$

in which the importance of the relationship between Z_T and Z_E is clearly seen. It is clearly insufficient to restrict Z_E to contain only a damping term, corresponding to Z_E being real. Both real and imaginary parts of Z_E will be important but the degree of importance will depend upon the relative magnitudes of the real and imaginary parts of Z_T. An interesting study on the application of complex control methods to the Salter Duck has been conducted by Nebel (1992) and describes both the strengths and pitfalls of the method.

The optimal results in Section 3.5 show the best that can be achieved, whereas the output power is dependent upon the form of the power take-off. Thus the maximum Eq. (3.44) is dependent upon the form of F_{ext} in Eq. (3.40) and a different expression may not produce as much output power. In terms of the control vector J identified earlier, Eq. (3.40) is equivalent to the form

$$J = \left(Re\{Z_E\}, \mathrm{Im}\{Z_E\}\right) \tag{3.46}$$

and the process adopted is equivalent to the following optimisation problem: determine P_{max} from

$$\mathcal{P}_{max} = \max\left\{\mathcal{P}_{ext}\left(\omega; J\right)\right\} \tag{3.47}$$

for unknown J, with information from Eqs. (3.40), (3.42) and (3.46). Equivalently, and more usually, this would be formulated using the capture width $L(\omega; J)$ from Eq. (3.5), since this does not depend upon the incident wave amplitude.

Suppose that the system is tuned to provide the maximum mean output power at a chosen frequency ω_T, dependent upon the wave climate. This enables J to be determined, denote it by $J_0(\omega_T)$ and consider as fixed. The *Bandwidth* is then defined by the function $\mathcal{L}\left(\omega; J_0(\omega_T)\right)$ and this provides the variation of the capture width with ω for a system tuned to ω_T. It does not mean that the maximum value of L occurs at ω_T, since the dependency upon ω in the hydrodynamic coefficients may be dominant.

Examples of bandwidth curves are shown in Fig. 3.7 and are taken from a numerical study of the Bristol Cylinder by Thomas and Ó Gallachóir (1993). This device belongs to the *Third Generation* category and power is extracted in two modes of motion; in the present model the power take-off is affected via taut extensible cables. A full description of device, together with the modelling approach and an experimental study is given by Davis et al. (1981), with ample references contained therein. The calculations were performed for a device of $6\,m$ radius, operating in $42\,m$ of water and with a clearance above the cylinder of

$3\,m$. A two-dimensional model was employed and so the curves denote hydrodynamic efficiency.

Three curves are shown, indicating tuning for waves of period $8\,s$, $10\,s$ and $12\,s$, and the frequency measure is the non-dimensional quantity $v = \omega^2 h / g$. The maximum hydrodynamic efficiency curve is also shown and this is seen to be achieved at each tuning frequency. It is clear that the bandwidth narrows as the tuning period increases, with a corresponding increase in wavelength. This is because this device performs best and very efficiently when the particle paths are circular, as for deep-water waves, and becomes less efficient as the eccentricity of the elliptical paths increases. This is confirmed by the $8\,s$ bandwidth curve being the widest and the $12\,s$ curve being the narrowest. Thus the bandwidth variation parameter is an important consideration in this example and this is a general conclusion. The characteristics of the bandwidth are dependent mainly upon the mode of motion and geometry of the device and some devices may be broad-banded whereas others will be narrow-banded. However, as shown very clearly by this example, it is possible for a device to possess a broad bandwidth in some potion of the frequency spectrum and a narrow bandwidth in another part.

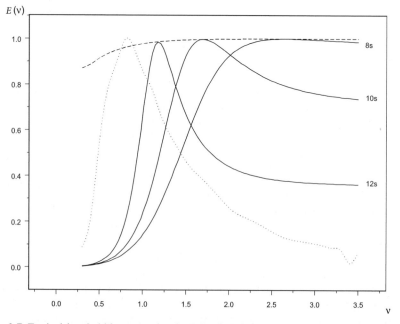

Fig. 3.7. Typical bandwidth curves for the Bristol Cylinder, showing the variation of hydrodynamic efficiency E with non-dimensional frequency v for three tuning frequencies (solid lines). The maximum efficiency curve (broken line) and the South Uist resource curve from Fig. 3.2 (dotted line) are also shown

3.6.2 Design

A normalised form of the South Uist spectrum presented in Fig. 3.2 is also shown as a dotted line in Fig. 3.7, with the normalisation achieved by dividing the power spectrum density by the maximum power level in the spectrum. It is included to show how the target resource interacts with the modelling approach and to help identify where the device must operate within a particular spectrum. Thus the performance of a device must in some way be targeted at a representative site spectrum or spectra; identifiable mechanisms to meet this requirement include tuning to an appropriate frequency or some methodology of influencing the device bandwidth. These are approaches based upon the idea of control discussed above but control does not provide the only mechanism since any flexibility in the geometrical structure of the device, summarized by the vector G, may also possess a potential influence. Including the geometrical parameters incorporates the concept of design, albeit only from a hydrodynamic perspective and this is an acknowledged limitation of the present approach. Device design is an evolutionary process and will change necessarily with gains in relevant knowledge and expertise, whether they are from improved mathematical models, novel laboratory experimental studies or the availability of improved construction materials. However, some convergence of ideas is necessary because not all devices can be built at demonstration or prototype stage and an assessment of design and performance characteristics will be required.

It is possible to make some preliminary comments about two particular hydrodynamic aspects of device design. The first is that it is generally accepted that WECs that are good absorbers of wave energy are also good wave generators, i.e. if the body is forced to move in its prescribed mode of motion, with the water initially at rest, then a uni-directional wavetrain will be generated. An accompanying constraint is that the amplitude of the forced body motion should be comparable to the amplitude of the generated wave for an efficient wave generator, consistent with the employment of modelling within the diffraction regime.

The second point is that most first and second generation wave energy devices have either been strongly site specific or site independent. Usually the distinction is dependent upon whether the device is shore mounted or a small offshore device. A typical shore-mounted, first generation device is an OWC that has been constructed to utilise a natural shore site such as a gully, whereas the general offshore second-generation device is anticipated as working well over a range of sea conditions and sites. The difference between the two cases is associated essentially with the bandwidth: site specific devices usually have a narrow bandwidth, whereas offshore devices are intended to have, but may not possess, a much broader bandwidth. This is seen in Fig. 3.2, which illustrates the difference in the magnitude of the maximum power level and the spectral width between the offshore and onshore resource at Pico. However, as potential offshore sites are identified by criteria established from wave resource assessment studies, then even seemingly site-independent WECs will require change to optimise the power generation. These changes will be dictated by factors such as body response amplitudes to dominant energy frequencies and constraints imposed by power take-off mechanisms. It is important to possess models that can respond to such demands.

Consider a device operating in a single sea state, characterised by the energy density $S_f(f)$ or the power density $P(f)$. The total power in the incident spectrum is $P_\infty = \int_0^\infty P(f) df = \rho g \int_0^\infty S_f(f) c_g(f) df$. In assessing the power absorbing capability of a particular device at a given site, there must be a recognition that only a portion of this power spectrum may be considered attainable and thus targeted. This introduces the targeted power in the spectrum PT defined by

$$P_T = P[f_1, f_2] = \rho g \int_{f_1}^{f_2} S_f(f) c_g(f) df . \qquad (3.48)$$

The physical frequency f (in Hz) may not be the best measure of frequency for modelling purposes, as shown in the text accompanying Fig. 3.7, and the most common alternative is the radian frequency ω. Other useful options are the non-dimensional frequency measures $v = \omega^2 h/g$ and $\omega^2 D/g$, with the former being particularly appropriate in finite depth or shallow-water applications and the latter in deep water; this being due to the appropriate choice of length-scale chosen for non-dimensionalisation in each case. To cover all possibilities, write P_T in the form

$$P_T = P[\mu_1, \mu_2] = \int_{\mu_1}^{\mu_2} P_\mu(\mu) d\mu \qquad P_\mu(\mu) = \rho g c_g S_\mu(\mu) \frac{d\mu}{df}, \qquad (3.49)$$

and which encompasses all such applications by allowing μ to be a generic measure of frequency with the physical measure corresponding to $\mu = f$. An example is given by comparison of the South Uist spectrum shown in Figures 3.2 and 3.7; Figure 3.7 shows the range $0.3 \leq v \leq 3.5$ and this corresponds to the range $0.0042 \leq f \leq 0.144$ in Fig. 3.2. The targeted part of the spectrum has a mean power level of $43.2\, kW/m$, in contrast to the value of $47.8\, kW/m$ quoted earlier for the full spectrum.

The simplest strategy to widen the bandwidth of a device with fixed stipulated geometry is to maximise the capture width; this is achieved by defining the control parameter vector J_1 by

$$\mathcal{L}_1(J_1) = Max\left\{\frac{1}{\mathcal{L}_T} \int_{\mu_1}^{\mu_2} \mathcal{L}(\mu; J(\mu)) d\mu\right\}, \qquad \mathcal{L}_T = \int_{\mu_1}^{\mu_2} \mathcal{L}_{max}(\mu) d\mu . \quad (3.50)$$

As the denominator L_T corresponds to the area under the broken line in Fig. 3.7, the maximum value of $\mathcal{L}_1(J)$ is unity and the resultant bandwidth curve from this measure is $\mathcal{L}(\mu, J_1)$. The optimisation process will generally require numerical procedures and the complexity will depend upon the number of control parameters contained within J. However, the definition of L_T is associated with a constant scaling factor and needs to be calculated only once. Such an approach corresponds to a de-tuning as the strategy for parameter choice attempts to move away from the influence of resonance.

The principal shortcoming of Eq. (3.50) is that although it addresses perceived deficiencies in bandwidth, it does not ensure that any improvement in bandwidth occurs in a particular part of the frequency range. This is illustrated in Fig. 3.7, where the dotted line shows a normalised representation of the resource and it is clear that any improvement in bandwidth should match the region where the resource is greatest. To address this deficiency, consider the measure

$$\mathcal{L}_2(J_2) = Max\left\{\frac{1}{P_T \mathcal{L}_T}\int_{\mu_1}^{\mu_2} P_\mu(\mu)\mathcal{L}(\mu;J(\mu))\,d\mu\right\}, \tag{3.51}$$

with the optimizing value J_2 and corresponding bandwidth function $\mathcal{L}(\mu,J_2)$. It is clear that the power density provides a weighting that moves the bandwidth towards the targeted regions of highest power. Note that the scaling factor $P_T\mathcal{L}_T$ is optional, as it does not contain variable parameters but can be useful from a practical numerical perspective.

Both Eqs. (3.50) and (3.51) describe a de-tuning approach to improve bandwidth for a stipulated body geometry but neither takes body parameters into account except in the calculation of the hydrodynamic coefficients. This deficiency can be overcome by including the body parameter vector G in the set of unknowns and thus Eq. (3.51) is extended to produce the final measure

$$\mathcal{L}_3(G_3,J_3) = Max\left\{\frac{1}{P_T}\int_{\mu_1}^{\mu_2} P_\mu(\mu)\mathcal{L}(\mu;G(\mu),J(\mu))\,d\mu\right\}, \tag{3.52}$$

with accompanying bandwidth $\mathcal{L}(\mu,G_3,J_3)$. A variant on this is

$$\hat{\mathcal{L}}_3(G_3,J_3) = Max\left\{\frac{1}{P_T}\int_{\mu_1}^{\mu_2} P_\mu(\mu)\hat{\mathcal{L}}(\mu;G(\mu),J(\mu))\,d\mu\right\}, \tag{3.53}$$

where $\hat{\mathcal{L}}$ is the non-dimensional capture width as in Eq. (3.34) and its importance identifies the presence of the physical width of the device in the capture process. The quantity L_T does not appear in Eqs. (3.52) or (53), since this will depend upon G; it may involve considerable effort in calculation and not provide the best strategy.

The measures in Eqs. (3.50) – (3.52) were first employed by Thomas and Ó Gallachóir (1993) in attempt to provide a design model for the Bristol Cylinder. Figure 3.8 shows the equivalent two-dimensional forms of the L_1, L_2 and L_3 measures for the Bristol Cylinder, for which a measure of fundamental performance is provided by the bandwidth curves in Fig. 3.7. The dotted line shows the power density of the resource in Figures 3.2 and 3.7; this may be considered to be the target. As expected, an increasing improvement of fit is provided by L_1, L_2 and L_3 respectively, with the same fixed geometry of Fig. 3.7 being employed in L_1 and L_2.

The improvement obtained by targeting the resource is clear and illustrates an important aspect of design, albeit from a hydrodynamic perspective. Newman (1979) also employed a particular form of Eq. (3.52) in his paper on the optimal performance of elongated bodies discussed in Section 3.5.4; the result that the body length should be equivalent to the dominant wavelength of the incident waves is based upon a selected set of calculations equivalent to a less rigorous optimisation process than that of Eq. (3.52).

A key requirement in these approaches is that an accurate hydrodynamic model is available to determine the capture width $\mathcal{L}(v, G, J)$. It is thus necessary to be able to determine the hydrodynamic coefficients to an acceptable degree of accuracy for the numerical optimisation process to be feasible. Such an approach also requires the numerical evaluation to derivatives and corresponds usually to a requirement of at least one order of magnitude of accuracy better than if the coefficients were employed for the equation of motion alone. Such considerations place stringent conditions upon the numerical methods described in Chapter 5. In addition, when performing a numerical investigation to determine optimal values of G and J, there is a need to impose constraints representative of the permissible physical parameter range. These strategies must also be implemented numerically but this does have one advantage: it is then relatively easy to bound any of the design parameters into specified ranges.

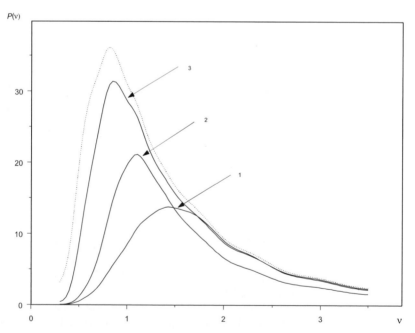

Fig. 3.8. The variation of captured power density $P(v)$ with non-dimensional frequency v for the Bristol Cylinder following implementation of optimisation strategies Eqs. (3.50) – (3.53) (solid lines) with the target South Uist resource (dotted line)

The approach described above targets the device to a particular power spectrum but is not complete for design. There may be many varying sea states that need to be considered and any optimisation process should address this issue. It may well provide an important filtering process in identifying whether or not the device can perform across a variety of sea-states.

3.6.3 Control Strategies

Suppose that the device has been constructed and installed, subject to whatever set of criteria which have been utilised. The question now arises: how should it be controlled to operate most efficiently? It is salient to make some preliminary remarks before attempting to address this difficult question.

The methodologies described in the previous two subsections are based upon an approach embedded in the frequency domain. This relies upon a representation of the typical spectra in Fig. 3.2 in terms of the series of harmonic functions in Eqs. (3.3) and (3.4), together with all accompanying ensuing approximations and caveats. However, the typical physical wave field is shown in Fig. 3.3 and the wave record suggests that frequency domain analysis, via Eq. (3.3), may not provide an acceptable method of controlling the device when relatively large individual waves occur in what may be considered to be at most a moderate sea. Another important consideration is the design and operation of the power take-off mechanism, which may not operate in a harmonic manner and require an appropriate description as a function of time that cannot be incorporated into Eq. (3.17).

These issues may seem to reduce to a comparison between the single-frequency equation of motion in Eq. (3.17) and its time-dependent form Eq. (3.22), or perhaps to a more general comparison of the relative strengths of frequency and time domain modelling. The time-dependent form will provide flexibility in the operation of the power take-off but it remains an equation that is valid only within the confines of linear wave theory and employs coefficients that are obtained using frequency domain modelling. In practice, the relative merits are more concerned with intent: frequency domain models are very useful in determining fundamental device properties and in design studies but control can only be enforced with a time-domain model.

The discussion in Section 3.6.1 considers those aspects of control associated with regular waves and a harmonic power take-off system. Attention is now directed towards a particular aspect of hydrodynamic control, introduced by Budal and Falnes (1980) for point absorbers and known as *Latching*. Consider the heaving buoy shown in Fig. 3.4, with horizontal dimensions sufficiently small to be considered as a point absorber and tuned optimally for maximum power in a regular sea. The tuning condition, capture width and displacement amplitude are given by Eqs. (3.30), (3.32) and (3.33) respectively. For a small device, the Froude-Krylov approximation is valid and the exciting force X may be obtained by integrating the pressure due to incident wave over the device. This gives a force that is in phase with the incident waves and with magnitude proportional to the radius of the device. The general tuning condition demands that the body velocity is in phase with the exciting force, which means that the body displacement lags the surface wave elevation by $\pi/2$. As the radius decreases, the displacement amplitude will increase to a value that cannot be achieved without breaking modelling or physical constraints. In addition, the constrained optimisation approach leading to Eq. (3.37) requires an increase in damping that may not be desirable or achievable within the power take-off system.

Latching was conceived to recognise and address these concerns while retaining the small-volume attractiveness of point absorbers. It is easiest described schematically and Fig. 3.9 is very similar to the original figure presented by Budal and Falnes.

The two curves in Fig. 3.9 show the displacement of the device (solid curve) and a representative incident wave motion (dotted line); this is only intended as a schematic and the exact phase relationship between the two motions is given by the optimal criterion in Eq. (3.30). This figure demonstrates a generic latching strategy that immediately halts the motion of the device at some point in the cycle by an appropriate controlling force $F_{ext}(t)$. As the instantaneous power is $F_{ext}(t)U(t)$, the device cannot generate power when it is fixed and will not do so until released. This occurs when it is considered appropriate to generate power again. This approach will avoid very large displacement amplitudes but the resulting body motion is no longer harmonic and so the time dependent equation Eq. (3.22) must be used to describe the motion. Greenhow and White (1997) employed this approach for a heaving point absorber device in regular waves to confirm the original concepts.

The latching control strategy is sub-optimal and the method of implementation, such as the location of fix-and-release points, belongs to the user. Although the method is easiest explained for a regular wavetrain, it is essentially an output control upon the time-dependent form of the equation of motion. It is thus applicable

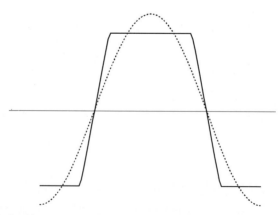

Fig. 3.9. Generic latching strategy, showing desired displacement (heavy solid line) against sinusoidal time dependency (broken line)

in any incident sea and it is straightforward to adapt the concept illustrated in Fig. 3.9 to the time series in Fig. 3.3. However, its use depends upon an accurate calculation of the hydrodynamic coefficients and an efficient and accurate algorithm to evaluate the convolution integral. Possible strategies for heaving point absorbers in irregular seas are considered by Barbarit, Duclos and Clément (2004) and for motions of a more general device by Barbarit and Clément (2006). The second paper discusses evaluation of the convolution integral and on extending such techniques to more general applications. A general review on control strategies and their applications is given by Falnes (2001).

3.7 The Oscillating Water Column

OWCs represent the class of devices that have received the greatest collective development, with perhaps the two best known sites being on the islands of Islay off the west coast of Scotland and Pico in the Azores. Floating or sea-bed devices tend to have been developed on an individual basis with research effort targeted at particular needs, whereas OWCs operate using similar technologies and possess similar requirements. In addition the great majority of OWCs is shore-mounted and this holds for devices either installed or in the prototype and development stages and whether of stand-alone design or integrated into a breakwater. Attention is directed towards the shore-mounted devices. Note that the two mentioned examples (LIMPET in Islay and the Pico plant) are described in Chapter 7. A brief description of the principles is presented in this section.

A generic shore-mounted OWC is shown in cross-section in Fig. 3.10. Incident waves induce a time-varying pressure field in the air chamber inside the device and power is extracted via a turbine and generator. The turbine is driven by the varying differential pressure field across it and includes both inhaling and exhaling phases in the wave cycle. A self-rectifying turbine, which rotates in the same direction independent of the direction of flow, is usually employed. Before considering the available modelling techniques for power extraction, it is appropriate to discuss the key features of the device shown in Fig. 3.10 from a hydrodynamic and aerodynamic perspective.

The shape and thickness of the front wall must ensure that the device survives severe weather conditions, particularly those forces associated with wave slamming. Water flowing past the lip may induce vortex shedding and hence rather turbulent flow within the vicinity of the front wall, though this may be mitigated by a circular lip. As the device is shore-mounted, the incident wave-field is likely to be at best weakly nonlinear and the same remark will apply to the pressure field within the chamber. The forced air-flow within the chamber is unlikely to be smooth or laminar and there is the possibility of water droplets suspended within the air. It is clear that although the global operating principle of an OWC is simple, in the sense that it is straightforward to describe, the construction of a good model to described the working principle of an OWC is likely to be a far from trivial task.

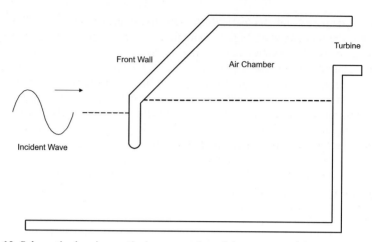

Fig. 3.10. Schematic showing vertical cross-section of shore-mounted OWC

Modelling the hydrodynamic behaviour alone, for the phenomena identified above, is difficult and requires advance numerical techniques. The influence of nonlinearity is considered in the absence of viscosity by Mingham et al. (2003), viscosity and turbulence are included via a k-ε model in a study by Alves and Sarmento (2006). Both of these papers describe two-dimensional studies and this approach is common in shore-mounted OWC modelling, in recognition of the considerable increase in complexity in moving to three-dimensions but also acknowledging that two-dimensional models can provide an important input in the appropriate circumstances. One point to note is that the LIMPET device on the Scottish island of Islay, described in Chapter 7, does not conform to the schematic in Fig. 3.10 and slopes uniformly. For the sake of convenience, Fig. 3.10 is taken to provide the generic form.

A major difficulty is that experimental studies on OWCs are not easy to perform: the hydrodynamic and pneumatic flows require different model scales and the influence of vortex shedding and viscous effects is difficult to infer from small-scale experiments. This makes mathematical and numerical modelling a particularly valuable tool in the development of OWCs. For convenience, the existing models are separated again into those appropriate for the frequency and time domains.

3.7.1 Frequency Domain Modelling

In spite of the reservations identified above concerning the applicability of linear wave theory, most of the modelling of OWCs involving power take-off has assumed that linear wave theory is valid. This should not be interpreted as a severe deficiency in preliminary modelling; the device must perform most efficiently in small waves and there are many phenomena that cannot be described adequately without an increase in the number of parameters.

The schematic shown in Fig. 3.11 is usually employed for two-dimensional studies and this can be compared readily with the more realistic configuration in Fig. 3.10. There are two main differences, allowing for the turbine to be placed in the roof or in the rear wall. In practice the front wall must possess a considerable thickness, associated with the survivability of the device and with vortex shedding and turbulence, whereas the idealised model has a wall of negligible thickness. The second difference concerns the sloped portion of the upper wall in the installed device and the idealised model retains a vertical wall at the front of the device. These differences are not easy to assess; the practical wall thickness will probably diminish the power incident to the device and the sloped wall may improve the internal air movement. In a preliminary model, both approximations are deemed acceptable.

The fundamental model is due to Evans (1982), who introduced the concept of an oscillating pressure patch on the water surface and equivalent to the interior water surface of an OWC. The original work was for an array of OWCs rather than just an individual device but the limitation to a single device is enforced here. Water wave theory assumes that the pressure on the air-water interface is equal to the air pressure p_a and this is taken usually to be constant. With reference to Fig. 3.11, air outside the chamber will be at atmospheric pressure p_a whereas the surface pressure inside the chamber will take a different value and will change due to the interior conditions in the chamber. If $p_c(t)$ is the air pressure in the chamber, then for harmonic time dependency write

$$p_c(t) = p_a + p(t), \quad p(t) = Re\{\mathbb{P}e^{-i\omega t}\}, \qquad (3.54)$$

where $p(t)$ is the difference between the chamber pressure and the external atmospheric pressure. The rate of change of the volume inside the chamber is due the

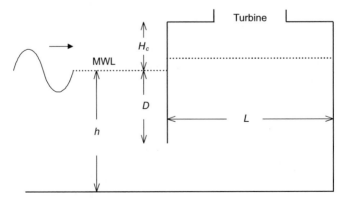

Fig. 3.11. Schematic showing model approximation to Fig. 3.10

change in surface level inside the chamber and can be represented as a volume flux $Q(t)$, measured positive in an upwards direction. In linear theory it can be written as

$$Q(t) = Q_S(t) + Q_R(t) = Re\{(Q_S + Q_R)e^{-i\omega t}\}, \tag{3.55}$$

associated with the scattered and radiated wave fields respectively and directly analogous to the decomposition in Eq. (3.8) for the forces on a floating body. The analogy is extended by writing $Q_R(t)$ as

$$Q_R(t) = -\{A(\omega)\dot{p}(t) + B(\omega)p(t)\}, \tag{3.56}$$

where A and B behave very much like the added-mass and damping coefficients encountered previously and so the same notation is employed. Using the representation

$$Q_R = -\overline{Z}_R(\omega)P, \quad Z_R(\omega) = B(\omega) + i\omega A(\omega) \tag{3.57}$$

identifies a further analogy with the floating body analysis, this time with Eq. (3.15). As with the floating body, calculation of A and B will depend upon the geometry of the OWC and numerical determination is usually required. At this point it is possible to consider the mean power generated via the optimal approach of Section 3.5; the optimal result for P_{opt} is directly analogous to the floating body result of Eq. (3.30), with U replaced by Q_S.

The control approach of Section 3.6 requires the power take-off mechanism to be stipulated. Evans assumed that the pressure across the turbine is related to the volume flux of the air inside the chamber and that the two quantities are related by a complex constant of proportionality α. i.e.

$$Q_S + Q_R = \alpha P, \tag{3.58}$$

The mean rate of working of the turbine is $P_{turb} = \langle p_c(t)Q(t)\rangle$; utilising the forms in Eqs. (3.54) – (3.58) enables this expression to be written as

$$P_{turb} = \frac{|Q_S|^2}{8B}\left\{1 - \frac{|Z_R - \overline{\alpha}|^2}{|Z_R + \alpha|^2}\right\}. \tag{3.59}$$

The conjugate control methods of Section 3.6.1 may now be implemented and the maximum output power P_{max} occurs when $\alpha = \overline{Z}_R$ giving

$$P_{max} = \max\{P_{turb}\} = \frac{|Q_S|^2}{8B} \quad \text{when} \quad P = \frac{Q_S}{2B}. \tag{3.60}$$

This is in agreement with the optimal result in Eq. (3.30) for the floating body and the results in Eqs. (3.44) and (3.45) for the maximum power generated by the external force. The importance of the damping B and the diffraction flow Q_S are clear, thus it is essential to be able to determine these quantities accurately and the necessity for efficient and accurate numerical methods is identified again. General numerical methods of industrial standard and employing the boundary element technique would be required to determine the volume fluxes and capture width for a three-dimensional model of the shore-mounted OWC. Applications are given by Lee and Newman (1996) and Brito-Melo et al. (1999), with the latter paper concentrating upon the requisite diffraction flow calculations referenced to the Pico plant.

Attention is now directed at the two-dimensional application in Fig. 3.11. This requires a finite depth model rather than an infinite depth one and the extension from the original work of Evans was provided by Smith (1983). It is often assumed that the oscillating pressure and volume flux will be in phase; thus the constant of proportionality α is real and the complex conjugate control leading from Eqs. (3.59) to (3.60) cannot be enforced, with a corresponding diminution in output power. The resulting hydrodynamic efficiency at the non-dimensional frequency $v(=\omega^2 h/g)$ is given by

$$\mathcal{E}(v) = 4\frac{h}{L}\frac{\alpha_0}{\sqrt{v}}\bigg/\left[\left(\frac{h}{L}\frac{\alpha_0}{\sqrt{v}}+\tilde{B}\right)^2+\tilde{A}^2\right], \tag{3.61}$$

dependent upon the frequency and geometry; α_0 is real and a non-dimensional form of α and $\tilde{A}, \tilde{B}$ are non-dimensional forms of the coefficients in Eq. (3.56). This expression possess a maximum value of

$$\mathcal{E}_{max}(v) = \frac{2}{1+\sqrt{\tilde{A}^2+\tilde{B}^2}}, \quad \text{when } \alpha_0 = \frac{L\sqrt{v}}{h}\sqrt{\tilde{A}^2+\tilde{B}^2} \tag{3.62}$$

and this is a maximum value rather than an optimal one, since it depends upon a particular choice of the power take-off parameter. A physical implementation requires the values of the hydrodynamic coefficients $\tilde{A}$ and $\tilde{B}$, these can be obtained accurately and efficiently by the method of Evans and Porter (1995).

In terms of the design approach described in Section 3.6.2, Eq. (3.60) is obtained from Eq. (3.59) by specifying the geometrical parameter vector $\mathbf{G}=(D,L)$ and varying the complex control parameter $\mathbf{J}=(\alpha)$. If the device is tuned to the frequency v_T for fixed $\mathbf{G}$, then α is obtained from the condition in Eq. (3.60) and the corresponding bandwidth curves $\mathcal{E}(v;\mathbf{J}_0(v_T))$ are obtained from Eq. (3.59).

For this two-dimensional geometry, Evans et al. (1995) combined the techniques introduced by Thomas and O Gallachoir (1993), outlined in Section 3.6.2, with the rapid and accurate numerical technique of Evans and Porter (1995) to maximise the function

$$\mathcal{E}_3(\mathbf{G}_3,\mathbf{J}_3) = Max\left\{\frac{1}{P_T}\int_{v_1}^{v_2}P_v(v)\mathcal{E}\big(v;\mathbf{G}(v),\mathbf{J}(v)\big)\,dv\right\} \tag{3.63}$$

for the Pico power spectrum shown in Fig. 3.2. (The subscript $_3$ is used in this expression to denote a direct analogy with the capture width / hydrodynamic efficiency function of Eq. 3.52). A constrained numerical optimisation scheme was implemented to ensure that the parameters remained within acceptable bounds and the findings suggested that the device could achieve a hydrodynamic efficiency of just over 90 % for the approach adopted. Of greater interest are the predicted dimensions of the optimal device values in comparison with those previously chosen for the Pico plant. The chamber length L was very similar for the predicted and chosen values but there was a slight difference in the values of the barrier depth D, with the numerical approach suggesting a value slightly greater than that chosen. This parameter helps to target the requirements of the spectrum.

The hydrodynamic efficiency value quoted above is derived from a model that matches the hydrodynamics reasonably well but does not provide an accurate description of the chamber aerodynamics or the power take-off mechanism. A discussion of these failings is given by Sarmento and Falcão (1985), who also identify the difficulty of performing scale experiments with OWCs.

To obtain a better understanding of the aerodynamics in the chamber, and hence of the hydrodynamic-aerodynamic coupling, follow the approach outlined originally by Sarmento, Gato and Falcão (1990) and employ the notation of Eqs. (3.54) and (3.55). Note that the aerodynamics of air turbines and their application to OWCs are assessed in detail in Chapter 6. The chamber is taken to possess an arbitrary width W for convenience but the modelling is genuinely two-dimensional and the choice $W=1$ is permissible. Denote the volume of air in the chamber and the water volume contained within the chamber and above the barrier by $V_c(t)$ and $V_w(t)$ respectively. With reference to Fig. 3.11, the still water values of these quantities are LH_cW and LDW; they are related at all times by

$$V_w + V_c = L(D+H_c)W .$$ (3.64)

From Eqs. (3.55) and (3.64)

$$\frac{dV_w}{dt} = Q(t) = Q_S(t) + Q_R(t) = -\frac{dV_c}{dt} .$$ (3.65)

If $\rho_c(t)$ and ρ_a are the density of the air inside and outside the chamber respectively, then the adiabatic gas law gives

$$\frac{\rho_c(t)}{\rho_a} - 1 = \frac{1}{\gamma}\left(\frac{p_c(t)}{p_a}-1\right) = \frac{p(t)}{\gamma} ,$$ (3.66)

where γ is the usual ratio of specific heats and Eq. (3.66) is valid under the assumption that both sides of the identity are small, i.e. the pressure and density inside the chamber do not vary much from their ambient values. The air mass in the chamber $M_c(t)$ can only change by a non-zero flux across the turbine, measured positive in an outward direction, and thus

$$\frac{dM_c}{dt} = \frac{d}{dt}(\rho_c V_c) = -\rho_a Q_t(t), \qquad (3.67)$$

where $Q_t(t)$ is the volume flux at the turbine. Now combine Eqs. (3.64) – (3.67) to yield $Q_t(t)$ in the form

$$Q_t(t) = \frac{\rho_c(t)}{\rho_a} Q(t) - \left[L(D+H_c)W - V_w(t) \right] \frac{1}{\gamma p_a} \frac{dp}{dt}. \qquad (3.68)$$

This relationship couples the hydrodynamic and aerodynamic domains and is weakly nonlinear; a linearised form can be obtained easily by replacing $\rho_c(t)$ by ρ_a and $V_w(t)$ by LDW in Eq. (3.68),

$$Q_t(t) = Q(t) - \frac{LH_cW}{\gamma p_a} \frac{dp}{dt}. \qquad (3.69)$$

In the absence of aerodynamic considerations, $Q_t(t)$ and $Q(t)$ would be the same and this is the approximation applied in the original Evans (1982) model. Both Eqs. (3.68) and (3.69) show clearly how compressibility in the aerodynamic domain can play an important role and so demands inclusion in the modelling process.

The instantaneous power at the turbine is $p(t)Q_t(t)$ and analogous to Eq. (3.58) it is assumed that

$$Q_t(t) = \alpha p(t), \qquad (3.70)$$

with α being real. For known $Q(t)$ the differential pressure $p(t)$ can be determined from the particular integral solution of Eq. (3.69) and hence the instantaneous power $P_{aero}(t) = \alpha p^2(t)$ can be determined. The addition of air compressibility introduces another device parameter, the chamber height H_c, into the geometrical design vector G. Weber and Thomas (2000) optimised the quantity

$$\mathcal{E}_{aero}(G_{aero}, J_{aero}) = Max\left\{ \frac{1}{WP_T} \langle P_{aero}(G, J) \rangle \right\} \qquad (3.71)$$

with parameter vectors $G = (D, L, H_c)$ and $J = (\alpha)$ to assess the influence of compressibility. This work was completed for an OWC of fixed width W ($\neq 1$) rather than one of unit width, to enable comparisons to be made with the dimensions of the Pico plant. The findings confirmed the importance of air compressibility, identified earlier by Falcão and Sarmento (1985), and although the hydrodynamic efficiency did not diminish appreciably relative to the hydrodynamic model the parameter values at the maximum efficiency were appreciably different. It may appear from a comparison of Eqs. (3.63) and (3.71) that similar quantities are not being optimised but it is straightforward to show that the approach adopted for Eq. (3.71) is both analogous to and extends Eq. (3.63). Instead of employing a power density from Fig. 3.2, Eq. (3.71) utilises the time series approach of Fig. 3.3 and Eq. (3.2) with averaging over an appropriate time period. In a sense this is a hybrid method, in which a time series is employed but the difficulties of instantaneous control are not addressed.

An improved model of the power take-off chain requires that a better measure of output power be employed rather than just the aerodynamic power at the turbine and the final approach utilises the mechanical output power. With the exception of the electrical generator this almost corresponds to a complete system analysis. The requirement is to link the hydrodynamic-aerodynamic coupling described above to a description of the mechanical output from the turbine, in a representation possessing physical parameters that may be incorporated into the optimisation process. Once more the input is taken from the specialist literature on the installation of turbines into OWCs and the following model is considered appropriate by Falcão and Justino (1999) for a turbine of Wells type.

The performance of the turbine, subject to justifiable approximation, can be characterised in terms of the non-dimensional flow Φ, pressure Ψ and power Π defined by

$$\Phi = \frac{Q_t}{ND_0^3}, \quad \Psi = \frac{p}{\rho_a N^2 D_0^2}, \quad \Pi = \frac{P_{mech}}{\rho_a N^3 D_0^5}, \tag{3.72}$$

where N and D_0 are the rotational speed and outer diameter of the turbine respectively. These quantities are related by

$$\Phi = f_\Phi(\Psi), \quad \Pi = f_\Pi(\Psi), \tag{3.73}$$

and the unknown functions f_Φ and f_Π are dependent upon the turbine geometry. For the Wells turbine developed for the Pico plant, the relationship f_Φ between Φ and Ψ is linear, confirming Eq. (3.70). The power – pressure relationship f_Π is shown in Fig. 3.12 and this curve possesses an interesting property: following almost linear growth, the power reaches a peak and then decreases rapidly as the turbine stalls. The free parameters to determine the instantaneous mechanical power P_{mech} are now N and D_0, which replace α of the previous formulations.

This extended turbine model has been utilised by Weber and Thomas (2001) to optimise the quantity

$$\mathcal{E}_{mech}(\boldsymbol{G}_{mech}, \boldsymbol{J}_{mech}) = Max\left\{ \frac{1}{WP_T} \langle P_{mech}(\boldsymbol{G}, \boldsymbol{J}) \rangle \right\} \tag{3.74}$$

with parameter vectors $\boldsymbol{G} = (D, L, H_c, D_0)$ and $\boldsymbol{J} = (N)$ for the Pico spectrum in Fig. 3.2. The initial work in Eq. (3.63) was an attempt to match the OWC geometry to the shape of the spectrum and is essentially a weighted optimisation problem, as can be seen clearly from the integrand in Eq. (3.63). This is less evident from Eq. (3.74) and there is an important difference, since Eq. (3.74) attempts to match the device parameters to Fig. 3.2, subject to the constraint imposed by

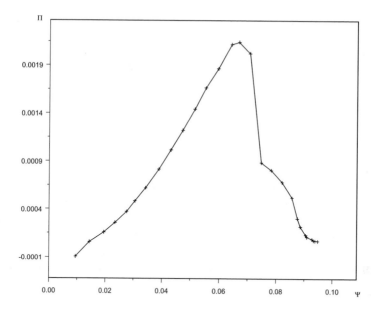

Fig. 3.12. The variation of the non-dimensional turbine power Π with non-dimensional pressure Ψ for the Pico turbine: measured values (×) and interpolation (solid line)

Eqs. (3.72), (3.73), and Fig. 3.12. The resulting output efficiency is much reduced, moving from around 90% for the hydrodynamic model of Eq. (3.63) to about half that value for the complete optimisation in Eq. (3.74). There are two particular aspects worthy of mention; the first is that the turbine power curve (Fig. 3.12) must match the wave climate in some sense and the second is that complete optimisation must replace partial optimisation in the design process. This work was continued to a target involving multiple sea-states and subsequent local control by Weber and Thomas (2003).

3.7.2 Time Domain Modelling and Control

The frequency domain modelling employs an approach beginning with the hydrodynamics and moving towards the mean mechanical output via an inclusion of the aerodynamics. One of its most important contributions is to provide a reasonable measure of determining the dimensions of a device. Time domain modelling usually assumes that the hydrodynamic problem has been resolved, that the OWC has been constructed and must now be controlled to provide the maximum output within the operating circumstances and range of sea-states. This topic is not within the remit of this review, which is intended to focus upon the hydrodynamic aspects of conversion; power take-off consideration for an OWC is a very specialist topic and is discussed in detail in Chapter 6. However, the work presented thus far shows how the frequency and time domains interact and a short discussion of those aspects of turbine control applicable to the hydrodynamics is beneficial.

The difference between the frequency and time domain models has already been discussed in Section 3.6 and in the present context, this is epitomised by the information supplied in spectral form (Fig. 3.2) or as a time series (Fig. 3.3). Reference has already been made to Sarmento et al. (1990) and Falcão and Justino (1999) for models of the power take-off but both are control papers involving time domain models with realistic constraints, such as the suitability of by-pass valves to prevent large chamber pressures.

Two other important issues associated with control are identified by Falcão (2002) and Perdigão and Sarmento (2003). The paper by Falcão considers the control of the Wells turbine once the OWC has been constructed, primarily by employing the rotational speed N as the controlling variable and is in keeping with the control modelling described herein. Perdigão and Sarmento consider control of a variable pitch turbine, in contrast to the fixed-pitch assumed so far, and show that appreciable improvements in efficiency can be obtained by this change of design. Both of these papers have implications for future models based upon the hydrodynamic approach.

3.8 Discussion

This brief review has attempted to provide a broad overview of the hydrodynamic theory that underlies the conversion process of the energy possessed by ocean waves into a form more useful to mankind. There are many shortcomings undoubtedly and there are certainly omissions; some are deliberate, others due to the enormous breadth of the field and the accompanying limitation of space. Any review must to certain extent, unintentionally or otherwise, reflect the interests and blinkered perspective of the author and any such perceived deficiencies are acknowledged.

It is hoped that this contribution will encourage those readers whose enthusiasm has been fired by the necessity and certainty of energy extraction from ocean waves will find in this article a good starting point to acquire a working expertise in the discipline. It is proposed that a good basis on wave mechanics and hydrodynamics in the maritime environment can be obtained from the text books in the reference list, followed by an introduction to wave energy via the more advanced texts of Falnes (2002) and the conference proceedings edited by Count (1980). With such an armoury, the other articles cited in this review and the excellent complimentary chapters in this book can be tackled with confidence.

References

Alves M, Sarmento AJNA (2006) Nonlinear and viscous diffraction response of OWC wave-power plants. Proc 6th Eur Wave Tidal Energ Conf. Glasgow, Scotland, pp 11–17

Barbarit A, Clément AH (2006) Optimal latching control of a wave energy device in regular and irregular waves. Appl Ocean Res 28:77–91

Barbarit A, Duclos G, Clément AH (2004) Comparison of latching control strategies for a heaving wave energy device in random sea. Appl Ocean Res 26:227–238

Brito-Melo A, Sarmento AJNA, Clément AH, Delhommeau G (1999) A 3-D boundary element code for the analysis of OWC wave-power plants. Proc 9th Int Offshore Polar Eng Conf. Best, France, pp 188–195

Brooke J (2003) Wave energy conversion. Elsevier

Budal K, Falnes J (1975) A resonant point absorber of ocean-wave power. Nature 257:478–479 [with Corrigendum in Nature 257:626]

Budal K, Falnes J (1980) Interaction point absorbers with controlled motion. In: Count BM (ed) Power from sea waves. Academic Press, pp 381–399

Chaplin RV, Folley MS (1998) An investigation into the effect of extreme waves on the design of wave energy converters. Proc 3rd Eur Wave Energ Conf. Patras, Greece, pp 314–317

Chaplin RV, French MJ (1980) Aspects of the French flexible bag device. In: Count BM (ed) Power from sea waves. Academic Press, pp 401–412

Count BM (ed) (1980) Power from sea waves. Academic Press

Crabb JA (1980) Synthesis of a directional wave climate. In: Count BM (ed) Power from sea waves. Academic Press, pp 41–74

Cummins WE (1962) The impulse response function and ship motions. Schiffstechnik 9:101–109

Davis JP, Finney R, Evans DV, Thomas GP, Askew WH, Shaw TL (1981) Some hydrodynamic characteristics of the Bristol Cylinder. Proc 2nd Int Sym Wave Tidal Energ. Cambridge, England, pp 249–260

Dean RG, Dalrymple RA (1991) Water wave mechanics for engineers and scientists. World Scientific Publishing

Evans DV (1976) A theory for wave power absorption by oscillating bodies. J Fluid Mech 77(1):1–25

Evans DV (1979) Some theoretical aspects of three-dimensional wave-energy absorbers. Proc 1st Symp Wave Energ Util. Gothenburg, Sweden

Evans DV (1980) Some analytic results for two and three dimensional wave-energy absorbers. In: Count BM (ed) Power from sea waves. Academic Press, pp 213–248

Evans DV (1981a) Power from water waves. Ann Rev Fluid Mech 13:157–187

Evans DV (1981b) Maximum wave-power absorption under motion constraints. Appl Ocean Res 3:200–203

Evans DV (1982) Wave-power absorption by systems of oscillating surface pressure distributions. J Fluid Mech 114:481–499

Evans DV, Ó Gallachoir BP, Porter R, Thomas GP (1995) On the optimal design of an Oscillating Water Column device. Proc 2nd Wave Energ Conf. Lisbon, Portugal, pp 172–178

Evans DV, Porter R (1995) Hydrodynamic characteristics of an oscillating water column. Appl Ocean Res 17:155–164

Evans DV, Thomas GP (1981) A hydrodynamical model of the Lancaster flexible bag wave energy device. Proc 2nd Int Sym Wave Tidal Energ. Cambridge, pp 129–141

Falcao AF de O (2002) Control of a oscillating water column wave power plant for maximum production. Appl Ocean Res 24:73–82

Falcão AF de O, Justino PAP (1999) OWC wave energy devices with air flow control. Ocean Eng 26:1275–1295

Falnes J (1980) Radiation impedence matrix and optimum power absorption for interacting oscillators in surface waves. Appl Ocean Res 2:75–80

Falnes J (2001) Optimum control of oscillation of wave-energy converters. Proc 11th Int Offshore Polar Eng Conf. Stavanger, Norway, pp 567–574

Falnes J (2002) Ocean waves and oscillating systems. Cambridge Univ Press

Faltinsen OM (1990) Sea loads on ships and offshore structures. Cambridge Univ Press

Goda Y (2000) Random seas and design of maritime structures, 2nd edn. World Scientific Publishing

Greenhow M, White SP (1997) Optimal heave motion of some axisymmetric wave energy devices in sinusoidal waves. Appl Ocean Res 10:141–159

Jefferys ER (1980) Device characterization. In: Count BM (ed) Power from sea waves. Academic Press, pp 413–438

Justino PAP, Clément AH (2003) Hydrodynamic performance for small arrays of submerged spheres. Proc 5th Eur Wave Energ Conf. Cork, Ireland, pp 266–273

Lee C-H, Newman JN, Nielsen FG (1996) Wave interactions with an oscillating water column. Proc 6th Int Offshore Polar Eng Conf. Los Angeles, USA, pp 82–90

Linton CM, McIver P (2001) Mathematical techniques for wave/structure interactions. Chapman and Hall/CRC

MacIver P (1994) Some hydrodynamic aspects of arrays of wave-energy devices. Appl Ocean Res 16:61–69

Mavrakos SA, McIver P (1997) Comparison of methods for computing hydrodynamic characteristics of arrays of wave power devices. Appl Ocean Res 19:283–291

Mei CC (1976) Power extraction from water waves. J Ship Res 20.63–66

Mingham CG, Qian L, Causon DM, Ingram DM, Folley M, Whittaker TJT (2003) A two-fluid numerical model of the Limpet OWC. Proc 5th Eur Wave Energ Conf. Cork, Ireland, pp 119–125

Nebel P (1992) Maximising the efficiency of wave-energy plant using complex control. J Syst Control Eng. Proc Inst Mech Eng 206:225–236

Newman JN (1962) The exciting forces on fixed bodies in waves. J Ship Res 6:10–17

Newman JN (1976) The interaction of stationary vessels with waves. Proc 11th Symp Naval Hydrodyn. London

Newman JN (1977) Marine Hydrodynamics. MIT Press

Newman JN (1979) Absorption of wave energy by elongated bodies. Appl Ocean Res 1:189–196

Perdigão J, Sarmento AJNA (2003) Overall-efficiency optimisation in OWC devices. Appl Ocean Res 25:157–166

Pizer DJ (1993) Maximum wave-power absorption of point absorbers under motion constraints. Appl Ocean Res 15:227–234

Pizer D, Retzler C, Henderson RM, Cowieson FL, Shaw MG, Dickens B, Hart R (2006) Pelamis WEC – Recent advances in the numerical and experimental modelling programme. Proc 6th Eur Wave Tidal Energ Conf. Glasgow, Scotland, pp 373–378

Pontes MT, Oliveira-Pires H (1992) Assessment of shoreline wave energy resource. Proc 2nd Int Offshore Polar Eng Conf San Francisco, USA, pp 557–562

Randløv P (1996) Final report and annexes to the Offshore wave energy converters (OWEC-1) project. EU Contract No. JOU2-CT93-0394. Danish Wave Power aps

Sarpkaya T, Isaacson M (1981) Mechanics of wave forces on offshore structures. Van Nostrand Reinhold

Sarmento AJNA, Falcão AF de O (1985) Wave generation by an oscillating surface-pressure and its application in wave-energy extraction. J Fluid Mech 150:467–485

Sarmento AJNA, Gato LMC, Falcão AF de O (1990) Turbine-controlled wave energy absorption by oscillating water column devices. Ocean Eng 17:481–497

Smith C (1983) Some problems in linear wave waves. PhD Thesis, Univ of Bristol

Thomas GP, Evans DV (1981) Arrays of three-dimensional wave-energy absorbers. J Fluid Mech 108:67–88 X Thomas GP, Ó Gallachóir BP (1993) An assessment of design parameters for the Bristol Cylinder. Proc 1st Eur Wave Energ Sym. Edinburgh, Scotland, pp 139–144

Tucker MJ, Pitt EG (2001) Waves in ocean engineering. Elsevier

Weber JW, Thomas GP (2000) Optimisation of the hydrodynamic-aerodynamic coupling for an Oscillating Water Column wave energy device. Proc 4th Eur Wave Energ Conf. Aalborg, Denmark, pp 251–259

Weber JW, Thomas GP (2001) An investigation into the importance of air chamber design of an oscillating water column wave energy device. Proc 11th Int Offshore Polar Eng Conf. Stavanger, Norway, pp 581–588

Weber JW, Thomas GP (2003) Some aspects of the design optimisation of an OWC with regard to multiple sea-states and combined object functions. Proc 5th Eur Wave Energ Conf. Cork, Ireland, pp 141–148

Wehausen JV (1971) The motion of floating bodies. Ann Rev Fluid Mech 3:7–268

4 The Wave Energy Resource

Stephen Barstow, Gunnar Mørk

Fugro OCEANOR,
Trondheim, Norway

Denis Mollison

Heriot-Watt University,
Edinburgh, Scotland, UK

João Cruz

Garrad Hassan and Partners Ltd
Bristol, England, UK

On an average day, about 1 *TWh* of wave energy enters the coastal waters of the British Isles. It is tempting to call this amount huge - it is about the same as the total energy of the terrible Indian Ocean tsunami of the 26th of December 2004. It brings home the scale of human energy demands to realise that this is also about the same amount of energy which is used in electricity in the British Isles on an average day.

The same approximate equivalence holds at a world scale: the total wave energy resource is of the same order of magnitude as world electricity consumption ($\sim 2\,TW$). The exploitable limit is probably at most about 10–25 % of the resource; thus ocean wave energy is potentially a significant contributor to human energy demands, not a panacea. Its key advantages are that it comes in a high quality form - mechanical energy of oscillation - and that it travels very long distances with little loss, so that small inputs over a large ocean can accumulate and be harvested at or near the ocean's edge. Other advantages include the point absorber effect, whereby devices can extract energy from a fraction of a wavelength on either side; this makes small devices, with capacities of the order of 1 *MW*, relatively attractive.

4.1 The Resource and its Origin

Wave energy is created by wind, as a by-product of the atmosphere's redistribution of solar energy. The rate of energy input to waves is typically .01 to .1 W/m^2. This is a small fraction of the gross solar energy input, which averages 350 W/m^2, but waves can build up over oceanic distances to energy densities averaging over 100 *kW/m* (note that the typical measure is power per metre width of wave front). Because of its origin from oceanic winds, the highest average levels of wave power are found on the lee side of temperate zone oceans.

Estimation of the resource is crucial when selecting suitable sites. In section 4.5.4 details on how available global wave model data, global satellite altimetry and measured buoy data were integrated in the EU supported WorldWaves project to derive a global wave and wind database with enhanced quality are given. Essentially, 10-years of 6 hourly wave and wind parameter data have been derived for 10,000 offshore grid points worldwide on a 0.5° grid. Based on these data sets, a number of global maps were produced. These are used in the following sections when discussing the global wave energy resource.

Firstly, Fig. 4.1 shows the 10-year mean annual wave power for all global points in the WorldWaves database. This map shows clearly that the most energy-rich areas of the global oceans are in the mid to high latitude temperate storm belts of both hemispheres, in particular between 40 and 60°. This figure gives, however, a slightly wrong impression of the relative energy levels in both hemispheres. The much higher resources in the southern hemisphere, where seasonal variations are much lower (compare Figures 4.2 and 4.3 showing January and July means), are seen more clearly in Fig. 4.4, where the mean wave power for all grid points is plotted as a function of latitude both annually and for the months of January and July.

On an annual basis, the highest levels in the Northern Hemisphere are off the west coast of the British Isles (see also Fig. 4.5) and Iceland and Greenland, with somewhat lower energy levels in the Pacific off the western seaboard of the US and Canada. Not surprisingly, the highest overall 50–100 *m* depth offshore energy levels in the Southern Hemisphere are located off Southern Chile, South Africa and the entire south and south west coasts of Australia and New Zealand. In equatorial waters, wave power levels of 15–20 *kW/m* on an annual basis over parts of all the ocean basins are observed, the highest coastal resources being off Northern Peru and Ecuador, although there may be significant El-Nino induced inter-annual variability in this area. Finally, in medium latitudes, Western Australia comes out best, with California also having a relatively high resource for the latitude.

Where in the world can the highest wave power regime be found? This turns out to be in the Southern Ocean at or around 48°S, 90°E which is about 1,400 *km* east of Kerguelen Island. Here, the annual average exceeds 140 *kW/m*. In the Northern Hemisphere, the maximum is found at 57°N, 21°W, some 400 *km* west of Rockall. Here, a "modest" by comparison 90 *kW/m* can be found. The maximum global monthly power level is practically the same in the two hemispheres, at or just above 200 *kW/m*. Note that these locations are given as example and are clearly unsuitable for wave energy projects: the mere distance to shore (and depth) is enough to make them unattractive. It is estimated that the first wave energy farms will be deployed in the 50–60 *m* depth region, i.e., in the limit of the transition from nearshore to the more attractive offshore wave climates.

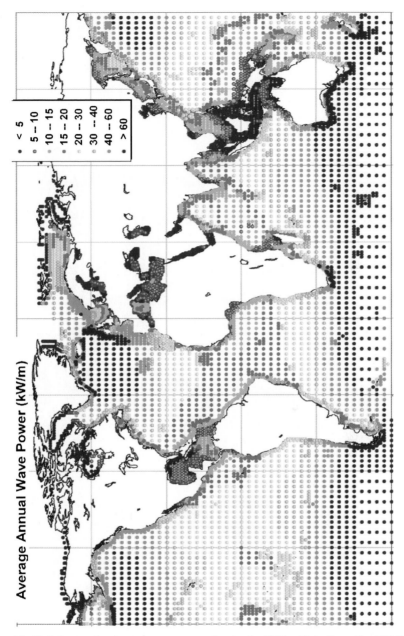

Fig. 4.1. Global annual mean wave power estimates in kW/m (data from the ECMWF WAM model archive; calibrated and corrected by Fugro OCEANOR against a global buoy and Topex satellite altimeter database)

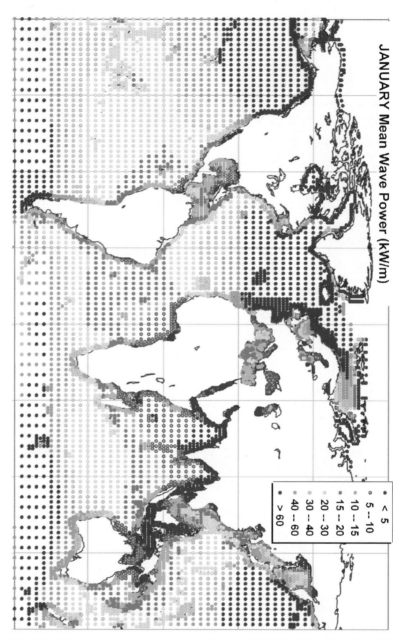

Fig. 4.2. Global January mean wave power estimates in kW/m (data from the ECMWF WAM model archive; calibrated and corrected by Fugro OCEANOR against a global buoy and Topex satellite altimeter database)

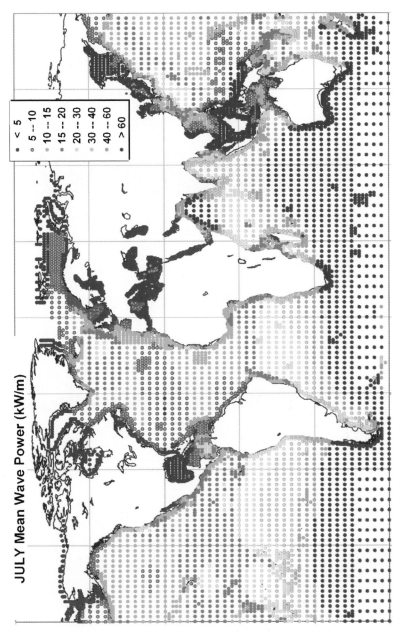

Fig. 4.3. Global July mean wave power estimates in kW/m (data from the ECMWF WAM model archive; calibrated and corrected by Fugro OCEANOR against a global buoy and Topex satellite altimeter database)

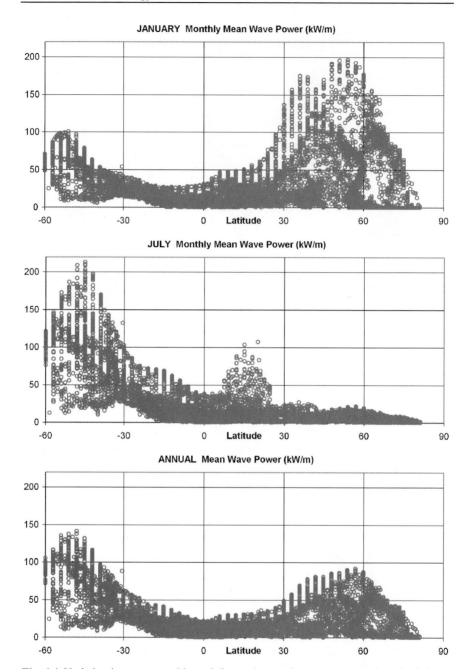

Fig. 4.4. Variation in mean monthly and (bottom) annual wave power against the latitude for all global grid points

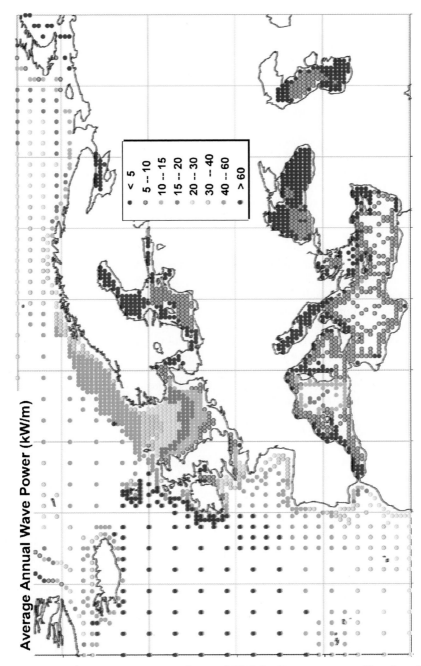

Fig. 4.5. Annual mean wave power estimates (*kW/m*) for European waters (data from the ECMWF WAM model archive; calibrated and corrected by Fugro OCEANOR against a global buoy and Topex satellite altimeter database)

Figures 4.2 and 4.3 give an indication of the seasonality of the global resources. However, if the ratio of the minimum monthly power level against the annual mean is evaluated, a clearer idea of how the resource varies over the year is obtained (Fig. 4.6). Figure 4.7 presents this data in a different way showing the variation of the same statistic against latitude. There is a dramatic difference between hemispheres as far as the stability of the wave energy resources is concerned with large summer to winter changes in the north, the largest being in ice impacted waters.

There are also large seasonal changes in the monsoon areas of South East Asia due to the seasonal switch in wind direction in these areas. This is particularly large along seasonally downwind and upwind coastlines, or indeed two sides of an oceanic island, due to the influence of fetch on wind wave growth. In the South China Sea, the seasonality is large due to the fact that there is little swell to smooth out the seasonal wind sea changes. This is also a dominant feature of the Arabian Sea which, along, with the Bay of Bengal, are the only sea areas in the Northern Hemisphere with larger summer than winter energy levels, due to a combination of stronger summer monsoon winds and higher swell influx at that time of year (i.e., the Southern Ocean winter). Another interesting feature is the area of the southern Indian Ocean to the south and west of Australia which shows regionally surprisingly large seasonality. This is due to the fact that this area roughly corresponds to the location, in January, of the Indian Ocean Anticyclone. This anticyclone has a remarkably large seasonal displacement towards the west during winter, leading to a large seasonal change in storm frequency in the aforementioned area.

In coastal areas, it can be seen that there are particularly stable wave energy resources off Chile, Namibia, Eastern Australia, Sierra Leone and Liberia, the Pacific coasts of Mexico and most of the South Pacific island nations, although these nations, particularly the low lying ones, are nowadays concerned with wave power for other reasons.

Finally, Fig. 4.8 plots the relationship between the extreme and mean significant wave height. One can consider, in very simple terms, that the lower this ratio is, the more feasible a wave energy project might be as the extreme conditions relate to design and to a certain extent operational costs and the mean represents the resource or the income (see also Hagerman, 1985, who introduced a similar ratio, his Figure of Merit). The change in this ratio from deep water to shallow water is particularly interesting, and is discussed further in section 4.5.

In terms of resource, there are many attractive areas in the globe when looking for a suitable site to locate a wave energy farm, as Figs. 4.1 to 4.8 suggest. The challenge is to choose a location which provides not only the adequate resource but also all the necessary conditions to ensure the continuous and reliable operation of the wave energy converters. A key driver is the proximity to shore, linked also with the water depth, as it influences not only costs (e.g. underwater power cables) but mostly the possibility of safely recovering the devices and conducting maintenance in a local shipyard. All of these variables (resource, cost, O&M) need to be carefully measured and given considerable thought.

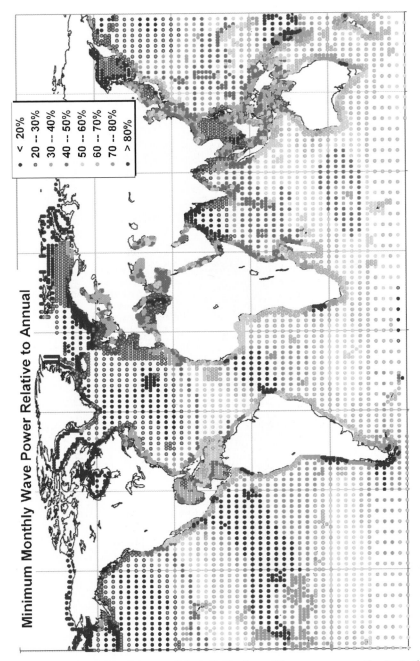

Fig. 4.6. The ratio of the minimum of the individual monthly wave power estimates to the annual value gives an indication of the seasonal variability of the resource (data from the ECMWF WAM model archive; calibrated and corrected by Fugro OCEANOR against a global buoy and Topex satellite altimeter database)

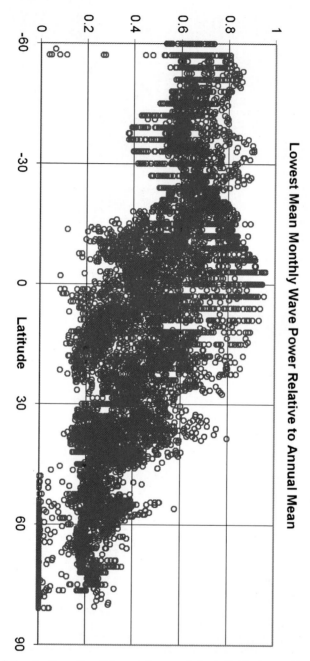

Fig. 4.7. Variation in the ratio of the minimum of the individual monthly wave power estimates to the annual value against latitude shows clearly the much lower seasonality in the far south. Note that the grid points with zero are from areas impacted by ice in the northern hemisphere

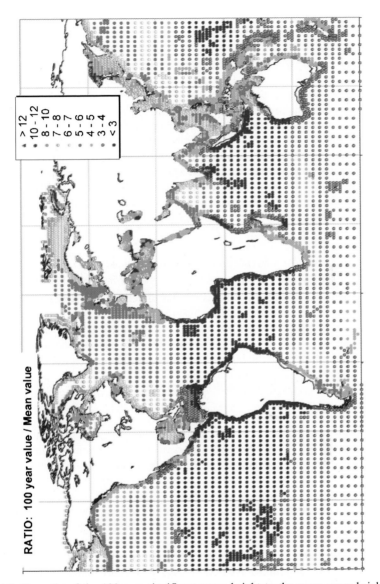

Fig. 4.8. The ratio of the 100-year significant wave height to the mean wave height, inspired by Hagerman's Figure of Merit (1985), roughly reflects the ratio of design costs for a wave energy plant against its income (the resource). The lower this ratio, the better it is. The 100-year value is calculated by a simple extrapolation of the 10-year significant wave height distribution at each location fitted by a 3-parameter Weibull distribution. Note that this ratio may be underestimated in areas impacted by hurricanes, cyclones, typhoons and other extreme events with long return period as 10-years is then not sufficiently long for reliable extreme estimates using this method (data from the ECMWF WAM model archive; calibrated and corrected by Fugro OCEANOR against a global buoy and Topex satellite altimeter database)

4.2 Sea States and their Energy

Deep water surface waves are oscillations of the sea surface layer under gravity and, to a good approximation, they consist of the linear superposition of a large number of simple components. This basic component is a sinusoidal wavetrain, with period T, which appears to travel at its phase velocity, $U=gT/2\pi$.

In fact, the water particles are not travelling: in a simple sinusoidal wave they oscillate in circles, whose amplitude a_d falls off exponentially with depth d ($a_d=a\ \exp(-2\pi/L)$).

The energy of the wavetrain, per unit area, is $E=\rho gH^2$, where ρ is the density of the water and H the root mean square wave height ($H^2=a^2/2$). More crucially for the present purposes, the oscillations move energy in the direction of the wavetrain and this energy flux, i.e. power per unit width, is E multiplied by the group velocity $U/2$, resulting in cH^2T, where the constant c is given by $\rho g^2/4\pi$ ($\cong 7.87\,kW/m^3s$).

Following the linear superposition assumption, a general sea state can therefore be described by its directional spectral density, $S(f,\theta)$, which specifies how the energy is distributed over frequencies (f) and directions (θ). Note that to specify the instantaneous sea state the phase is also required.

The power in a general sea state is the integral of the energy density multiplied by the group velocity. The spectral moments (m_n) are particularly useful to quantify the power; m_n denotes the nth moment of the frequency spectrum, given by

$$m_n = \int f^n S(f)df .\qquad (4.1)$$

A selection of spectral parameters is typically used to characterise a sea state. In an effort to standardise the nomenclature the spectral approach is used throughout this chapter. But firstly some short notes on alternative nomenclatures / definitions are presented. To understand the original definitions it is necessary to recall a typical plot of a wave record (surface elevation vs. time). In a pre-computational environment to process such a plot meant subtracting the mean water level and searching for the point where the surface profile crosses the zero line upwards. Such a point identifies the start of a new wave. The next zero-upcrossing point would then define the end of this wave and the start of the next one (Goda, 2000). Extreme individual values could be easily extracted, such as the maximum wave height (H_{max}) and maximum wave period (T_{max}) in a record, but once the waves are identified and counted, other parameters could also be derived. In terms of wave height, the mean wave, the highest one-tenth ($H_{1/10}$) and the highest one-third ($H_{1/3}$) wave are the most interesting from a statistical point of view.

The significant wave height (H_s), a standard measure in the offshore industry, is by definition $H_{1/3}$: the waves in a record were counted and selected, in descending order starting from the highest wave until one-third of the total number of waves was reached. The mean value is $H_{1/3}$. The same procedure could be applied to the wave period, resulting in $T_{1/3}$, the significant wave period.

When trying to link Eq. (4.1) to H_s, it was found that $H_{1/3}$ is approximately $4\sqrt{m_0}$, which therefore corresponds to the significant wave height when using a spectral approach (H_{m0}). Note that the relation between H_{m0} and the root mean square wave height (H_{rms}) is also clear ($H_{m0} = 4H_{rms}$), which could suggest that the use of H_{rms} is particularly suitable to keep the physical meaning (but it is particularly unsuited when comparing it to the standard significant wave height). Overall the spectral approach is mathematically consistent, as Table 4.1 suggests. Every parameter can thus be obtained from Eq. (4.1). Knowing the relation between H_{m0} and H_{rms}, the wave energy flux can be therefore given by $\dfrac{c}{16}H_{m0}^2 T_{-10}$, in W/m, for deep water waves ($c = \rho g^2/4\pi, \cong 7.87\,kW/m^3 s.$).

To estimate the directional content of a given sea state is also of vital importance to both the characterisation of a specific site and to predict the performance of a wave energy converter. In order to fully define the directional spectra $S(f,\theta)$ several methods can be applied.

Two basic approaches are possible, following either stochastic or deterministic methods. The methods have specific limitations and potential, and differ considerably in terms of the degree of parameterisation and on the computational effort required, for the same frequency and directional resolution. An overview of the main methods is given is subsections 4.2.1 and 4.2.2, following Benoit et al. (1997).

Stochastic methods are based on the random phase assumption. The cross-spectrum between each pair of signals is calculated, and the directional spectrum is derived by inverting the relation with the cross-spectra, using one of the available techniques. The form of the directional spreading function, which if convoluted with the frequency spectrum results in the directional spectrum, is always assumed, but major differences exist when fitting the data to the model: Fourier expansions use a deterministic fit, parametrical models use fixed form functions and the latest generation of methods, like the Maximum Likelihood Method (MLM) or Maximum Entropy Method (MEM), fit statistically to the data. The latter class is by far the best option when modelling real data following a stochastic approach, particularly when evaluating data from directional wave buoys or any other single-point system.

Table 4.1. Definition of the key spectral parameters

Spectral Nomenclature	Definition	Description
H_{m0}	$4\sqrt{m_0}$	Significant wave height (H_s)
T_{-10}	m_{-1}/m_0	Energy period (T_e)
T_{02}	$\sqrt{m_0/m_2}$	Zero-upcrossing period (T_z)

Deterministic methods keep the phase information but only allow one or two directions per frequency. The cross-spectrum is not evaluated and the Fourier coefficients from each individual signal are used. The number of methods is much smaller than the stochastic equivalents and only two are typically used. The approach might prove useful in cases where reflective structures are present.

4.2.1 Directional Spectra Estimation Using Stochastic Methods

In the stochastic approach it is assumed that the wave field can be described by

$$\eta(x,y,t) = \iint \sqrt{2S(f,\theta)dfd\theta} \cos\left[k(x\cos\theta + y\sin\theta) - 2\pi ft + \varphi\right], \quad (4.2)$$

where $\eta(x,y,t)$ is the surface elevation, $S(f,\theta)$ the directional spectrum, k the wave number, θ the wave direction, f the wave frequency and φ the phase function. Stochastic methods carry the assumption that φ is randomly disturbed over the entire $[0,2\pi]$ range with uniform probability density, which in turn means that all wave components are treated as independent of each other. When analysing situations where phase-locked waves are present (e.g.: in the vicinity of a reflective formation) such assumption might be biased. All these methods follow a two step approach:

1. Computation of the cross-spectra between each pair of signals, usually by means of the Fast Fourier Transform;
2. Calculation of the directional spreading function, using the relation between the cross-spectra and the directional spectra and one of the methods below described.

The relation mentioned in 2. can be given by

$$G_{ij}(f) = E(f) \int_0^{2\pi} H_i(f,\theta)H_j^*(f,\theta)e^{-i\vec{k}\cdot(\vec{x}_j - \vec{x}_i)}D(f,\theta)d\theta, \quad (4.3)$$

where $G_{ij}(f)$ is the cross-spectra, $E(f)$ the frequency spectra derived from the surface elevation signal, $H_m(f,\theta)$ the transfer function between the surface elevation and another wave signal (e.g.: pressure, horizontal velocity, etc.)[1], and $D(f,\theta)$ the directional spreading function.

To solve the system of equations that Eq. (4.3) represents, the first step is to calculate the cross-spectra of the original signals[2]. The output can be rearranged in

---

[1] '*' denotes the complex conjugate.
[2] Given the complex nature of the cross-spectra, the real part is typically called coincident or co-spectrum and imaginary one quadrature or quad-spectrum.

terms of Fourier Coefficients. For single point measurement systems with three independent signals (e.g: Waverider buoy, ADCP), only four coefficients can be calculated, irrespective of the method followed to calculate $D(f,\theta)$. These methods differ considerably in several assumptions and computing potential, and will therefore lead to different results. $D(f,\theta)$ has some inherent proprieties, namely

$$\int_0^{2\pi} D(f,\theta)d\theta = 1, \tag{4.4}$$

which follows from $E(f) = \int_0^{2\pi} S(f,\theta)d\theta$. In addition: $D(f,0) = D(f,2\pi)$ and the spreading function is also positive over the $[0,2\pi]$ interval, a condition that some less precise methods fail to obey, producing erroneous results.

There is a vast amount of stochastic methods available for calculating $D(f,\theta)$ in order to solve Eq. (4.3). Some of them can be grouped in classes like Fourier Series Decomposition or Parametrical methods.

For the first class, which can be understood as an introduction, two different methods were reviewed.

Truncated Fourier Series Decomposition Method (TFSM)

The Fourier series expansion of $D(f,\theta)$ is given by

$$D(f,\theta) = \frac{a_0}{2\pi} + \frac{1}{\pi}\sum_{n=1}^{\infty} a_n \cos n\theta + b_n \sin n\theta, \tag{4.5}$$

with

$$a_0 = \int_0^{2\pi} D(f,\theta)d\theta = 1. \tag{4.6}$$

In the simplest form, the series is reduced to a finite number of terms (N). If $N=2$ (single point measuring systems) only (a_1,b_1) and (a_2,b_2) are calculated from Eq. (4) it is possible to obtain

$$D(f,\theta) = \frac{1}{\pi}\left(\frac{1}{2} + a_1\cos\theta + b_1\sin\theta + a_2\cos 2\theta + b_2\sin 2\theta\right). \tag{4.7}$$

Although computationally efficient, the method yields the serious drawback of outputting negative values for the directional spreading function in some occasions, which is inherently flawed (Kuik et al., 1988). Refinements of this method include a weighting function that will ensure that an everywhere positive $D(f,\theta)$ (e.g.: Longuet-Hinggins et al., 1963). Note that Eqs. (4.5) and (4.7) have a deterministic nature but the methods are intrinsically stochastic, given the random phase assumption.

Direct Fourier Transform Method (DFTM)

A refinement of the TFS method is available in Kinsman (1963), following earlier work from Barber (1961). Again a Fourier expansion is made, but of greater complexity, resulting in

$$D(f,\theta) = \alpha \left\{ 1 + 2\sum_{n=1}^{j} \left[C'_{ij} \cos\left(kX_n \cos\theta + kY_n \sin\theta\right) + \right. \right.$$
$$\left. \left. + Q'_{ij} \sin\left(kX_n \cos\theta + kY_n \sin\theta\right) \right] \right\}, \tag{4.8}$$

where C'_{ij} and Q'_{ij} are the normalised components of the co and quad-spectra, respectively, and X_n and Y_n the spatial lags linked with the covariance function (recall that the cross-spectrum is the Fourier transform of the covariance function). α is a constant factor used to allow the unit integral condition over $[0,2\pi]$. The TFS and DFT methods produce similar (but not equal) results which are usually of lower quality when compared with methods which statistically fit to the data when solving the system of equations given by Eq. (4.3).

In conclusion, Fourier Series decomposition methods will lead to directional spectra estimations which typically have low spatial resolution. More recent approaches allow more accurate estimates, but it is still interesting to compare some parameters, like the mean direction, derived from all methods.

Parametrical methods use fitting techniques, either deterministic or statistically based, to derive $D(f,\theta)$. The common trend among them is the direct assumption (*a priori*) of a given shape for the directional energy distribution (Kuik et al., 1988). The models can also be unimodal or bimodal, where a linear combination of unimodal models is used. Cosine spreading functions are typically implemented (e.g: Mitsuyasu et al. 1975).

To avoid the Fourier decomposition and statistically fit the model to the data, Isobe (1990) initially developed a Maximum Likelihood Fitting method (MLF) that aimed to maximise a likelihood function linked with the Fourier coefficients of the signals. Some limitations, like the inability to detect two peaks at the same frequency, and mostly the availability of more efficient techniques lead directly to alternatives, neglecting this class of methods.

The evolution of the estimation techniques can be clearly predicted by looking at the first two classes of methods: to overcome the initial limitations that contribute to the rough estimates and sometime negative distributions the parametrical methods were derived. Although many shapes can be assumed, the number of available Fourier coefficients limits the number of shape parameters, which is a major weakness of the approach. Sand (1984) suggested a deterministic approach that would be further studied by Schäffer and Hyllested (1994), but a review on some of the newer generation stochastic methods, like the Maximum Likelihood and Maximum Entropy methods, is firstly conducted.

Extended Maximum Likelihood Method (EMLM)

Following on the early work of Capon et al. (1967), the Maximum Likelihood method takes a linear combination of the cross-spectra,

$$D'(f,\theta) = \frac{1}{E(f)} \sum_{i,j} \alpha_{ij}(f,\theta) G_{ij}(f),$$ (4.9)

with

$$D'(f,\theta) = \int_0^{2\pi} D(f,\theta) w(\theta,\theta') d\theta'.$$ (4.10)

The EMLM spreading function estimate can then be considered as the convolution product of the true $D(f,\theta)$ with the window function $w(\theta,\theta')$. Naturally, the estimate and the real $D(f,\theta)$ will be more and more similar as $w(\theta,\theta') \to \delta(\theta,\theta')$, where $\delta(\theta,\theta')$ is the Dirac function. The window function is given by

$$w(\theta,\theta') = \sum_{i,j} \alpha_{ij}(f,\theta) H_i(f,\theta') H_j^*(f,\theta').$$ (4.11)

Isobe et al. (1984) proved that the best $D'(f,\theta)$ is

$$D'(f,\theta) = \frac{\kappa}{\sum_{i,j} H_i(f,\theta) G_{ij}^{-1}(f) H_j^*(f,\theta)},$$ (4.12)

where κ satisfies the unit integral condition over $[0,2\pi]$. One possible limitation associated with the method was pointed by Benoit and Teisson (1994) when testing numerical and laboratorial data, and is associated with its tendency to produce broader directional peaks. However such tendency is harder to quantify when real sea data is being used.

Iterated Maximum Likelihood Method (IMLM)

An alternative procedure to derive the MLM estimate was introduced by Pawka (1983), in an attempt to match the cross-spectra from the initial MLM estimate with the one from the original signals. This is directly linked with the influence of $w(\theta,\theta')$, as Eqs. (4.10) and (4.11) demonstrate, and follows from Eq. (4.12).

In the iterated version of the method a quantity $\varepsilon^i(f,\theta)$ is added to the derived spreading function in the previous iteration. $\varepsilon^i(f,\theta)$ can be given by

$$\varepsilon^i(f,\theta) = \frac{|\lambda|^{\beta+1}}{\lambda\gamma}, \qquad (4.13)$$

with

$$\lambda = D'(f,\theta) - \Delta_{MLM}^{i-1}(f,\theta), \qquad (4.14)$$

where $\Delta_{MLM}^{i-1}(f,\theta)$ stands for the MLM estimate computed from the cross-spectra based on $D'(f,\theta)$ (and not the original one). β and γ are control parameters that support the convergence of the iterative algorithm (as a rough guide, the order of magnitude of β and γ is 1 and 10, respectively). The iterative process is halted after a fixed number of iterations.

In Benoit et al. (1997) it is mentioned that the IMLM should produce a more precise estimate than the EMLM for single-point measurement systems. The tendency for producing more narrow banded spectra was confirmed when using data from Waverider buoys (see 4.2.3).

Extended Maximum Entropy Principle (EMEP)

Hashimoto et al. (1994) have extended the Maximum Entropy Principle (MEP) for arrays of measuring devices, with both by the original MEP and the EMEP producing identical results for single-point measuring systems such as a directional wave buoy. The EMEP algorithm follows:

$$D'(f,\theta) = \frac{1}{c} e^{\sum_{n=1}^{N} a_n \cos n\theta + b_n \sin n\theta}, \qquad (4.15)$$

where $c = \int_0^{2\pi} e^{\sum_{n=1}^{N} a_n \cos n\theta + b_n \sin n\theta} d\theta$. The coefficients a_n and b_n are unknown and there is a total of M equations that can be used to solve the problem, with M being the non-zero components of the cross-spectra. Each equation can be written to illustrate the difference between the measured and the model cross-spectra, obtained by direct introduction of Eq. (4.15) in Eq. (4.3). This difference, as for the MLM, is not necessarily null, so a total of M residuals (ε_m, $m = 1,..., M$) will appear. The optimal estimate is the one which minimises the sum of the squares of the residuals, a non-linear problem that can be solved using Newton's method. The optimal number of coefficients (N) is obtained by minimising the Akaike Information Criteria (AIC; Akaike, 1973):

$$AIC = M\left(\ln 2\pi + 1 + \ln \sigma^2 + 4N + 2\right), \qquad (4.16)$$

where σ^2 is the estimated variance of ε_m. The EMEP is therefore a powerful stochastic method, as the number of harmonic components given by Eq. (4.15) is adapted to the data. In addition, statistical variability is also taken into account. Hashimoto et al. (1994) proved that the MEP and EMEP methods yield similar outputs for single-point measuring systems and in the process also reported the similarities with the outputs of the much more computationally intensive Bayesian Directional Method.

Bayesian Directional Method (BDM)

The Bayesian Directional Method (BDM) is the only stochastic method which does not assume any shape with regard to the directional spreading function. Such nature ensures that the method is highly adaptive, a desirable characteristic, but it also contributes to a difficult numerical implementation and a partly unsuitability for single-point measuring systems, as the computational time required is very high when compared to the conventional alternatives and the increase in accuracy is often minimum. Nevertheless, it is able to detect a broad variety of directional spectra, like uni, bi and trimodal, symmetrical peaks, etc.

The method was adapted from Bayesian techniques used in probability theory by Hashimoto et al. (1987). The $[0,2\pi]$ range is sub-divided into N segments, and the BDM estimate is a piecewise-constant function over each of the N segments. It is defined recursively as:

$$\begin{cases} x_n = \ln D'(f,\theta_n) \\ D'(f,\theta_n) = \sum_{n=1}^{N} e^{x_n} I_n(\theta) \end{cases}, \qquad (4.17)$$

where

$$\begin{cases} \theta_n = (n-1/2)\Delta\theta = (2n-1)\pi/N \\ I_n(\theta) = \begin{cases} 1, & \text{if } (n-1)\Delta\theta \le \theta \le n\Delta\theta . \\ 0, & \text{otherwise} \end{cases} \end{cases} \qquad (4.18)$$

The high value of N (typically between 40 and 90) and the statistical treatment that the errors in the estimated cross-spectra suffer lead to a very complete (and computationally intensive) approach. To add to the complexity, the non-linear system of equations is closed by a smoothness condition applied to the estimated directional spreading function:

$$\sum_{n=1}^{N}(x_{n+1} - 2x_n + x_{n-1})^2 \to 0, \qquad (4.19)$$

with a parameter (or hyperparameter) linking the Eq. (4.19) and the system of equations given by the introduction of Eq. (4.17) in Eq. (4.3). This parameter corresponds to the one which minimises the Akaike Bayesian Information Criteria (ABIC; Akaike, 1973).

4.2.2 Directional Spectra Estimation Using Deterministic Methods

Deterministic methods have one very distinctive factor from stochastic approaches: the phase information is retained from the original data set, and not assumed to be randomly disturbed over the entire $[0, 2\pi]$ range with uniform probability density.

Furthermore, the cross-spectrum is not evaluated and the Fourier coefficients from each individual signal are used. The wave field is decomposed into wave components of a given direction, amplitude and phase (deterministic decomposition), and so the free surface elevation follows:

$$\eta(x,y,t) = \sum_{m=1}^{M} \sum_{n=1}^{N} a_{mn} \cos\left[k_m \left(x\cos\theta_n + y\sin\theta_n \right) - \omega_n t + \varphi_{mn} \right], \qquad (4.20)$$

where m and n denote the frequency and directional bins, respectively. Deterministic methods assume a high value for M but a low value for N (i.e: $N=1$ or $N=2$), to ensure that the amplitude and the phase of the original signals can be determined from the Fourier coefficients. Such characteristic intrinsically limits the ability of the methods to model a wide range of directional spectra shapes, in a similar way to what occurs with parametric methods (in common they have the high level of parameterisation).

Deterministic methods are far scarcer than stochastic methods. Sand (1984) presents a review on the basic principles and Schäffer and Hyllested (1994) provide the two approaches that are commonly used: SDD and DDD, single and double directional decomposition, respectively.

Single Direction Decomposition (SDD)

The inputs for the SDD and DDD approaches are the same: the surface elevation and two horizontal velocities, (u,v). The Fourier transform of each signal is derived, resulting in C_η, C_u/F and C_v/F, respectively. F is the transfer function between the surface elevation and the horizontal velocity signals at a certain depth, as given by linear wave theory.

The SDD method has one main assumption: a three-dimensional sea can be represented by a sum of components of different frequencies. It is straightforward to implement this by considering the vector $\vec{u}_f(t)$, with

$$\begin{cases} u_f = \Re\{C_u e^{i\omega t}\} \\ u_v = \Re\{C_v e^{i\omega t}\} \end{cases}, \qquad (4.21)$$

where $\Re$ denotes the real part, ω is the angular frequency and t is time. The norm of $\vec{u}_f(t)$ has two maxima and two minima, which appear for

$$\tan 2\omega t = \frac{2\left(A_u B_u + A_v B_v\right)}{A_u^2 - B_u^2 + A_v^2 - B_v^2}, \qquad (4.22)$$

where the A and B coefficients are given from $C_u = A_u - iB_u$ and $C_v = A_v - iB_v$. Considering that the two solutions which give the maxima are ωt and $\omega t + \pi$, the correct direction of propagation is chosen from the two by evaluating which leads to a positive η_f, with

$$\eta_f(t) = \Re\{C_\eta e^{i\omega t}\}. \tag{4.23}$$

Note that the amplitude and phase of the wave are linked with C_η, so the velocities are used only for deriving the wave direction from Eq. (4.22).

Double Direction Decomposition (DDD)

The DDD method assumes that two components (directions) exist per frequency. Such assumption allows it to be more suitable for situations where reflections are relevant. Knowing that the subscripts 1 and 2 refer the wave directions, the following decomposition can be obtained:

$$\begin{cases} C_\eta = C_{\eta_1} + C_{\eta_2} \\ C_u = C_{\eta_1}\cos\theta_1 + C_{\eta_2}\cos\theta_2 \\ C_v = C_{\eta_1}\sin\theta_1 + C_{\eta_2}\sin\theta_2 \end{cases} \tag{4.24}$$

Using the expansion $C_j = A_j - iB_j$ for $j = \eta, u, v$, a non-linear system of equations can be derived (6 unknowns). The optimal solution is of the form

$$\cos\frac{\theta_2 - \theta_1}{2} = \pm\frac{A_u B_v - B_u A_v}{\sqrt{(A_\eta B_v - B_\eta A_v)^2 + (B_\eta A_u - A_\eta B_u)^2}}, \tag{4.25}$$

which is reintroduced in

$$\left(\cos\frac{\theta_2 + \theta_1}{2}, \sin\frac{\theta_2 + \theta_1}{2}\right) = \frac{\cos\dfrac{\theta_2 - \theta_1}{2}}{A_u B_v - B_u A_v}\left(A_\eta B_v - B_\eta A_v, B_\eta A_u - A_\eta B_u\right), \tag{4.26}$$

to obtain the final solution (θ_1, θ_2).

To conclude, and in both the SDD and DDD methods, the estimated directional spectra follows Schäffer and Hyllested (1994), in which $S(f,\theta)$ was obtained by the convolution of the raw directional spectra with a smoothing window. Extra care needs to be taken as for some frequencies no solution can be found, hence the energy is neglected (5 to 10% in typical applications). A later study (Hawkes et al., 1997) showed that, for a variety of synthesised sea states that were generated in large wave basins, and despite the fundamental differences between these methods and the stochastic ones, the results were very similar to those obtained with the Maximum Entropy Principle.

4.2.3 Case Study Using Waverider Data

For single-point measuring systems, such as a directional Waverider buoy, valid options are the EMLM, IMLM, EMEP, SDD and DDD methods. Using the recorded displacements in three directions, sensitivity studies relating the major outputs (H_s, dominant direction, etc.) and the evaluation of the corresponding wave roses (polar plots) are valuable to quantify the relative importance of the inputs (like the frequency and directional resolutions, number of iterations, etc.). Figure 4.9 shows the estimated directional spectrum using selected methods for which the computational burden is comparable. Data from a single half-hour file is presented (the sampling frequency of the buoy is equal to 1.28 Hz). The dominant direction is sensitive to the method (oscillating between 280 and 290 compass bearings), but clearly the major differences are between the DFTM method and all

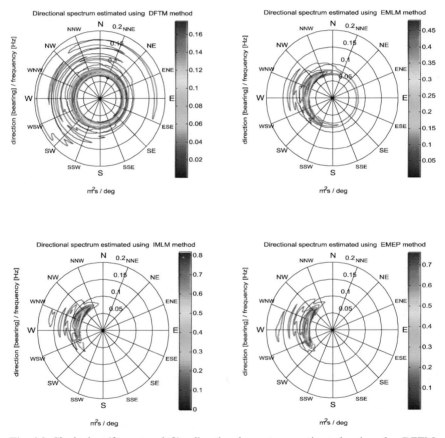

Fig. 4.9. Clockwise (from top left): directional spectrum estimated using the DFTM, ELML, IMLM and EMEP methods (30 *min* of data from a Directional Waverider buoy; $H_{m0} = 4.2\,m$)

the others. The insufficiency of such method in producing a narrow banded spectrum is also clear from Fig. 4.9. The EMLM still produces some fully circular contours (meaning, erroneously, that waves with a certain frequency would come from all directions, with low values of energy density) but greatly minimises the problems of the DFTM algorithm. The low spatial resolution of these methods is also reflected in the differences in the scale of the plots (being broader implies that the maximum values of energy density are naturally reduced). Finally the IMLM and EMEP outputs are similar in shape, though the dominant direction differs by 10 deg (IMLM: 290 compass bearings; EMEP: 280 compass bearings). It is emphasised that certain sea states might be more or less suitable for a specific estimation technique, but as a general rule the latter methods are both more stable and more precise. Several computational packages have been developed to derive the directional spectrum. Johnson (2002) presented the DIWASP toolbox (DIrectional WAve Spectrum) in a MATLAB environment. More recently, Cruz et al. (2007) developed a similar toolbox by emulating other wave analysis packages, applying the methodology to single point measurement systems and implementing a quality check procedure to handle corrupted data sets. Additional estimation methods, such as the wavelet method suggested by Donelan and Krogstad (2005), are to be assessed in futures versions.

4.3 Wave Growth, Travel and Decay

Once created, the sea state is well described as a linear, energy conserving process. However, the generation of waves is a strongly nonlinear process. Energy is input, through air pressure and shear stress, mostly to the high frequency end of the spectrum, and subsequently transferred to lower frequencies through nonlinear wave-wave interactions (Hasselmann 1962, 1963). Also, the waves created are not unidirectional, even in a steady wind, but have a characteristic directionally spread spectrum, centred on the wind direction.

   If a steady wind blows over a large expanse of sea, it gradually builds waves up until they reach a state limited by steepness, in which the rate of energy input is balanced by losses through breaking ('white horses'). The limiting state is well described by the simple scaling law formulae of Pierson and Moskowitz (1964)

$$E \infty U^4, \int_0^f S(y)dy = \exp(-k(fU)^{-4}), \qquad (4.27)$$

which depend only on the wind speed U. Later empirical work of Mitsuyasu (1975) proposed a formula for the limiting spectrum's directional spread. In this limiting state, the phase velocity is approximately equal to the wind speed, and the steepness (defined by a height to mean wavelength ratio) takes a fixed value. The Froude scale says that the rate of energy input is proportional to U^3 (the usual dependence of wind power on velocity), while the fetch (distance) required for the build-up is proportional to wavelength, and thus to U^2. Therefore the power in a fully developed wind sea is proportional to U^5 (for more detail, see Mollison, 1986).

When the wind speed decreases, waves spread out; their steepness thus decreases, and they cease to lose energy at a significant rate, as long as they remain in deep water. Such waves, separated from the conditions that created them, are known as swell. A typical oceanic sea state may consist of several spectral components: a locally generated wind sea, plus one or more swell components. The latter will typically have narrower spectra as regards both frequency and direction. These various components, to a good approximation, behave in a simple additive fashion, as regards both water movements and energy flows.

4.4 Wave Climate Estimation

Because of its variability on all time scales, a wave climate is not easy to measure. Wave buoys, scalar or directional, can give good estimates of the sea state from a 20 or 30 minute sample, but are too expensive to maintain for long term wave climate estimation, except at a small number of key reference sites. Wave climate estimates have therefore come to rely largely on computer hindcast wind-wave models; for example, the WAM model (WAMDI Group, 1988) now provides global coverage, and has been used, for example, to produce a wave energy atlas (WERATLAS) for European waters (Pontes et al., 1996). Figure 4.10 (from Mollison, 1986) shows estimates of the long-term average resource for Western Europe, made using an earlier wind-wave model. Members of the WERATLAS project team have continued to work over the last decade in the development of what became known as WorldWaves (Barstow et al., 2003). This is an integrated MATLAB package for calculating time series from directional wave spectra and associated statistics anywhere worldwide both in offshore and shallow waters. In a typical project, long term predictions would be made with the models and this would be validated with a few months of in-situ (coastal) buoy measurements. A comparison of synchronous wave model predictions would then allow any model errors to be removed over the longer term.

Normally, data is required for a period of at least 10 years, to allow for seasonal, year-to-year and longer-term variability (such as the El-Nino/Southern Oscillation in the Pacific). For sites in depths of less than $100\,m$, specific consideration of local bathymetry is necessary; and various numerical shallow water models are available that can be used to calculate the nearshore wave climate, starting from data for a deep water reference site (see, e.g., Southgate 1987). For instance, WorldWaves can now provide up to 50-years of wave energy time series at a coastal site, integrating long-term wave model hindcast data with multi-mission satellite altimetry (used for validation and calibration purposes) to specify the incoming deep water waves, long-term buoy data in a few sites, global coastline and bathymetric data and the state-of-the-art SWAN coastal wave model for transferring the offshore climate to the target coastal location.

Fig. 4.10. Early wave power estimates for Western Europe (Mollison, 1986)

In the following subsections a review of the main sources of wave data for mapping the wave energy resource is given. Although there are many different types of measurement principle and various types of numerical wave models, only the most relevant for wave energy applications are described. A full review was published recently in Kahma et al. (2005).

4.4.1 Buoys

Buoys have been used for measuring waves since the early 1960s. The most successful has been the Datawell Waverider which measures its own acceleration on a gravity stabilised platform. This sensor was refined in order to also measure the tilt in two orthogonal directions, then allowing for the first time the full directional spectra. The Hippy 40, as it became known, has been used since the early 1980s. In the mid-1980s, the seminal Wave Direction Calibration Project (WADIC) was carried out at the Ekofisk field in the central North Sea (Allender et al., 1989). All seven of the commercially available wave buoys were compared with regard to a number of platform based wave instrumentation including an array of downlooking laser profilers, current meter triplets, pressure transducers, among other sensors. All but one of the buoys used the Hippy sensor. The OCEANOR Wavescan buoy was particularly successful in WADIC and continues to collect directional wave data today, 23 years after its introduction. It is an example of a so-called heave/pitch/roll buoy. Datawell also paved the way for more reasonably priced directional measurements through its Directional Waverider, launched in the late 1980s. This buoy measures the heave and sway accelerations (outputting the corresponding displacements) rather than the heave, pitch and roll as in the buoys which preceded it. This meant that a more compact spherical hull could be used as measurement platform. Barstow and Kollstad (1991) showed that the Directional Waverider provided reliable data when compared with a Wavescan buoy.

Although the Hippy is still in use, notably in the NOAA network in the USA, most manufacturers of directional buoys have phased it out in preference for more robust sensors without moving parts. Examples are the Seatex Motion Reference Unit (MRU) and the OCEANOR Seasense. The latter is now used in all the buoys shown in Figs. 4.11 and 4.12 (see also Barstow et al., 2005), providing both heave/pitch/roll output for the Wavescan buoy and heave/displacement for the Seawatch buoys.

In offshore waters around the world, long-term buoy wave measurement networks are still relatively few and far between. Networks with directional measurements are even scarcer, even with the clear notion that directional information is essential for a number of applications (e.g.: forecasts, costal defence, etc). Some of the most important networks are:

1. The NOAA-NDBC buoy networks in the US (covering East and West coasts of the USA, the Gulf of Mexico and the Hawaiian Islands), in addition to the more recent Canadian network. Unfortunately, only one of the deep water buoys currently measures directional spectra.
2. The Indian National Data Buoy Programme which is currently probably the largest national program with offshore directional buoys used as standard.
3. National networks in Spain, Greece, France and Italy although most buoys are rather close to the coast and hence quite site specific.
4. Long term measurements carried out in Norwegian waters and the North Sea for the offshore industry.

Fig. 4.11. The Wavescan heave/pitch/roll metocean data buoy has been used for directional wave measurements since 1984 and there are today more buoys than ever in operation, mainly in deep ocean, high current conditions. This buoy has, for example, collected data in the Agulhas current for one year, possibly the toughest environment in the world

Fig. 4.12. The Seawatch buoy (left) and the Seawatch Mini mk. II (right) are used mainly for coastal monitoring. These buoys are examples of heave/displacement buoys. The mini buoy would normally be used for wave energy resource mapping where other metocean parameters such as wind data were not considered to be required

4.4.2 Satellite Altimeters

The back-scattered signal from satellite altimeters, when properly interpreted, can provide significant wave height measurements close to the accuracy of a buoy, from an orbit of typically 1,000 km (see, e.g., Krogstad and Barstow, 1999). Measurements are made each second, whilst the satellite flies over a repeat net of ground tracks at about 6 km/s. This provides enormous amounts of wave data worldwide, and with, at present, a steady flow of new data from three or more operational satellites, millions of new observations are becoming available each month. Global long-term satellite altimeter measurements have been performed during 1985-1989 by the US Navy's Geosat and the Geosat-Follow-on mission from 2000 by ESA's ERS-1 (from 1991 to 1996), ERS-2 (1996 to 2003), EnviSat (launched in 2002), and most importantly for our purposes due to its longevity, the US/French Topex/Poseidon mission from 1992 to 2005. The Topex-Follow-on mission (Jason) has provided data since 2002. Full resolution altimeter significant wave height and wind speed data for most of these satellites can be found on the World Wave Atlas.

The altimeter is basically a radar which sends a pulse down to the ocean surface at nadir. The significant wave height is obtained from the slope of the leading edge of the return pulse, while the total backscatter gives us the wind speed.

Each satellite altimeter has to be validated and calibrated in order to remove altimeter-dependent biases on significant wave height and this is generally and most reliably done by comparison with long-term offshore buoy data. When comparing temporally varying significant wave height data from the buoy with spatially varying data from the satellite it is important that the buoy measurments are from areas where the gradients in significant wave height are rather small, which means in practice moorings far from coasts as wave conditions often vary rather strongly near to coasts both due to geographical sheltering effects, fetch limited wave growth in offshore blowing winds and shallow water effects.

Emphasis is given here to the Topex satellite altimeter as this is the most successful mission to date. Altimeter data from both the ERS missions as well as Envisat have routinely been assimilated into the ECMWF wave models (see 4.4.3) and are therefore not independent of the model data. Algorithms for the correct interpretation of the back-scattered radar return pulse from satellite altimeters have been gradually improved (see, e.g., Krogstad and Barstow, 1999; Queffeulou, 2004). The accuracy of the Topex altimeter can be seen in the satellite – buoy comparison shown Fig. 4.13. All Topex altimeter data globally for 1992 to 2002 have been analysed applying the bias corrections described above as well as an automatic data control, removing, for example, unphysical along-track variations in wave height. The different satellites fly on different exact repeat orbits, returning to exactly the same ground track on each repeat. The repeat time is 10 days for Topex and Jason, 17 days for Geosat and GFO and 35 days for the ESA satellites. For this reason, the ESA satellites have a much denser coverage spatially, but longer return period to the same location.

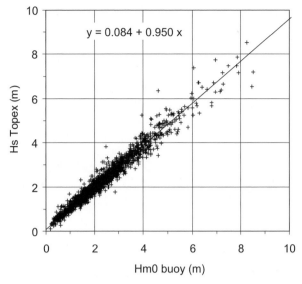

Fig. 4.13. Comparison of significant wave height between the Topex Side B altimeter for 1999–2002 and 13 NOAA buoys; coincident data for Side B. The best fit regression is used to calibrate the satellite wave heights before use

4.4.3 Global Wave Models

Many meteorological centres today run wave models regionally and globally and dedicated long-term hindcasts have also been performed. The wave models simulate the growth, decay and propagation of ocean waves based on input winds over the area in question. In the aforementioned WERATLAS project, it was confirmed that the ECMWF (European Centre for Medium-Range Weather Forecasts) wave model was the best available at that time. ECMWF have maintained that lead (see Fig. 4.14). The reasons for this are related both to the high quality of the ECMWF wind fields and, amongst other things, the assimilation of, in particular, satellite altimeter and also synthetic aperture radar wave data in the models. The WAM wave model (Komen et al., 1994) has been operational at ECMWF since 1992. Further details of the model itself with references can be found at the ECMWF website. The development of the WorldWaves package, through various EU projects, also adopted the ECMWF data.

The current global wave model data base adopted by the WorldWaves project are 0.5° lat/long resolution data from ECMWF's operational global WAM model covering the period December 1996 to present (updated monthly). In addition, for the Mediterranean, Baltic and Black Seas, data are sourced from the finer resolution Mediterranean model. In order to reduce the amount of data, only 1.5 data points are retained in the open ocean areas.

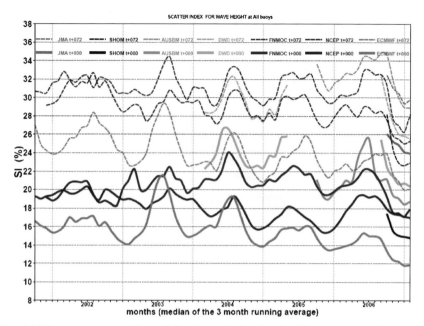

Fig. 4.14. Long-term comparison of the Scatter Index for significant wave height, SI, between model and a multi-buoy data set. The superiority of the ECMWF analysis and forecasts is evident here (the SI is the standard deviation of the difference between the model and buoy normalized by the buoy mean). Figure courtesy of Peter Janssen at ECMWF

In addition, ECMWF have also completed a long-term global ocean wind and wave hindcast (ERA-40 data) which can provide data back to 1957, and, when combined with the operational data up to 50-year time series are available. These data sets are therefore often used in projects requiring longer-term data. The ERA-40 data are, however, run on a coarser 1.5° grid, reducing their usefulness in enclosed sea areas.

The output from the WAM model is the full directional spectrum each 6 hours. When the model data are used as input to a shallow water model (SWAN in World-Waves), the best results are obtained. These sets are, however, voluminous, and, instead, the wave parameters in Table 4.2 are the default data currently available globally under WorldWaves. Based on the significant wave height, energy period and direction for wind sea and swell separately together with the peak wave period, the full directional wave spectrum is then synthesised when carrying out offshore to nearshore transformations. The full directional spectra files from ECMWF are available from mid-1998 from the operational model at the 0.5° resolution and for the entire ERA-40 hindcast period from mid-1957 to mid-2002 with 1.5° resolution. These data sets are frequently used in projects requiring the highest accuracy.

It is important to understand how the separation of the WAM directional spectra into wind sea and swell parts is carried out. First, the windsea part of the spectrum is defined by

$$1.2 \, (28/c) \, U^* \cos(\theta - \theta_w) \geq 1, \qquad (4.28)$$

Table 4.2 Wave and wind parameters available from the global archives

Parameter long name and units	Alternative short name
Significant wave height, *m*	H_{m0} or H_s
Mean wave direction, *deg*	
Peak period of 1d spectra, *s*	T_p
Energy period, *s*	T_{m-10}, T_{-10} or T_e
Significant wave height for the wind waves, *m*	$H_{m0\,ws}$ or $H_{m0\,w}$
Mean direction of wind waves, *deg*	M_{dirws}
Mean period of wind waves, *s*	$T_{m-10\,ws}$ or $T_{m-10\,w}$
Significant wave height for the swell, *m*	$H_{m0\,sw}$ or $H_{m0\,s}$
Mean direction for the swell, *deg*	M_{dirsw}
Mean period for the swell, *s*	$T_{m-10\,sw}$ or $T_{m-10\,s}$
Wind speed equivalent to 10 *m* height, 10 *min.* average, *m/s*	W_{sp} or U_{10}
Wind direction Eq. to 10 *m* height, 10 *min.* average, *deg*	θ_w

where $c = c(f)$ is the phase speed of the waves with frequency f and U^* is the friction velocity. If τ is the surface stress then

$$\tau = U^{*2} = C_d\, U_{10}^2, \tag{4.29}$$

where C_d is the drag coefficient and U_{10} is the 10 *m* wind speed. θ is the mean propagation direction of the waves and θ_w is the direction of the wind. What is not wind sea is then classified as swell.

In some parts of the world (e.g., the Pacific coasts of South America), two or more swells can frequently coexist. In this case, the three parameters describing the swell spectrum will poorly describe the swell energy-direction spectrum. In such areas, the full directional spectra data must then be used when trying to predict coastal wave conditions. At any offshore location in the Pacific in the northern winter (October to March), crossing northerly and southerly swells will frequently occur. Using parameter data for predicting coastal wave energy resources will for this reason either over- or under-predict, often significantly, even if the models were perfect.

4.4.4 Model Validation and Calibration

The satellite altimeter data represent a high quality, independent and globally covering wave and wind data set which can be used for the direct validation of wave model data, even though this is limited to significant wave height and wind speed. The correlation between satellite and buoy data were confirmed using long-term data sets from the NOAA wave buoy network. The WAM model significant wave heights are compared first against 13 NOAA buoys and then against satellite altimeter data sampled from the area around each of the same buoys. From the resultant scatter plots for each of the 13 locations, various statistics were calculated. The results in the form of the bias and the scatter index are presented in the maps in Figs. 4.15 and 4.16. As it can be seen, the results are more or less identical independent of whether one chooses the buoy or satellite data for the validation. This gives us the confidence to use the Topex data globally as a reliable reference, as if a worldwide buoy network was available.

The wave model data is available on a regular 0.5° lat-long grid. The satellite data are, on the other hand, available about each 6 km each N-days (where N is the exact repeat period for each satellite) along ground tracks globally. In the World-Waves project, a global offshore wave model data set was created using the corrected (against buoy data) Topex data from 1996 to 2002 to validate and subsequently calibrate the ECMWF model data. This was found to be worthwhile as there was typically a systematic bias on the raw model data. Removing that bias significantly improves the data quality, particularly in enclosed seas such as the Mediterranean. Not all grid points are close enough to a Topex satellite track to be useful for validation purposes (i.e., considered to be derived from the same wave climate population). Thus, a sub-set of calibration points were identified. Of a total of about 10,000 wave model grid points globally about 5,700 are calibration points (in oceanic areas, practically all points are validation points as there are not strong wave climate gradients in such areas) and a larger spatial separation between model and satellite data for calibration is acceptable. Near coastal waters, the selection of calibration points and satellite data extraction locations used for validation has to be done manually as it is not necessarily the closest point which is the best choice in such areas with strong wave gradients.

The results of the global model validation are presented in Fig. 4.17 (correlation coefficient). Although showing a largely very high correlation, the model tends to underestimate the significant wave height in a systematic way. Using the regression coefficients to calibrate the model data removes the bias and gives more accurate offshore boundary conditions for coastal wave modelling. Since 2002, significant improvements have been made to the ECMWF models and the bias is now much reduced. The global validation is currently being repeated using the Jason altimeter data (this satellite flew the same ground tracks as Topex and the same validation points can therefore be used). The validation can be further refined by using data from all satellite altimeter missions. This is particularly advantageous in areas with relatively strong wave climate gradients.

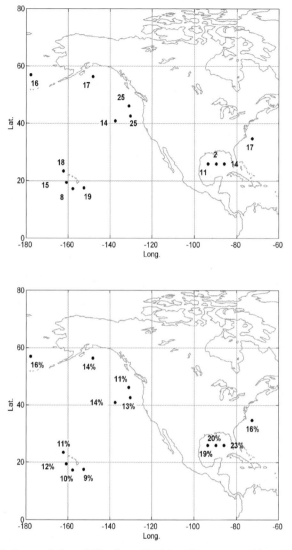

Fig. 4.15. Validation statistics of H_{m0} from WAM against NOAA buoy data for 1996 to 2002. Top: mean difference in *cm*; Bottom: residual Scatter Index in %

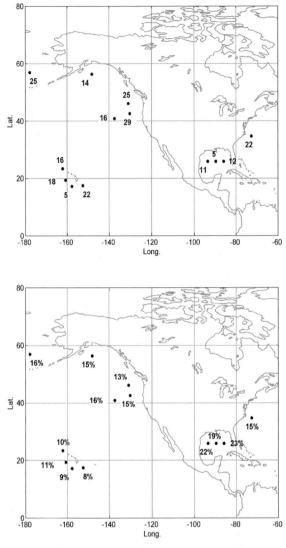

Fig. 4.16. Validation statistics of H_{m0} from WAM against calibrated Topex data for the locations of the NOAA buoys in Fig. 4.16 for 1996 to 2002. Top: mean difference in cm; Bottom: residual Scatter Index in %

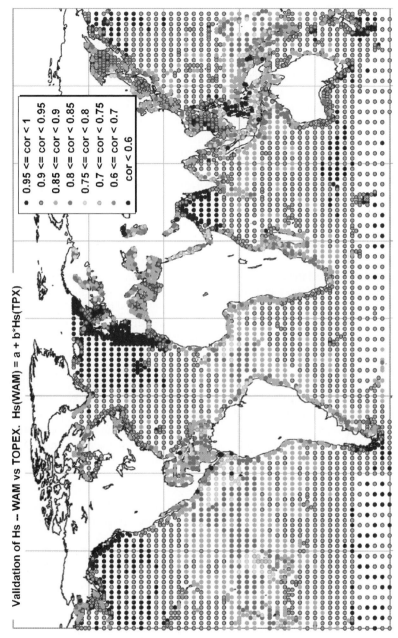

Fig. 4.17. Global map showing the correlation coefficient between the ECMWF WAM operational model data and Topex/Poseidon data for all global validation points for December 1996 to 2002

The ECMWF model was then finally calibrated on a global basis for inclusion in the WorldWaves software package. This was done by adjusting significant wave heights and, analogously, the wind speed on a point-by-point basis for all global grid points. For oceanic points, wave periods were adjusted to conserve the wave steepness in some enclosed seas where wind sea dominates. Directional wave spectra can also be calibrated on a case-by-case basis.

Such an approach, combining an 'Atlas' data base for offshore reference sites with computer tools for calculating the resource at specific locations, was pioneered for the evaluation of the UK wave energy resource by Whittaker et al. (1992). This two-level methodological approach has major advantages of flexibility and long life, in that the components – both data base and computer tools – can be updated individually as relative technological advances are made.

4.5 Wave Energy in Shallow Water

Once waves enter shallow water, roughly defined as water of depth less than half a wavelength, they change in a number of respects. For a start, the basic solutions of the equations for surface waves under gravity are different, because of the lower fixed boundary to water movement. In the extreme, where the wavelength is much greater than the water depth, the group velocity is no longer half the phase velocity but equal to it. This means that waves are no longer dispersive, but travel as solutions: the classic example of this is a tsunami, where waves with a wavelength of 100 km or more can be created by a sudden fall or rise in the sea floor. On a much more modest scale, the same behaviour can be seen when an ocean swell runs up a gently sloping beach. As ocean waves enter shallow water, they slow down; for a wave approaching at an angle to the depth contours this has the effect of changing its direction; the wave spectrum consequently becomes narrower directionally, with principal direction close to perpendicular to the depth contours.

In slowing down, the wavelength decreases and height increases, leading to breaking. Waves lose energy both through breaking and through bottom friction, which can be significant in depths from 50 m to around 100 m (Mollison, 1983).

In WorldWaves, the offshore model data sets described in section 4.4 are used as boundary conditions to the state-of-the-art SWAN model (Booij et al., 1999) for calculating coastal wave conditions anywhere worldwide. SWAN then performs the offshore to nearshore transformation for each time step, generating a new time series of wave parameters and/or full directional spectra at the nearshore target location(s) of interest. Geographic tools are available for quickly selecting any geographical area worldwide for the analysis, setting up a computational area with nesting and selecting the offshore boundary grid points and target nearshore locations.

SWAN is a third-generation spectral wave model, developed for the calculation of the propagation of random waves from deep to shallow water, accounting for the following physics:

- Wave propagation in time and space, including the effects of shoaling, refraction due to current and depth, frequency shifting due to currents and the effects of non-stationary depth;
- Wave generation by wind;
- Triad and quadruplet non-linear wave-wave interactions;
- Whitecapping, bottom friction, and depth-induced breaking;
- Wave induced setup;
- Transmission through and reflection from obstacles.
- Wave diffraction

A typical project will involve the following stages:

- Select and recalibrate the offshore model points using validation data from all satellite altimeters providing data near to the grid points to be used in the offshore to nearshore transformation. Calibrate the full offshore directional spectra.
- Check and update the bathymetric data from admiralty charts.
- Set up the computational grid with suitable choices of grid meshes (both in the outer and nested grids) and other relevant SWAN parameters controlling model physics and numerics.
- Run SWAN for the time period of interest, taking account of water level variations in shallow target locations where tides and storm surge can significantly impact the resource and design conditions.
- Quality control the nearshore time series data.
- Calculate the required wave statistics.

In some cases, one might have satellite data near to the target location which could be used to check the offshore to nearshore transformation. If not, it is wise to carry out short term buoy measurements at the target location to validate the model results (note that the SWAN simulation can be updated for exactly the same time period as the measurements to give the possibility of direct validation). If bias is found between the SWAN simulations and the measurements, this information can be used, for example, to calibrate the long-term model predictions.

As waves move from deep water into shallower water, the ratio of maximum to mean wave energy can decrease considerably. In a simulation where the wave transformation across the shelf and towards the coast of Southern Australia was modelled using the SWAN wave model the maximum storm wave energy is reduced relative to the mean by about a factor of 3 from 50 to $6\,m$ water depth. It also became clear that the extremes are reduced by a much large amount than the normal (average) sea states.

4.6 Discussion

In this chapter a detailed account of the origin and methods to estimate the wave energy resource were given. Resource assessment studies are essential when evaluating the possible locations for a wave energy project, and site specific measurements and surveys are necessary prior to any final decision on where to install a wave farm. As demonstrated, for a first selection of suitable areas numerical models play an important role and their validation is critical.

In an operational environment, the understanding of the incoming wave field and its correlation with the performance of the wave energy converter(s) is crucial, hence the emphasis given to the several methods to estimate the directional spectrum. Wave measurement systems that are capable of delivering such information, such as surface buoys or bottom-mounted acoustic Doppler current profilers (ADCPs), are commercially available but it is likely that a new wave energy industry will have additional specifications for such devices like, e.g., increased sampling frequency, longer autonomy, etc.

This contribution aims to show to the readers the areas in the globe where wave energy projects are more likely to be installed and to present the considerable amount of experience that can be emulated from the offshore industry and from oceanography. Those interested in extending their knowledge in this field can find suitable references in the articles mentioned throughout the chapter and in key textbooks such as Goda (2000) and Tucker and Pitt (2001). A vital area (forecasting) should also be approached by such readers, as it strongly affects the O&M aspects of a wave energy farm.

References

Akaike H (1973) Information theory and an extension of the maximum likelihood principle. In: Petrov, Csaki (eds) Proc 2nd Int Symp Inform Theory. Akademiai Kiado, Budapest, pp 267–281

Allender J, Audunson T, Barstow S, Bjerken S, Krogstad HE, Steinbakke P, Vartdal L, Borgman L, Graham C (1989) The WADIC Project: A comprehensive field evaluation of directional wave instrumentation. Ocean Eng 161:505–536

Barber NF (1961) The directional resolving power of an array of wave detectors. Ocean Wave Spectra. Prentice-Hall, pp 137–150

Barstow SF, Aasen SE, Mathisen JP (2005) WaveSense marks the first 20-years service for the Wavescan buoy. Sea Technol, pp 53–57

Barstow SF, Kollstad T (1991) Field trials of the Directional Waverider. In: Proc 1st Int Off Pol Eng Conf (ISOPE), pp 55–63

Barstow SF, Mørk G, Lønseth L, Schjølberg P, Machado U, Athanassoulis G, Belibassakis K, Gerostathis T, Stefanakos CN, Spaan G (2003) WorldWaves: Fusion of data from many sources in a user-friendly software package for timely calculation of wave statistics in global coastal waters. Proc 13th ISOPE Conf. Oahu, Hawaii, USA

Benoit M (1992) Practical comparative performance survey of methods used for estimating directional wave spectra from heave-pitch-roll data. In: Proc 23rd Int Conf Coastal Eng (ASCE). Venice, Italy, pp 62–75

Benoit M, Frigaard P, Schäffer HA (1997) Analysing Multidirectional Wave Spectra: A Tentative Classification of Available Methods. In: Proc IAHR Sem Multidirectional Waves Interact Struct. 27th IAHR Congress. San Francisco, USA, pp 131–158

Booij N, Ris RC, Holthuijsen LH (1999) A third-generation wave model for coastal regions. 1. Model description and validation. J Geophys Res 104(C4):7649–7666

Capon J, Greenfield RJ, Kolker RJ (1967) Multidimensional maximum-likelihood processing of a large aperture seismic array. Proc IEEE 55:92–211

Cruz J, Mackay E, Martins T (2007) Advances in Wave Resource Estimation: Measurements and Data Processing. Proc 7th Eur Wave Tidal Energy Conf. Porto, Portugal

Donelan M, Krogstad H (2005) The Wavelet Directional Method. In: Khama et al. (eds) Measuring and analysing the directional spectra of ocean waves. EU Cost Action 714.71–80

Goda Y (2000) Random seas and design of maritime structures, 2nd edn. World Scientific Publishing

Hagerman G (1985) Oceanographic design criteria and site selection for ocean wave energy conversion. In: Proc IUTAM Sym Lisbon 1985. Hydrodynamics of Ocean Wave Energy Utilization. Springer Verlag

Hashimoto N, Kobune K, Kameyama Y (1987) Estimation of directional spectrum using the Bayesian approach and its application to field data analysis. Rep Port Harbour Res Inst, Vol. 26

Hashimoto N, Nagai T, Asai T (1994) Extension of Maximum Entropy Principle for directional wave spectrum estimation. In: Proc 24th Int Conf Coastal Eng (ASCE). Kobe, Japan, pp 232–246

Hasselmann K (1962) On the nonlinear energy transfer in a gravity-wave spectrum. 1. General theory. J Fluid Mech 12:481–500

Hasselmann K (1963) On the nonlinear energy transfer in a gravity-wave spectrum. 2. Conservation laws, wave-particle correspondence, irreversibility. J Fluid Mech 15:273–281

Hawkes PJ, Ewing JA, Harford CM, Klopman G, Stansberg CT, Benoit M, Briggs MJ, Frigaard P, Hiraishi T, Miles M, Santas J, Schäffer HA (1997) Comparative Analyses of Multidirectional Wave Basin Data. In: Proc IAHR Sem Multidirectional Waves Interact Struct, 27th IAHR Congress. San Francisco, USA, pp 25–88

Isobe M, Kondo K, Horikawa K (1984) Extension of MLM for estimating directional wave spectrum. In: Proc Symp Descript Model Directional Seas, Paper A-6. Lingby, Denmark

Isobe M (1990) Estimation of directional spectrum expressed in a standard form. In: Proc 22nd Int Conf Coastal Eng (ASCE), pp 467–483

Johnson D (2002) DIWASP, a directional wave spectra toolbox for MATLAB: User Manual. Research Report WP-1601-DJ (V1.1). Centre for Water Research, University of Western Australia

Kahma K, Hauser D, Krogstad HE, Lehner S, Monbaliu JAJ, Wyatt L (2005) Measuring and Analysing the Directional Spectra of Ocean Waves. EU COST Action 714. EUR 21367. ISBN 92-898-0003-8

Kinsman B (1965) Wind Waves. Prentice-Hall, pp 460–471

Komen GJ, Cavaleri L, Donelan M, Hasselmann K, Hasselmann S, Janssen PAEM (1994) Dynamics and Modelling of Ocean Waves. Cambridge Univ Press

Krogstad HE, Barstow SF (1999) Satellite Wave Measurements for Coastal Engineering Applications. Coastal Eng 37:283–307

Kuik AJ, van Vledder G, Holthuijsen LH (1988) A Method for the Routine Analysis of Pitch-and-Roll Buoy Wave Data. J Phys Oceanogr 18:1020–1034

Longuet-Higgins MS, Cartwright DE, Smith ND (1963) A variational technique for extracting directional spectra from multicomponent wave data. J Phys Oceanogr 10:944–952

Longuet-Higgins MS (1985) Accelerations in steep gravity waves. J Phys Oceanogr 15:1570–1579

Lygre A, Krogstad H (1986) Maximum Entropy Estimation of the Directional Distribution in Ocean Wave Spectra. J Phys Oceanogr 16:2052–2060

Mitsuyasu H, Tasai, F, Suhara T, Mizuno S, Ohkuso M, Honda T, Rikishi K (1975) Observations of the directional spectrum of ocean waves using a cloverleaf buoy. J Phys Oceanogr 5:750–760

Mollison D (1983) Wave energy losses in intermediate depths. Appl Ocean Res 5:234–237

Mollison D (1986) Wave climate and the wave power resource. In: Evans DV, Falcao Af de O (eds) Hydrodynamics of Ocean Wave–Energy Utilization. Springer–Verlag, Heidelberg, pp 133–156

Mollison D (1994) Assessing the wave energy resource. In: Barnett V, Turkmann KF (eds) Statistics for the Environment 2: Water Related Issues. Wiley, pp 205–221

Pawka SS (1983) Island shadows in wave directional spectra. J Geophys Res 88(C4):2579–2591

Pierson WJ, Moskowitz L (1964) A proposed spectral form for fully developed wind seas based on the similarity theory of SA Kitaigorodskii. J Geophys Res 69:5181–5190

Pontes MT et al. (1996) An atlas of the wave-energy resource in Europe. J Offshore Mech Arctic Eng 118:307–309

Queffeulou P (2004) Long term validation of wave height measurements from altimeters. Marine Geod 27:495–510

Sand SE (1984) Deterministic Decomposition of Pitch-and-Roll Buoy Measurements. Costal Eng 8:242–263

Schäffer HA, Hyllested P (1994) Analysis of Multidirectional Waves Using Deterministic Decomposition. In: Proc Int Symp: Waves – Physical and Numerical Modelling. Univ British Columbia, Vancouver, Canada, pp 911–920

Southgate HN (1987) Wave prediction in deep water and at the coastline. Report SR 114. HR Wallingford, UK

Tucker MJ, Pitt EG (2001) Waves in Ocean Eng. Elsevier Science Ltd, London

WAMDI Group (1988) The WAM Model – a third-generation ocean wave prediction model. J Phys Oceanogr 18:1775–1810

Whittaker TJT et al. (1992) The UK's Shoreline and Nearshore Wave Energy Resource. UK Dept of Trade & Industry, ETSU WV 1683

Woolf DK, Cotton PD, Challenor PG (2003) Measurement of the offshore wave climate around the British Isles by Satellite Altimeter. Philos Trans R Soc Lond Ser A, pp 27–31

5 Numerical and Experimental Modelling of WECs

The design of a wave energy converter relies heavily on results from numerical simulations and experiments with scale models. Such results allow not only fundamental design changes but also the optimisation of selected configurations. For ongoing development, and particularly at an early stage, numerical models give the flexibility of assessing a large number of versions at a relatively low cost. Physical models are then tested in wave tanks to validate the numerical simulations and to investigate phenomena which are not evidenced by the computational packages. This chapter provides an overview on the numerical techniques that have been used to model the hydrodynamics of wave energy converters (WECs), details on wavemaker and wave tank design, guidelines on experimental techniques and finally a case study related to one of the most studied concepts which reached the full-scale prototype stage.

5.1 Fundamentals of Numerical Modelling

João Cruz

Garrad Hassan and Partners Ltd
Bristol
England, UK

When designing a wave energy converter, and at several stages of development, numerical modelling is pivotal. In this section only the hydrodynamic numerical modelling is considered. It is critical at an early stage, as it allows several iterations of the same concept to be tested in the fastest way possible, but it is equally critical in later stages, when envisaging new generations of machines and/or trying to optimise control routines.

Chapter 3 already focused the differences between working in the frequency or in the time domain. Basically frequency domain solutions of the equations of motion rely on the assumption that the incident waves are the result of the superposition of single harmonic waves. Linear wave theory is used (i.e.: body motions are assumed small when compared with the wavelength) and thus the problem can be split into two other: the diffraction problem, where the body is fixed and subject to an incoming wave field, and the radiation problem, where the body is forced to move in otherwise undisturbed fluid. The velocity potential is obtained the sum of the diffraction potential and all the radiation potentials, which can be associated with the wave exciting forces and moments and with the hydrodynamic coefficients (added mass and damping), respectively. With such results the motions of the body can be derived, and these are usually expressed in a non-dimensional form through the response amplitude operator (RAO). Additional constrains can be introduced by external mass, damping of stiffness matrices (e.g.: to assess the influence of different mooring arrangements or of different power take-off settings).

When non-linear effects are judged to be significant, time domain solutions need to be implemented. There are several ways to derive such models, but in the majority of cases the non-linear analysis is based on direct pressure integration over the body surface at each time step of the simulation (McCabe, 2004). Simplifications, like reducing the body surface to a mean wetted-surface, can be implemented, leading to a considerable reduction in the computational time that is required to run the simulations at the expense of the maximum possible accuracy. The main difference to the frequency domain approach is therefore the possibility of adding non-linear effects in the equations of motion, which are typically linked with convolution integrals that take into account effects that persevere after the motion of the body stops (hence such integrals are sometimes referred to as 'memory functions').

To this date the frequency domain approach as been used in a much larger number of applications than the time domain equivalent. The fact that this book is dedicated to an overview of the several stages of development of wave energy converters lead to the choice of emphasising such approach in this section, as frequency domain models are particularly useful to those who are new to the field and are simultaneously valid tools to the more experienced readers. Firstly an introduction to panel methods is given, while in 5.1.2 details regarding specific studies involving several wave energy converters are addressed.

5.1.1 Introduction to Panel Methods

Panel methods, also referred to as Boundary Element Methods (BEM) in a wider engineering perspective, are computational methods used to solve partial differential equations which can be expressed as integral equations. Typically, BEM are applicable to problems where the Green function can be calculated. A thorough review on panel methods in computational fluid dynamics is presented in Hess (1990). Relevance is given to aerodynamics, but the main assumptions (e.g.: potential flow) and principles are relevant to general fluid mechanics problems. A sub-chapter focusing exclusively in free surface applications is also presented. The two common problems given as examples are:

1. a ship at constant forward speed in an undisturbed wave field;
2. a fixed structure facing incoming regular waves.

Note that an extension of the case described in 2. also includes the problem of an oscillating body in an undisturbed media, which is particularly relevant in wave energy conversion.

In 1., Rankine type sources were originally used and both the submerged portion of the hull and the surrounding free surface were panelised. In 2. the singularities are more complex and only the body is discretised. Newman (1985) developed a practical technique to address such issues and later applied it to a variety of case studies. Many references can be found in the literature, but given the introductory nature of this section Newman's key communications are followed.

A review on the basic principles that rule the application of panel methods in marine hydrodynamics is given in Newman (1992). It is emphasised that many of the common problems in this subject, like wave resistance, motions of ships and offshore platforms, and wave structure / interaction can be addressed following potential flow theory, where viscous effects are not taken into account. The objective is therefore to solve the Laplace equation with restrictions imposed by boundary conditions. The domain is unbounded (with the solution being specified at infinity), so a numerical approach that arranges sources and (optionally) normal dipoles along the body surface can be considered to solve the hydrodynamic problem. Two different representations can be considered, following Lamb (1932): the potential or the source formulation. In the first one, Green's theorem is used, and the source strength is set equal to normal velocity, leaving the dipole moment, which is equal to the potential, unknown. On the other hand, the source formulation relies solely on source terms with unknown strength to describe the potential. In both cases, similar Fredholm integral equations can be solved.

The pioneer work of Hess and Smith (1964) is mentioned by Newman, in which the source formulation was used for three-dimensional bodies of arbitrary shape. For the first time, a linear system of N algebraic equations was derived by establishing boundary conditions at a collocation point on each of the N panels that were used to describe the fluid domain.

Hess and Smith (1964) also derived the analytical expressions for the potential and velocity induced by a unit density source distribution on a flat quadrilateral panel, avoiding numerical integration that could lead to erroneous results when the calculation point is in the vicinity (or on) the panel.

To conclude his keynote paper, Newman (1992) also points out the basic differences between the source and the potential formulation. It is mentioned that the computational effort required for both approaches is roughly equivalent. The differences manly involve:

1. issues linked with thin bodies, where normal dipoles prove to be more stable than sources;
2. the fluid velocity, that in the source formulation can be evaluated from the first derivatives of the Green function, whereas in the potential formulation the second derivatives are necessary. Nevertheless the latter is not robust when using flat panels to discretise a curved surface, given that the velocity field induced by the dipoles changes quickly over distances similar to the panel dimensions;

3. 'irregular frequencies', which are related to flawed solutions in problems involving bodies that pierce the free surface. It is a common problem of both approaches but more likely to appear in the source formulation (Yeung, 1982).

When choosing a method to solve a specific problem there are two main versions that can be followed: a low-order method, where flat panels are used to discretise the geometry and the velocity potential, and a high-order method, which uses curved panels, allowing (in theory) a more accurate description of the problem. The high-order method has inherent advantages and disadvantages when compared with the low-order equivalent. Lee et al. (1996a) and Maniar (1995) showed the increase in computational efficiency, i. e., the method converges faster to the same solution when the number of panels is increased in both. The possibility of using different inputs for the geometry, like an explicit representation, also contributes to an increase in accuracy. Another significant advantage relies on the continuity of the pressure and velocity on the body surface, which is relevant for structural design. The main disadvantage is linked with the lack of robustness that the method yields, failing to converge in some cases. Such issues can be particularly severe when a field point is in the vicinity of a panel or near sharp corners.

The concerns associated with the computational burden have been progressively loosing the initial importance as computers evolved. However such issues remain clear when developing a new code, particularly when studying complex problems. It is also clear that the pre-processing, linked with the calculation of the panel representation and relevant parameters, like areas and moments of inertia, and the solution of the linear system itself, are the steps which require the majority of the effort.

Newman and Lee (1992) performed a numerical sensitivity study on the influence that the discretisation has on the calculation of wave loads. The effects of the number of panels and their layout were investigated. Convergence tests were also performed. Such focus on accuracy was clear since the early simulations, but computational limitations were clear. A classic case is the one described in Eatock Taylor and Jeffreys (1985), where the hydrodynamic loads calculated are of 'uncertain accuracy'. The recent hardware developments allow much more detailed studies.

Typically, increasing the number of panels used in the geometric and hydrodynamic representations will lead to an increase in accuracy. One important exercise that should never be neglected when developing a code is the numerical verification of the results, ensuring that the solution is not divergent or convergent to the wrong solution. Naturally validation, i.e., the comparison with physically derived results, is also a key factor. The computational time required to solve the problem also increases with the number of panels, so an optimal ratio between accuracy and the number of panels can be derived. Equally relevant is the panel layout, which can be solely responsible for invalid solutions.

A few basic qualitative guidelines are pointed out by Newman and Lee (1992). These can be summarised in the following way:

1. near the free surface, short wavelengths demand a proportionately fine discretisation;

2. local singularities, induced by (e.g.) sharp corners, tend to require fine local discretisation;
3. discontinuities on the characteristic dimension of the panels should be avoided; ideally a cosine spreading (also referred to as spacing) function should be used for the panel layout (width of the panels is proportional to the cosine of equally-spaced increments along a circular arc);
4. problems involving complex geometries can require a high number of panels even for simple calculations (e.g.: volume).

Convergence tests are usually the answer to select the optimal discretisation. For representative wavelengths and for the same mesh layout, the number of panels is increased and the output evaluated. For a high enough value, the increase in the number of panels will not lead to a significant change in the solution.

The authors mention the word 'error' when comparing different numerical solutions, which according to many references is fundamentally wrong (Roache, 1998; Eça and Hoekstra, 2000). Recently several authors have conducted verification studies using numerical results related to different concepts (e.g.: Cruz and Payne, 2006; Sykes et al., 2007).

Newman and Lee (1992) also mention, using the low-order method (flat quadrilateral panels), a numerical 'error' of 0.1 % to 10 %, emphasising the need to validate all the results. The authors are directly associated with the development of a BEM code named WAMIT, at the Department of Ocean Engineering of the Massachusetts Institute of Technology (MIT). This code was initially verified through comparison with analytical solutions. Validation exercises were also conducted using experimental results. Together with these procedures, benchmarking with similar codes also has an important role to ensure that a code does not converge to the wrong answer. Examples of topics studied by this research group include wave loads on offshore platforms, time-domain ship motions, ship interactions in a channel, wave energy conversion and, on a more theoretical level (with implications to all fields), the development of a panel method based on B-splines. This high-order approach is justified by some fundamental differences, namely the possibility of describing more accurately the geometry and the velocity potential. Recent developments are presented in Newman and Lee (2002).

Other research groups have been actively involved in BEM code development. A particular strong one with regard to the study of wave energy conversion can be found at the École Centrale de Nantes (Laboratory of Fluid Mechanics). A complete suite of packages for several seakeeping problems has been under development since 1976 at ECN, resulting in:

1. AQUADYN, for general problems without forward speed;
2. AQUAPLUS, which assumes an encounter frequency for a moving vessel;
3. CUVE, which solves the problem of a vessel with internal tanks.

AQUADYN is a BEM code very similar to WAMIT, in particular to its low-order panel method solver. Several examples of the use of AQUADYN can be found in the literature (e.g.: Brito-Melo et al., 1998). Details about specific studies related to wave energy conversion involving AQUADYN and WAMIT, two of the most prominent BEM codes used in the field, are given in section 5.1.2.

5.1.2 Applications of Panel Methods to Wave Energy Conversion

It is fair to say that Salter's early work regarding wave energy absorption by different shapes, published in a wide audience journal like Nature (Salter, 1974), lead to similar studies in research groups spread worldwide. The first numerical simulations soon followed. A first attempt to numerical reproduce Salter's experiments was made by Katory (1976), in which inconsistent results were obtained (e.g.: the derived added mass matrix was not symmetric). Mynett et al. (1979) presented the first comprehensive numerical study with regard to cam shaped wave energy converters, following the experimental work performed by Salter on such shapes and the theoretical work of Mei (1976) and Evans (1976), where the principles of basic power take-off systems were described and characterised using linear wave theory. A modified hybrid element method, originally derived by Bai and Yeung (1976), was used. The forces, motions and the hydrodynamic efficiency of the device were assessed. The simulations were validated by direct comparison with the available experimental results, allowing the confirmation of the high efficiency of the cam shape in a broad band of wave frequencies. An interesting sensitivity study was also conducted, evaluating the impact of the change of shape, submergence ratio, water depth and the inclusion of a non-rigid support structure. Some key findings can be identified in Figures 5.1 and 5.2, which illustrate the relative influence of such parameters for constant water depth by plotting the efficiency (ε) as function off the non-dimensional frequency. Figure 5.1, where the optimal efficiency (ε_{opt}) is compared for selected configurations, shows the predominant influence of the submergence depth (s) with regard to other parameters like the angle θ, which partially defines the shape. Note that when $\theta = \pi/2$ and $s=0$ the theoretical limit for a semi-circle is reached, so $\varepsilon_{opt} = 0.5$ for all frequencies. Figure 5.2 shows a similar plot, now comparing the effect of the external damping ratio $\left(\hat{\lambda}_{22}/\hat{\lambda}_{33}\right)$, where the index '2' denotes heave and '3' pitch. The two curves per damping ratio correspond to two different values of the external carriage mass (which holds the support system). It is clear that as the ratio decreases so does the efficiency.

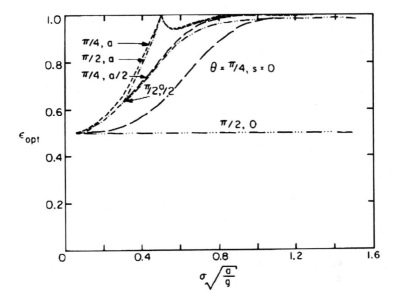

Fig. 5.1. ε_{opt} vs. non-dimensional frequency for different configurations (Mynett et al., 1979)

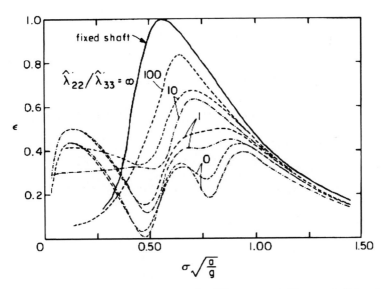

Fig. 5.2. ε_{opt} vs. non-dimensional frequency for different non-rigid supports (Mynett et al., 1979)

Mynett et al. (1979) therefore corresponds to the first numerical study concerning cam (or duck) shapes. In Standing (1980) numerical comparisons regarding the response amplitude operator in pitch and the capture width for a duck string were evaluated by means of a BEM code named NMIWAVE, from the National Maritime Institute, in a direct follow up of Mynett et al. (1979). Most of the subsequent work at the University of Edinburgh was experimental, with different models at different scales being tested in narrow and wide wave tanks, but Pizer (1992, 1993, 1994) applied a pure BEM approach to the duck geometry and in Cruz and Salter (2006) WAMIT was also applied to the same concept.

The use of pure BEM codes to study wave energy converters (WECs) was at first also linked with the study of Oscillating Water Column (OWC) plants. Brito-Melo et al. (1998, 2000a) modified the AQUADYN code originally developed at ECN (Nantes), producing a specific version dedicated to OWCs (AQUADYN-OWC). The major modification was associated with the supplementary radiation problem imposed by the oscillatory movement of the water in the inner chamber, which was solved by modifying the boundary condition through the pressure distribution. The study, conducted in the scope of the development of the Pico plant, showed an increasing level of depth: the initial configuration assumed an isolated structure surrounded by an infinite fluid domain (Fig. 5.3), whilst the final geometry included the neighbouring coastline and bathymetry (Fig. 5.4). Comparisons were made with a 1:35 scale model, validating the numerical results.

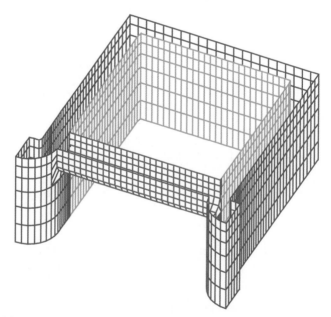

Fig. 5.3. Initial OWC configuration studied (Brito Melo et al., 1998)

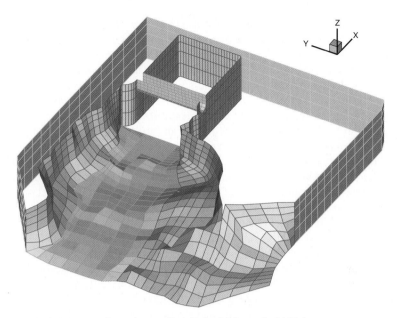

Fig. 5.4. Final OWC configuration studied (Brito Melo et al., 2000a)

Studies involving the integration of OWCs in breakwaters have also been conducted by the same wave power group at the Instituto Superior Técnico (Brito-
Melo, 2002b). Such approach required a number of changes in AQUADYN-
OWC, most of which due to the presence of the breakwater, and can be useful for
the modelling of new OWCs like the one to be integrated in the Porto breakwater
(Portugal) in the near future (Martins et al., 2005), or to numerically simulate experiments such as those described in Bocotti et al. (2007).

WAMIT has also been used, in its low-order option, to model OWCs. Lee et al.
(1996b) studied three different configurations: a moon pool in infinite water depth,
a bottom-mounted OWC and an OWC with extended walls (in the direction of
wave propagation). Two approaches were conducted to incorporate the inner free
surface effects. Firstly, the source code was modified to take into account a new
dynamic boundary condition. Secondly, a virtual surface was fitted to the inner
free surface, with predetermined velocity distributions ruling the movement. The
study lacks experimental validation but a partial verification exercise was performed, comparing the outputs of both approaches, which were found to be
closely correlated. Several numerical problems were identified, like the difficulty
in implementing the principles associated with resonance in a linear code, and the
influence of thin walls, which can lead to inaccuracies when representing the linear system of equations. Numerical sensitivity exercises were also conducted by
evaluating different discretisations of the geometry and by comparing the derived
values for the exciting force from direct pressure integration and from the Haskind
relation.

Delauré and Lewis (2003) applied WAMIT in the modelling of an OWC, following a similar approach to the second one employed by Lee et al. (1996b), where generalised modes of motion were used to model the inner free surface. The article follows up on a series of contributions from the same authors, where a review on similar applications, parametric studies and benchmarking with experimental results were presented (Delauré and Lewis, 2000a; 2000b; 2001). The agreement between numerical and experimental results was shown to be particularly good for small amplitude waves and for an 'open chamber' configuration (no external damping). One of the results confirms Newman's earlier work (Newman, 1992), by pointing out the differences between the results from the potential and the source formulation, with the latter being judged less suitable for problems involving thin wall structures such as OWC plants.

Returning to the previously mentioned work at the Wave Power Group of the University of Edinburgh, Pizer (1994) used a custom made BEM code, previously developed at the University of Strathclyde during the author's PhD studies, to compare numerical with experimental results from a solo duck (Skyner, 1987); see Chapter 2. In the process of verifying the code, selected analytical results, such as a floating hemisphere, were also used. Recently a WAMIT model was derived to compare results from its high-order module to both the low-order predictions from Pizer and the experimental results from Skyner. The radiation impedance matrix, the exciting force and the non-dimensional capture width were calculated. The results show, as expected, a better agreement with the previous numerical predictions than with the experimental results. Nevertheless the correlation with the latter is at least as good as the previous (i.e., when using the original numerical calculations). Examples are given in Figures 5.5 and 5.6 for the real part of the hydrodynamic impedance matrix and the modulus of the wave exciting force, respectively. Typically the WAMIT curves seem to be vaguely shifted in terms of frequency when compared to the experimental ones. This is particularly clear when trying to identify the maximum / minimum value on each plot, and could be partially linked partially with the discretisation procedure or, as indicated by Payne (2006) in a study of a different concept, to inaccuracies in the description of the mass matrix. Sensitivity studies show that the outputs are strongly influenced by changes in this matrix, particularly in the moments of inertia. In the duck case the differences are most likely due to the presence of the vertical flat mounting struts from which the model is connected to the test rig. With regard to the non-dimensional capture width, the influence of the control parameters on the location of the peak value is not only clear but expected. In addition Pizer (1992) pointed out that in further studies regarding conservation of energy, Nebel (1992) came across unaccounted losses, which could be linked with the properties of the flow or a physical problem with the test rig.

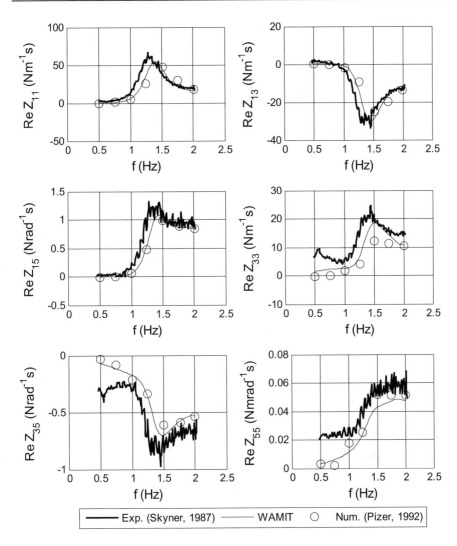

Fig. 5.5. Real part of the hydrodynamic impedance matrix – duck model results

Within the same research group, Payne (2006) used WAMIT to perform the hydrodynamic modelling of a sloped IPS buoy, comparing the results with those from two experimental models: a one degree-of-freedom model (Fig. 5.7) and a freely floating model (Fig. 5.8). The one degree-of-freedom version was developed by Lin (1999). WAMIT results, particularly in terms of the body motions, showed a shift in the frequency with regard to the experimental equivalents, a tendency that was linked with the influence of the discretisation of the inertia matrix. A numerical sensitivity study to quantify the influence of the radii of gyration

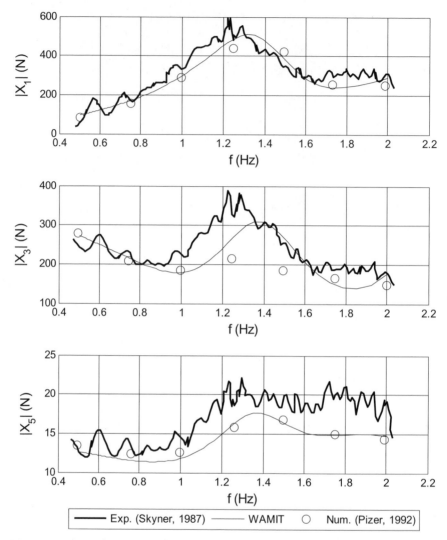

Fig. 5.6. Modulus of the exciting force (1- surge; 3- heave; 5 – pitch) – duck model results

was conducted to confirm that effect. The complexity of the model, namely the dynamometer that acts as the power take-off system and the inability to fully describe all the physical phenomena in a linear code can also be indicated as partially responsible for such discrepancies. An extensive review on the application of BEM codes to wave energy research, both in theoretical studies and when comparing numerical and experimental results, is also available in Payne (2006).

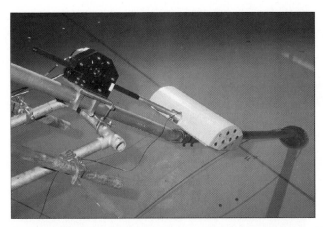

Fig. 5.7. One degree-of-freedom experimental model of the slopped IPS buoy (Lin, 1999)

To conclude, and to emphasise the importance of BEM modelling, particularly at the early stages of development, two examples related to full-scale concepts which will be addresses later in Chapter 7 can be given. Firstly, the Archimedes Wave Swing (AWS), for which the first numerical calculations were performed by Pinkster (1997), who derived the hydrodynamic coefficients for selected geometries. The AQUADYN code was also extensively applied to the AWS, allowing the recalculation of the hydrodynamic coefficients and also the exciting force for a wide range of configurations (Alves, 2002; Prado et al., 2005). Figure 5.9 shows one of the early numerical discretisations of the AWS pilot plant. Recently results from AQUADYN were used to estimate the wave profile directly above the full-scale pilot plant, which was installed in late 2004 offshore Póvoa de Varzim in Northern Portugal (Cruz and Sarmento, 2007). Starting from a library of hydrodynamic coefficients related to nine scenarios for different levels of tides and floater positions, the aim was to characterise the sea state at the actual pilot plant's location using the available pressure sensors. Two approaches were performed: a first one purely based in linear wave theory, neglecting the presence of the device, and a second one, based on the results from AQUADYN, which allowed a detailed quantification of the effects of the presence of the plant on the wave profile directly above it. Comparisons with a Datawell Waverider buoy located at a certain distance from the plant validated the methodology.

In a similar way, and also from an early stage, the Pelamis wave energy converter (WEC) has been developed using a variety of computer codes, of different scope and complexity. In the basis of all the developed tools is the computation of the hydrodynamic coefficients, exciting force and motions in several degrees-of-freedom using a linear BEM code named 'Pel_freq'. A detailed description of the complete software suite is given in Retzler et al. (2003), where validation exercises are described at several scales, though initial comparisons with results from a

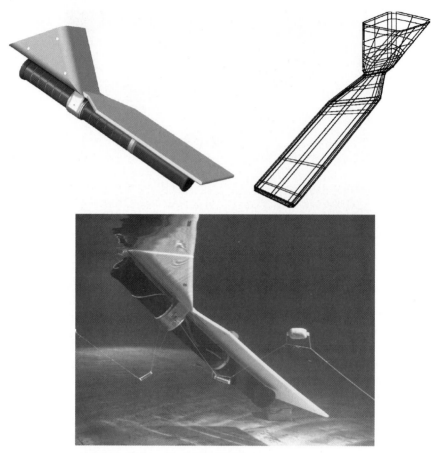

Fig. 5.8. Freely-floating model of the slopped IPS buoy (Payne, 2006): SolidWorks model (top left), MultiSurf model (top right) and experimental model (bottom)

1:35 scale model were already presented in Yemm et al. (2000). An updated version of the 2003 article is given in Pizer et al. (2005). It is emphasised that the outputs of the frequency domain code are extensively used as inputs in the time domain simulation also developed by Pelamis Wave Power (formerly Ocean Power Delivery) (linear and nonlinear), and in interfaces with other numerical tools for selected problems (e.g.: mooring load analysis). A case study related to the modelling of the Pelamis WEC, a concept already mentioned in Chapter 3 (and described in detail in Chapter 7), is given in section 5.4.

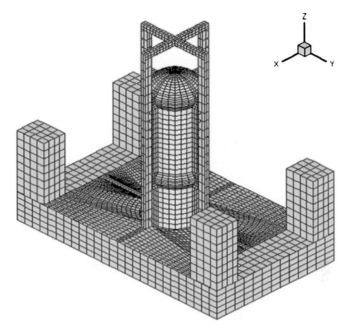

Fig. 5.9. AQUADYN's discretisation of the AWS geometry (Alves, 2002)

5.2 Wave Tank and Wavemaker Design

Matthew Rea

Edinburgh Designs Ltd
Edinburgh
Scotland, UK

Most modern tanks use two types of wavemakers. Flap paddles are used to produce deep water waves where the orbital particle motion decays exponentially with depth and there is negligible motion at the bottom. Typical applications are the modelling of floating structures in deep water and the investigation of the physics of ocean waves. Often the hinge of the paddle is mounted on a ledge some distance above the tank floor.

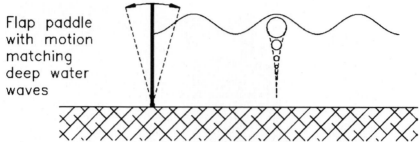

Flap paddle
with motion
matching
deep water
waves

In water that is deep compared to the wavelength
wave motion is confined to the surface layers

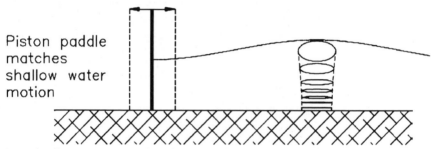

Piston paddle
matches
shallow water
motion

In shallow water the horizontal water particle
motion is almost constant at all depths

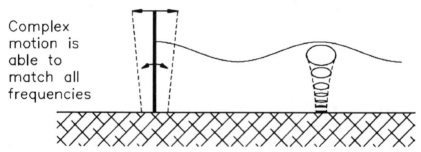

Complex
motion is
able to
match all
frequencies

Real mixed seas have a combination of wavelengths
requiring more complex paddle motion

Fig. 5.10. Description of the motion of a wavemaker

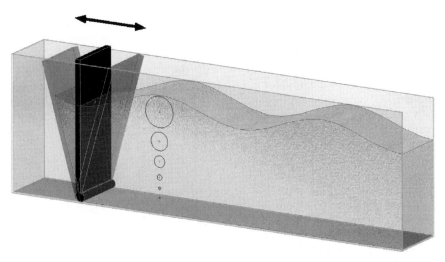

Fig. 5.11. Schematic of a flap paddle

Piston wavemakers are used to simulate shallow water scenarios, where the water depth is roughly smaller than half a wavelength. Here the orbital particle motion is compressed into an ellipse and there is significant horizontal motion on floor of the tank. This type of paddle is used to generate waves for modelling coastal structures, harbours and shore mounted wave energy devices.

Most early wave tanks were custom designs produced in the laboratory where they were used, hence there are many unique and innovative designs. These include displacement pistons, sliding wedges and other more complex machines like double hinged flaps. The design goal is to try to match paddle motion to the water motion and minimise the evanescent waves immediately in front of the paddle. These unwanted waves decay naturally but reducing their amplitude minimises the

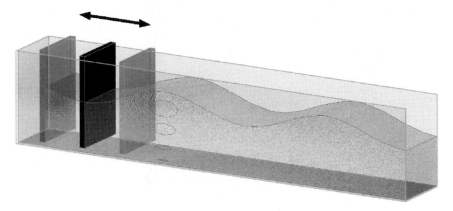

Fig. 5.12. Schematic of a piston paddle

unusable space in front of the paddle. All paddles will have an optimum frequency where the horizontal motion is very close to the motion of the water. This is the frequency where the inertia of the water, or added mass, is lowest. As the frequency is increased the mismatch between the paddle and the water motion causes the added mass to increase. This effect can be seen in a wave tank where a piston wavemaker generates high frequency waves, although the motion is small, the paddle moves a block of water that appears to be attached to it. It takes a few wavelengths for the natural wave to transform from this motion and travel down the tank. High frequencies do not require high power but can exert very high inertial loads on the structure. At low frequencies the volume displaced by the paddle limits the wave height. A piston will displace twice as much water as a flap, with the same stroke, so the wave will be approximately twice as big. Although the loads are low the design focus becomes the paddle stroke and preventing leakage round the structure (see Figures 5.13 and 5.14).

The type of research to be conducted in the tank determines the choice of tank size and wavemaker. First determine the sea state that is to be modelled; how deep is the water and what is amplitude and frequency range in the open ocean? The open full-scale sea wave spectra should be split into component wave fronts so that the amplitudes at different frequencies can be determined. The next step is to set the scale factor for the tank and models. There are many arguments that big is better however model scale is ultimately determined by the available budget. For most tanks this is in the range between 1/10 to 1/100 scale. The wavemaker type will depend on the relationship between the waves and the water depth. If the water depth is less than half the wavelength, or will be varied, a piston should be chosen.

5.2.1 Tank Width

The choice of tank width depends on the proposed model tests. The most straightforward tank is a single paddle in a narrow flume that represents a 2D slice, with the model fully blocking the width of the tank. This type of model is relatively easy to analyse because the waves and flow act in a plane. Visibility is excellent and models are readily accessible. It is a very good and economic tank for early investigations. A slightly wider tank with a single paddle can have a 3D model subjected to long-crested waves that pass round the sides so that 3D edge effects can be observed. The main difficulty is that as the width increases the frequency of the resonant cross wave becomes very close to the working frequency of the tank. For example a $0.7\,m$ deep tank $1.2\,m$ wide will have a cross wave of $0.78\,Hz$. The most realistic mixed seas have to be modelled in a wide tank with multiple individually controlled paddles. Software control of the paddles will allow a full range of waves and wave spectra to be generated. The width of the tank depends on model width and the angle of waves required on either side. For a line of paddles the angular spread is limited by the angle from the model to the tip of the line of straight paddles. One way round this is to build the bank of paddles in a curve (see Fig. 5.21).

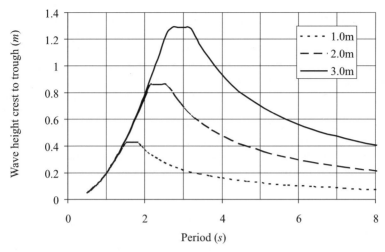

Fig. 5.14. Theoretical wave height for a 0.5, 0.75.1.0 *m* deep piston paddles with a stroke of +/– 0.5 x water depth

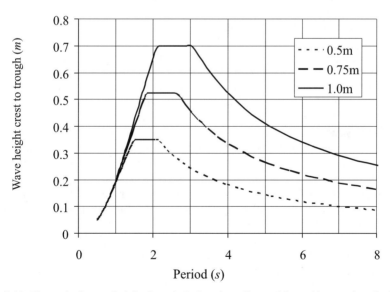

Fig. 5.13. Theoretical wave height for a 1, 2, 3 *m* deep flap paddles with a stroke of +/– 17 degrees

5.2.2 Tank Length

The tank has to have enough length to allow for three distinct areas. First there is the paddle and enough space for the evanescent waves to decay. Waves from a well-controlled paddle need to travel approximately twice the hinge depth of the paddle to become fully developed. The model zone depends on the size and motion of the model. Towing tanks are the extreme example where the length has to be sufficient to allow the carriage to accelerate, run and the slow down. For wide tanks the combination of width and length determines the angle of waves that approach the model. Finally there is the wave absorbing beach which has to be at least half the length of the design wavelength to achieve 90 % absorption.

5.2.3 Paddle Size

The angular motion of a flap paddle is determined by the quality of the control system. With position feedback it is reasonable to run up to +/–12 degrees. With force feedback or other 2nd order correction they will run well up to a displacement of +/–18 degrees. Piston paddles can move larger distances and are typically designed with a stroke of 50–100 % the water depth. A paddle for generating solitons will require a total travel distance of at least twice the water depth.

The first analysis of wave generation was published by Biesel and Suquet (1951) and provides solutions for relationship between wave height, stroke and force for hinged and piston wave generators. This was refined by Gilbert, Thompson and Brewer (Gilbert et al., 1971) who produced design charts that give engineering solutions for wavemaker design. The analysis is based on linear theory and takes no account of breaking waves. Higher frequency waves are limited by breaking; for regular waves the limiting steepness is 1 : 7 so the linear wave height curve is combined with the breaking wave limit. This tends to overestimate the size of the maximum breaking wave so a practical solution is to truncate the top 15 % of the curve. The paddle will create waves above this height but they will by unsuitable for research but useful for demonstrating the tank to visitors. Lower frequency waves are limited by the displacement of the paddle. As an approximate guide a flap paddle should extend about 35 % of the hinge depth above the waterline.

5.2.4 Multiple Paddles

A bank of individually controlled paddles can produce angled waves by setting a phase difference in the drive signal to each paddle. The most common layout is a rectangular tank with a straight line of absorbing paddles facing a beach on the opposite side. At first this seems a restricted arrangement but the hard sides can be used to reflect waves towards the model so the virtual angle that the paddles cover is greater than the physical width. Computer driven paddles are very versatile and can generate waves at 90 degrees to the paddles. 3D wave tanks are notoriously complex experimental environments and there is a strong argument for keeping a simple layout of one generating side, one absorbing side and two hard reflecting sides. Several large tanks have paddles along two sides in a L shape with beaches on the opposite sides. This is especially useful if there is current flow in the tank so waves can be run across and with the current flow. This arrangement leads to a complicated geometry where the two banks of paddles meet and waves get absorbed as they run along the beach which adversely affects the working area of the tank. Full computer control of the paddles allows the paddles to be laid out in any configuration leading to tanks with paddles arranged in a curve. Many coastal tanks have movable paddles that can be arranged around a model to provide waves from an appropriate direction.

Desired angle/frequency and the available budget determine the choice of paddle width. Multiple paddle wavemakers can generate angled waves up to a limit, which is determined by the paddle width and the wavelength. Normally this limit can be set where the apparent wavelength of the angled wave at the paddles is 2–4 times the paddle width. Near this limit the paddles generate a "ghost" wave at 90 degrees to the main wave. Figure 5.15 shows the operating envelope for various width paddles in 1 m deep water. Waves to the right-hand side of the curves are not possible. For example a bank of paddles, each 500 mm wide, will be able to generate a 1 Hz wave at 40 degrees but 700 mm paddles will not.

5.2.5 Drive and Control Systems

Early paddles used a crank to produce sinusoidal motion. An adjustable mechanical arm altered the stroke and the motor speed controlled the frequency. Some tanks had segmented paddles and angled waves could be produced by setting the phase of the cranks on a common drive shaft. This system could not be used for random waves and was time consuming to adjust. In the 1950s larger machines used a hydraulic drive with servo valve and electrical control system that could be directly driven with an analogue voltage. Most of the big naval towing tanks had direct servo hydraulic drives capable of generating long crested random waves.

With servo control it was possible to control the paddle motion from a signal generated in the control room. Single frequency waves were produced with a sine wave generator. Complex spectra were generated using a bank of adjustable filters to allow selected frequencies from a white noise source.

In the late sixties transistor amplifiers meant that direct drive electrical servo systems became possible. The size and reliability of electronic drives improved dramatically in the 1990s so that they are now competitive with hydraulic machines for all except the largest wave paddles. The control has become more sophisticated with specialised digital controllers available to correct for absorption of reflected waves and 2nd order harmonics.

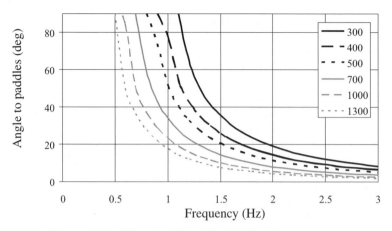

Fig. 5.15. Limiting angle for different width paddles in 1.0 *m* deep water

All modern tanks have wave generation software to drive the paddles. Data for the paddles is either pre-computed or generated in real time. A commonly used technique is to sum individual sine waves to create complex seas. Frequency, amplitude, angle and phase define a wave front. Summing individual wave fronts generates multi-spectral seas. Built-in functions allow regular sine waves, long crested multi spectral waves and mixed seas to be defined, each with a single line of text. Standard functions include the Pierson Moskowitz, Cosn, Cos2n, ISSC, Bretschneider, Neumann, Mitsuyasu and JONSWAP spectra, RMS merging, amplitude merging and freak waves.

5.2.6 Absorbing Wavemakers

Waves reflect off the surfaces of the model and from the sides of the tank. All tanks have resonant frequencies and often these lie within the working frequencies that are generated in the tank. A good beach will absorb much of the energy after it has passed the model but has little effect on cross-waves or models reflected from the model. This can be a major limitation on towing tanks where the productivity of the whole facility is determined by the settling time after a run has been completed. Active absorbing wavemakers dramatically increase the performance of a tank by prolonging the time that an experiment can run without the build up of spurious waves and also by decreasing the settling time between runs.

Traditional wavemakers work with a position feedback control system. This has the disadvantage that the swept volume of the paddle is dependent on the water level in front of the paddle. So the wave height generated is dependant on many factors including the size of an incoming wave or a poor quality beach.

During the first trials of the Duck wave energy converter Professor Stephen Salter found that wave height could vary by 30 % which made it very hard to measure the absorption of the device. Early experiments were unstable because waves were reflected back from the models and interacted with the wavemakers to create an uneven wave field. He overcame the problem by inventing a force feedback absorbing wavemaker that absorbed incoming waves by measuring the force on the front of the paddle and controlling the velocity (Salter, 1981). Now the absorption control is calculated by a digital controller so absorption is totally predictable and can be optimised for specific experimental conditions.

Other researchers have implemented wave absorption using different techniques such as measuring the incoming wave with a wavegauge mounted to the front of the paddle. This signal is brought into the paddle controller and the motion is modified to absorb and damp out the unwanted wave.

5.2.7 Absorbing Beaches

The wave, after it has passed the model, has to be absorbed. There are a wide variety of beach designs and the best summary is given in Ouslett and Datta (1986).

This survey assessed the performance of about 48 wave absorbers and several research papers. One factor that is common to many of the sloping beach designs is some form of innovative porosity mechanism, usually to channel the water flow caused by the wave advancing up the beach to be transferred back without affecting the wave. Similarly surface roughness is often used with the intention of tripping the wave over. The significant conclusions of the survey report are:

- A reflection of up to 10 % is to be expected even for well designed beaches and that the % reflection tends to increase with reduced wave height.
- It does not appear possible to attain reflection coefficients below 10 % for absorbers shorter than 0.5 to 0.75 of a wavelength.
- A porosity of 70 % in one case was shown to decrease the reflection coefficient by 2 %.
- Most beaches surveyed have a steepness of between 1:6 and 1:10 at the waterline.

Absorption, especially in a wide tank, is surprisingly difficult to define. It is dependant on amplitude, angle, and frequency. Many of the mechanisms that ultimately dissipate the energy rely on the Reynolds number so similar beaches will have different characteristics as the scale is altered. Another difficult with beaches is that they appear, in a tank, to be less effective than they are. A wave reflected from a beach that absorbs 90 % of the energy will be 31 % the height of the original wave.

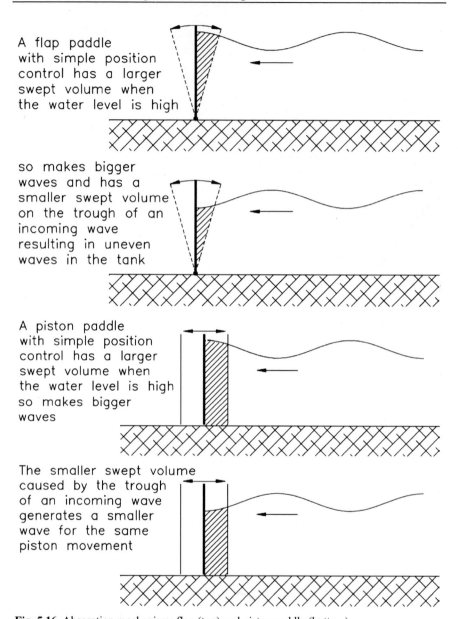

Fig. 5.16. Absorption mechanism: flap (top) and piston paddle (bottom)

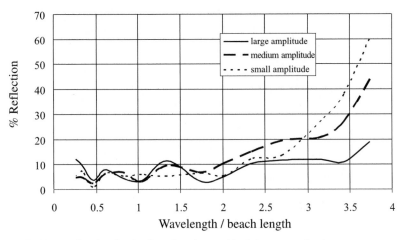

Fig. 5.17. Wave reflection for 6 *m* long beach tested in 3.0 *m* water depth

Sloping beaches do not have to run the full depth of the tank and can be sloped at up to 30 degrees without degrading performance. A typical wave tank beach will have a steep underwater section with a curved transition to a very gentle 6 degree slope meeting the waterline. Very little structure is required above the waterline as by this stage the wave has broken and its energy is dissipated. It is useful to allow water to run over the end of the beach so it does not cause back waves by surging back down the slope. Ripples caused by the wave breaking can affect smaller tanks. These can be reduced by covering the surface with an absorbent layer of foam or mesh material.

The loads on a beach can be high and it is particularly important to design for the up-thrust which can be just as high as the down-thrust. Beaches are also subjected to fully reversing cyclic loading so can fail in fatigue rather than by direct loading. It is very important to consider the mounting points where the entire structural load is transferred to the body of the tank.

Sloping beaches do not work so well in variable depth tanks. An alternative is to use mesh filled wedges. Multiple layers of plastic mesh dissipate the waves as they flow past and create eddies on the millions of sharp edges and ideally present the same impedance to waves as would unobstructed water in an infinitely long tank. The flow velocity varies for different waves so the foam density should increase progressively with depth and with distance down wave. There is a full description of the method of construction in Taylor et al. (2003).

5.2.8 Examples of Wave Tanks

To conclude this section a selection of photographs from different wave tanks is presented. Firstly, Fig. 5.18 shows an eight piston paddle arrangement in a wave tank at the University of Manchester, while Fig. 5.19 presents a similar tank but

Fig. 5.18. Wave tank at the University of Manchester (small displacement piston paddle)

with flap paddles at the University College London. Many more could be given as an example but two stand out by their unique character: the impressive 50 by 30 m (5 m deep) wave basin at the Ecole Centrale de Nantes (Fig. 5.20), which often receives several developers in the wave energy area, and the Edinburgh curved tank (Fig. 5.21), an uncommonly shaped fully-functional wave tank which replaced the Edinburgh wide tank in 2003. In contrast to the linear array of the majority of multi-directional wave tanks, the absorbing-wavemaker paddles are placed in a 90-degree arc in an attempt to improve the angular spread of the generated three-

Fig. 5.19. Regular waves at the University College London

Fig. 5.20. $50 \times 30 \times 5\,m$ wave basin at Ecole Centrale de Nantes

dimensional sea states, and to minimise cross-tank seiches. The lessons from this tank are expected to provide valuable input to the design and construction of a fully circular wave tank, as proposed in Salter (2001).

Fig. 5.21. Curved tank at the University of Edinburgh (Taylor et al., 2003)

5.3 Guidelines for Laboratory Testing of WECs

António Sarmento[1], Gareth Thomas[2]

[1]*Instituto Superior Técnico, Lisbon, Portugal*
[2]*Dept. of Applied Mathematics, University College Cork, Cork, Ireland*

This contribution is adapted from Sarmento A and Thomas G (1993), "Laboratory Testing of Wave Energy Devices", Wave Energy Converters Generic Technical Evaluation Study, Annex Report B1, Device Fundamentals/Hydrodynamics. C.E.C., Brussels.

5.3.1 Introduction

Tank testing, in both narrow and wide tanks, has played an important role in the progress of wave energy studies and is widely agreed to be essential for the calibration and validation of mathematical and numerical models. Most devices have been tested extensively either to validate a mathematical model or to supply vital information during the design process.

It must also be acknowledged that certain phenomena, of which device survivability is a good example, are not yet well understood from a theoretical viewpoint and good experimental programmes are vital to facilitate progress to be made in these important areas. Good laboratory experiments can also identify and isolate particular problems, often device specific, which are not addressed by contemporary theoretical models and thus provide an important input into the next generation of models.

There are however two fundamental characteristics of tank testing which do not have direct analogies in the modelling programmes. Wave tanks can be expensive to construct and this is especially so for wide tanks, with many wavemakers capable of generating multidirectional seas; in addition, they cannot easily be moved from one site to another and cannot be usefully employed without suitable wave-generating software and the availability of personnel with sufficient accumulated expertise to perform the required experiments. This means that established wave tanks are substantial investments in both materials and expertise and it is important to have the correct strategies for the maximum utilisation of such facilities.

Wave tanks have often been built for specific work programmes, but have been successfully used for purposes outside their original remit. So, for example, a facility originally intended for the testing of structures for the offshore oil industry can be used for studies on fish cages in the nearshore region and perhaps also for wave energy converters (WECs) in much deeper water offshore. However, despite the ability of wave tanks to perform a range of tasks, there are aspects of tank testing which are specific to WECs. A major one of these is that the standard testing practice for an offshore structure is to monitor the behaviour of a model under specified wave conditions and perhaps measure the pressures and forces on the structure or mooring systems; this can be an immensely difficult task, but for wave energy devices there is the added difficulty of including a simulator of the power take-off mechanism.

It is readily acknowledged that an Oscillating Water Column (OWC) was one of the first devices to pass successfully through all three stages of demonstration, prototype and full-scale generation. This situation has arisen because of the twin reasons that the power take-off mechanism, an air turbine, is sufficiently developed for immediate use and that OWCs can be built at shoreline sites with the relative ease of construction in a more benign environment than that of the proposed offshore WECs. A consequence of this progress is that OWCs have already been subjected to wide-ranging model testing programmes and this has not been the case for most WECs (one of the exceptions is presented as a case study in section 5.4). Recent experimental work has demonstrated that there are specific difficulties associated with the model testing of OWCs; furthermore some, but not all, of those problems which are presently being encountered by OWCs will be of direct relevance to offshore WECs at comparable stages of their development. These problems can be due to both device geometry and power take-off characteristics.

The purpose of this contribution is to outline progress to date and to identify those problems which have caused, or are likely to cause, greatest difficulty in testing programmes. Recommendations are then made with regard to the funding and laboratory practice requirements of future research programmes.

5.3.2 Laboratory Testing

Historical Perspective

Despite drawing heavily upon the theoretical expertise which originated in the fields of naval and commercial ship hydrodynamics, together with that of the offshore structures industry, the laboratory programmes for the testing of WECs have tended to develop in isolation of the better established industries. This may seem to be surprising, especially to the uninitiated to whom all structures designed to operate and survive in forty metres of water must seem to demand similar testing requirements.

This point seems even stranger upon further reflection, given that there are a number of large commercial and semi-state organisations which provide comprehensive testing facilities to the ship-building and offshore industries. In a similar manner wave energy has generally not utilised (though there are some exceptions) the facilities supported by national or European programmes.

There are three principal reasons for this relative degree of isolation and each is indicative of factors which have been, and continue to be, associated with the development of wave energy devices. The first is that funding has never been supplied to the wave energy community in a way which matched that of the offshore industry and even daily rates at commercial testing stations were prohibitively expensive, although in the long term this has probably not proved to be a disadvantage. The second is that the requirements of testing are different for the both cases; wave energy devices are inherently more complicated especially when the power take-off mechanism is also taken into account. Thirdly, wave energy requires wide tanks to act as development facilities, not just as testing facilities, and this demands considerable access to tank time.

The lack of suitable funding is exemplified by the construction one of the best known wide tank, constructed at the University of Edinburgh by Professor Stephen Salter in 1978 (see Chapter 2). When this tank was commissioned, some considerable surprise was expressed at the technological achievements which had been made, but astonishment greeted the cost for which the facility had been built - this was a small fraction of that which commercial organisations stated would be required. This success has epitomised much of the progress of wave energy, in which achievements have not been matched by funding. It must also be acknowledged that despite financial shortcomings the results from most experimental programmes, where these have been openly reported, appear to be acceptable.

Laboratory Practice

Most laboratory programmes have followed the same course as the parallel theoretical studies, in which the initial work has been in two dimensions (2-D) and then extended to three dimensions (3-D). The terminology Narrow Tank is usually reserved for experiments which investigate genuinely 2-D phenomena and Wide Tank refers to the 3-D case, which usually allows for the possibility of directional seas. Regular and irregular waves can be used in both cases. The relative ease of construction, combined with lower running costs, means that most institutions either possess or have ready access to a narrow tank.

Working in a narrow tank has many advantages and the use of 2-D models and investigations can often be readily justified on both scientific and engineering grounds. The best quality experiments are often carried out in narrow tanks for the very simple reason that specialist experimental equipment regularly operates better on the relatively small scale and the degree of control over the experimental conditions is generally very good. Much of the sophisticated experimental equipment, such as absorbing wavemakers (to act as wavemakers and/or beaches) and cylindrical wave gauges work best in narrow tanks and it must be recognised that the development of this equipment, again with an important input from Professor Stephen Salter, has been of enormous benefit to the wider community who conduct water wave experiments.

Moving to a wide tank introduces a number of difficulties which are not present in a narrow tank. One estimate has placed the cost per annum of running a wide tank at ten times that of a narrow tank, personnel excluded, and the increase in experimental difficulty is of a similar magnitude. There is a much greater level of uncertainty in a wide tank and wavemakers, beaches and wave probes; all have been causes of concern to experimentalists. As section 5.2 showed, such concerns have been progressively minimised over the years and there are now a considerable number of wave basins where large scale experiments can be conducted with confidence.

5.3.3 Shortcomings of Existing Practices

Fundamental Deficiencies

There have been many fundamental deficiencies in the testing programmes which have been completed to date. It is tempting to lay most of the blame upon the device teams, but to do so would be most unfair as they have been generally unwilling victims of circumstances beyond their control rather than the perpetrators of misdeeds. The point made earlier that the results of most experimental programmes seem to be acceptable is an important one and is mainly due to the enthusiasm and dedication of device teams.

The most obvious criticism of previous programmes is that insufficient funds were available and these were stretched as far as possible. This only tells part of the story as insufficient time is also a crucial factor and the way in which that time is used. Commercial tank testing is usually allocated a daily rate and this is entirely inappropriate for device development; the base rate should be measured in months rather than days and the device teams should be present to oversee tests whenever possible. This should not exclude the possibilities of sub-contracting device tests but there is a requirement that the sub-contractors are familiar with the expected behaviour of wave energy devices as well as being familiar with their tank facility. The concept of a wave tank as a development tool is an important one and needs an appropriate level of funding.

Monitoring

Much of the past experimental work has taken place under the cloak of secrecy as device teams have sought to hide their work from competitors. This approach is in many ways understandable, but it has not necessarily meant that experimental expertise is always employed. Indeed there are sometimes uncertainties in published results which are difficult to assess and lack of specialist experimental knowledge amongst device teams has been all too evident. There are two principal reasons for the laissez-faire approach. The first is that there are not yet standard practices for the testing of wave energy devices; the second is that there has been insufficient independent monitoring.

The question of establishing standard practices is a difficult one to deal with, particularly as experimental programmes have often been primarily used to confirm theories or concepts rather than be used as genuine device design tools. Testing has often been completed, in rather short time periods, by the device teams themselves without the benefit of expert advice. One crucial feature here is that research to date has been under funded for the level of progress attained and corners have been cut to match the meagre budgets.

The lack of suitable monitoring procedures is in many ways indicative of the fact that wave energy conversion is a rather new technology. Testing for the off-shore engineering industry is a tightly controlled process, with testing often carried out at considerable expense by independent specialist laboratories, with specified standards laid down by government regulations, insurance requirements or industry standards. Exacting standards are not required for device development, but it is important to establish standard laboratory practices for device testing; this would ensure confidence in experimental results and enable comparisons between the performances of different devices at model scale.

Scale Effects

One of the most commonly acknowledged difficulties of conducting experiments with wave energy devices is the presence of scale effects. This occurs because if only one experiment, or series of similar experiments using the same single facility, is chosen to investigate the behaviour of a device then the model scale chosen will not usually be appropriate for all of the phenomena which are associated with the hydrodynamic behaviour.

The initial testing of a device, in either a narrow or wide tank, usually utilises Froude scaling which is governed by the wave kinematics. Although this is a sensible approach to adopt, the range in magnitude of WECs can present problems. For instance, the horizontal dimension of a broad bandwidth terminator may be a hundred metres or more, whereas a point absorber type buoy might have a diameter of at most ten metres. The difficulty which arises is that small scale viscous effects, due to the laboratory scale chosen and which will not appear at full-scale, can corrupt the model tests and not permit simple comparison between model tests for devices which were carried out using the same facility.

There are also a number of phenomena which cannot be appropriately scaled in standard narrow or wide tank tests. In preliminary model tests these may seem to be relatively unimportant when compared to the determination of the basic hydro-dynamic behaviour and to a certain extent, this is true. However, all of the effects are associated with either real fluid or nonlinear effects and some of them possess a potentially catastrophic capability. These include nonlinear wave effects, which culminate in both engulfment and impact forces, vortex shedding from cables and structures, and turbulence.

Finite Channel Width Effects

The principal purpose of wide tank testing is to reproduce open sea conditions at model scale in order to monitor and test device performance. However, even for wide tanks the influence of the channel walls on the hydrodynamics can be appreciable and the behaviour which occurs in the tank can be more representative of a motion within a finite domain than the desired open sea conditions. The phenomenon has already been recognised at a fundamental level and the notation 21/2-D has been used to describe the modelling of 3-D models in relatively narrow wave flumes.

It was thought that the solution was simply to widen the tank, but theoretical work has shown that a wave tank can often provide a poor replacement for the open sea when a single vertical cylinder, or an array of such cylinders, is placed in a channel and subjected to a regular incident wave train even in comparatively wide tanks. The influence of the channel walls is considerable in many ways, particularly with regard to the pressure distribution over the cylinder, or cylinders', surface and to the reflected and transmitted waves in the tank.

It seems likely that similar conclusions will hold both for bodies which possess more general geometries and for irregular waves, although these have not been extensively studied, and such results have important implications for the laboratory testing of WECs. However, recent work suggests that the implications are very important and the testing of arrays in particular will require considerable care to isolate tank effects from interactions between the array members.

Lessons from OWC Testing

Detailed experiments using scale models of OWCs have identified many problems which need to be addressed. Some of these are of a generic nature and have been included above, but while most are presently specific to OWCs they may have wider applications in the long term.

Testing of offshore WECs requires that the model is placed in the working area of a wave tank, which may be quite small even for large tanks, and essentially this lies between the wavemakers and the beach (at a few selected wavelengths from both). All of the advanced testing for OWCs has been concerned with shore-mounted devices so that a wavemaker is present but the absorbing beaches are replaced by models of the coastline which includes the OWC. The OWC under test is usually strongly site specific and it becomes necessary to model the bathymetry in the vicinity of the site to a degree of acceptable scale and accuracy, but this will often require very small device models due to the limitations enforced by the physical dimensions of the tank. Very small scale models will not allow the hydrodynamic losses and wave breaking to be well represented by the model and this affects the capacity of the model tests to simulate the influence of significant wave height. A further difficulty is that the removal of beaches will most certainly lead to problems of unwanted reflections.

The minimisation of hydrodynamic losses does not generally require detailed simulation of the bathymetry, power take-off or control procedures. The essential requirement is for larger model scales and this means that different scale tests are required for different phenomena. A further example of this is the importance of wave breaking and impact tests, which should include wave breaking. Scale effects are recognised as being extremely important and require considerable study.

Power take-off mechanisms are not generally simulated in detail. A good model of a turbine is to use a device which dissipates the pneumatic energy of the air such that the flow versus pressure characteristic does not deviate much from that of the turbine to be used in the prototype. Such a simulating device could have a nonlinear characteristic as in the case of an orifice plate or an approximately linear characteristic as in the case of a rotating disk or a porous plug.

Although the three-dimensional nature of the physical modelling has already been stressed, there are certain aspects of OWCs which can be suitably modelled in the first instance by two-dimensional models. These include the impact forces mentioned previously and also the testing of control procedures, for which linear waves will suffice at an initial stage but will eventually require irregular waves.

5.3.4 Results and Conclusions

Funding

The cost of constructing major wave tank facilities, i.e. wide tanks, together with the funding levels required to maintain equipment and support personnel on an ongoing basis, is very considerable. If such facilities are to be financially justifiable then they must be able to regularly attract funding and not be subjected to long periods of enforced idleness. Commercial alternatives do exist nowadays, but a major step forward could still be achieved if a large scale wave tank was built to be benefit of the wave energy community, possibly with the support of the EU. This, along with funding for the early stage developers, would allow the appropriate testing of different concepts by the tank operating crew, providing independent validation and certification of the device. Narrow tanks are considerably less expensive to build, require little general maintenance and have low running costs; all device teams should have ready access to a narrow wave tank.

There are many advantages to the funding of centres of testing expertise, of which one should be associated with offshore devices and another with OWCs. An agreed common approach to testing would be required and this would be a major step forward. The host facility would provide an element of neutrality and thus ensure that test results of different devices could be fairly compared. There are two important points which must be addressed: the first is that the wave energy community must have universal confidence in the testing centres and the second is that the wider community must have an input mechanism into the management and policy of the testing centres

Testing Programmes

It is not possible at this stage to determine how an agreed testing programme would be constructed, although certain elements can be readily identified. The validation of linear theory for the prediction of device performance in regular waves is clearly the first task once the appropriate model scale of the testing programmes has been established. At the opposite end of the wave amplitude scale to linear waves are extreme waves and the device response to extreme wave should be tested to assess prospects for storm survivability. The testing procedures for the central region, involving irregular seas, will be more difficult to determine and cannot be done at this stage. One of the major reasons for this is that WECs will almost certainly become more site specific than they are at present and consequently they will be designed to operate best within certain sea conditions; this means that spectra will be device and site specific. An approach is to use generic spectra to study the basic device response to irregular waves and site specific spectra for details of device behaviour at the proposed site. However, this is a difficult subject to resolve; it will require general agreement within the wave energy community, but the task should be given an urgent priority.

Scale Effects

The importance of scale effects in laboratory experiments has already been identified. An attempt is made here to categorise the various fluid-structure interactions which occur and to suggest a suitable scales for experimental investigation of the mechanisms. The suggested scales are the minimum values which should be used and larger scales are often more desirable, but there is a balance to be drawn between financial cost, available facilities and meaningful results. The list below utilises the experience accumulated under several development programmes.

Offshore Device Behaviour

This does not require detailed knowledge of the local bathymetry; constant depth testing can be employed and device considerations will dominate the experimental regime. The appropriate scale is usually dependent upon tank size and wave-making capacity, but too small a scale can introduce small-scale viscous effects into the hydrodynamic interaction processes.

- Suggested Scales: *1 : 50* (First Choice), *1 : 100* (Second Choice)

Validation of Numerical Models / Optimisation

The design of a WEC requires the use of scale models to validate the numerical predictions carried out by frequency and time domain models. Furthermore, there are critical aspects like survivability that need to be addressed. The experiments should naturally include the highest level of detail possible, but it is recognised that at this stage not all important factors can be taken into account. The most noticeable example is the power take-off mechanism, which is either not modelled or at best a simulator is used. Another important question is whether the influence of nonlinear wave effects upon the capture width can be accurately assessed. If alternative configurations should be tested, this is the ideal scale to do so.

- Suggested Scales: *1 : 20* (First Choice), *1 : 33* (Second Choice)

Nonlinear and Hydrodynamics

There are a number of hydrodynamic mechanisms which are either difficult to model at small scale or are not yet understood from a theoretical viewpoint. Examples are engulfment and impact forces, viscous losses and turbulent effects. Almost all are potentially catastrophic, but most can be modelled using 2-D tests in the first instance and another common feature is that as large a scale as possible should be used.

- Suggested Scale: *1 : 7* or *1 : 5*

Component Testing

The ideal scale for component testing will strongly depend on the type of WEC, particularly with regard to the power take-off mechanism. An extreme example can be found in the experience from OWCs: for turbines the major issues concern the influence of turbulence and the importance of water particles in the air flow to the turbine. Additional blade problems can also arise and stall is one important factor. The modelling difficulties are great and the flows, for OWCs, are very complicated. Hydraulic systems will need testing at considerably larger scales than linear generators.

- Suggested Scales: *1 : 5, 1 : 2, 1 : 1*

5.3.5 Recommendations

As there are a number of recommendations and these are both of a generic and a technical nature, it has been decided that the list will be divided into a generic list concerning standard practices and a more technical list; all recommendations are important. Very serious consideration should be given to the generic list before further investment in the testing of wave energy devices is undertaken.

Generic

1. Access to wide tank facilities is still limited and strongly conditioned by financial motives; a standard facility should be supported by future wave energy programmes; sufficient funds should be available to enable device teams to conduct experiments at the specialist centres.
2. Such facility should focus the study of offshore devices and array interaction; additionally OWC R&D should continue to be encouraged, especially when integrated in costal defence mechanisms such as breakwaters.
3. Standard testing procedures should be established for all WECs, but this will require a high degree of agreement from device teams. This should include linear waves in regular seas to validate mathematical models and extreme waves to test storm survivability; agreed generic spectra should be used, in conjunction with site specific spectra as necessary.

Technical

4. The applicability of all aspects of 2-D and 3-D testing should be assessed, particularly with regard to losses, forces and impact pressures. This also should include the influence of flume width on both genuine 2-D tests and those best described as 21/2-D (i.e. 3-D tests in wave flumes).
5. The importance of scale effects in model tests is not fully understood. A detailed comparison of experimental tests should be undertaken with two different scale models, say 1 : 10 and 1 : 100 to assess this phenomenon.
6. A comparison should be made of the regular and irregular wave testing requirements with regard to the evaluation of losses, forces and impact pressures.
7. The influence of different power take-off simulators in model experiments should be considered.
8. Detailed experimental studies to monitor time-domain simulation models should be undertaken.

5.4 Case Study: Pelamis

Ross Henderson

Pelamis Wave Power Ltd
Edinburgh
Scotland, UK

The role of numerical and experimental modelling in the development of the Pelamis WEC is discussed in this section. A detailed description of the concept is given in Chapter 7, but a short summary is desirable prior to the presentation of the models that have been built by Pelamis Wave Power (formerly Ocean Power Delivery) over the last years. The Pelamis WEC is an offshore, floating, slack-moored wave energy converter consisting of a set of slender semi-submerged cylinders linked by two degree-of-freedom hinged joints. The power take-off (PTO) consists of hydraulic cylinders mounted at the joints, which pump fluid via control manifolds into high-pressure accumulators for short-term energy storage. Hydraulic motors use the smooth supply of high-pressure fluid from the accumulators to drive grid-connected electric generators. A response inclined to the horizontal can be induced by controlling of the PTO to give different levels of restraint in each joint axis. The inclined response offers an effective hydrostatic stiffness reduced from a vertical response, resulting in a natural frequency controllable through the PTO with minimal reactive power requirements.

The wave loading and response is limited in large seas by the inherent design characteristics of the machine. The machine's overall length is chosen to be comparable to the wavelengths for which maximum power capture is desired. In longer waves the segments move with smaller phase differences, thus relieving the load on the structure and the power systems. In small and moderate seas, the wave loading is dominated by the strong dynamic buoyancy force. In high waves the dynamic buoyancy limits as the cylinder sections become locally submerged, and the weak inertial force becomes more important. The rate of rise of wave loading with wave height therefore reduces as the wave height becomes comparable to the cylinder diameter. These two features – frequency and amplitude limits – protect the machine in long and high storm waves.

Numerical and experimental modelling has been at the core of the Pelamis development programme since its inception. The culmination of the numerical and experimental modelling programmes was achieved with the construction, installation and test of a full-scale prototype, 120 m long and 3.5 m in diameter, with a rated power of 750 kW, at the European Marine Energy Centre (EMEC) in Orkney, Scotland. The several stages of development are detailed in the following subsections.

5.4.1 Numerical Simulation

The PEL suite of software, developed by Pelamis Wave Power over a number of years, is used to model the hydrodynamics of the Pelamis and allow the analysis of results. It comprises three main programs of increasing computational complexity: Pel_freq, Pel_ltime, Pel_nltime; see Table 5.1. Approximate CPU times are shown for a set of 50 wave spectra representing the sea-states for an average year. Recently the possibility of using real seas spectra from single point measurement devices was also included in the simulation. Each of the main programs is fully configurable with respect to geometry (i.e. tube lengths & diameters, ballasting, roll-bias), control applied at the joints (linear impedance for **Pel_freq** but arbitrary for others), and waves (Figures 5.22 and 5.23).

Table 5.1 The PEL Suite main components

Program	Body Dynamics	Hydrodynamics	Control	CPU	Applications
Pel_freq linear frequency domain	linear	3D freq. dep. coef. 2D freq. dep. coef.	linear	1 sec	large parametric studies with simplified control
Pel_ltime linear time domain	linear	3D impulse response 3D freq. dep. coef. 2D freq. dep. coef.	Arbitrary non-linear	2 hrs	power absorption in small and moderate seas
Pel_nltime non-linear time domain	non-linear	2D freq. dep coef.	Arbitrary non-linear	4 days	survivability in large seas

A time-domain model of the Pelamis power take-off system was developed and included as an optional routine within **Pel_ltime** and **Pel_nltime**. It includes all effects associated with the real hydraulic system to enable accurate power prediction studies and allow detailed control algorithms to be developed within the context of the entire machine operating in representative conditions. The separation of control routines and the models of physical systems allows for the easy translation of control programmes between simulations and actual hardware.

Control subroutines sample joint angles from the existing hydrodynamics subroutines just as a real controller samples transducer signals. A control algorithm is then applied and the output is passed to the PTO model in a similar format to that of the real controller. The PTO subroutine then models the physical hydraulic system and provides the resulting applied joint moment to the rest of the program. Other useful signals such as chamber and accumulator pressures and flows are also output for analysis. Effects included: fluid compressibility, valve characteristics, delays, flow losses, friction, accumulation, generation characteristics.

A finite element model of the mooring system has been integrated into the time-domain simulations. This allows the effect of the mooring on both power capture and survivability to be examined and included in numerical studies. The model is generally definable to allow different mooring configurations to be examined.

In addition several auxiliary programs perform pre- and post-processing tasks in relation to the PEL suite. These are summarised in Table 5.2.

The PEL hydrodynamics simulation has been verified using tank test data, while the power take-off models were verified using laboratory test rigs. Offshore engineering consultants WS Atkins carried out an independent verification of the hydrodynamics modelling of the PEL suite.

In addition to the PEL suite, PWP has made use of the commercial simulation package 'Orcaflex'. Orcaflex is a marine dynamics program originally designed for static and dynamic analysis of flexible pipeline and cables. The OrcaFlex Pelamis model is built up from a combination of several types of component (buoys, lines, spring-dampers, winches) placed within an environment with seabed, waves, wind and current all specified by the user. Wave types available are Airy wave theory, Stokes' 5th order theory, Dean's stream function theory and Fenton's cnoidal theory, all of which have been tested on the Pelamis model.

Orcaflex was used in particular to analyse the mooring system under extreme events. For example, the sensitivity to wave height and period of the mooring system was examined using controlled test cases with a single Dean stream wave with small precursor. A maximum wave height of $28.6\,m$ is the maximum expected within the 100 year storm spectrum according to a HR Wallingford report in a sea state characterised by $H_{m0} = 15.4\,m$, $T_{02} = 14.6\,s$. Worst case snatching and extreme loading events were also simulated to test the mooring design.

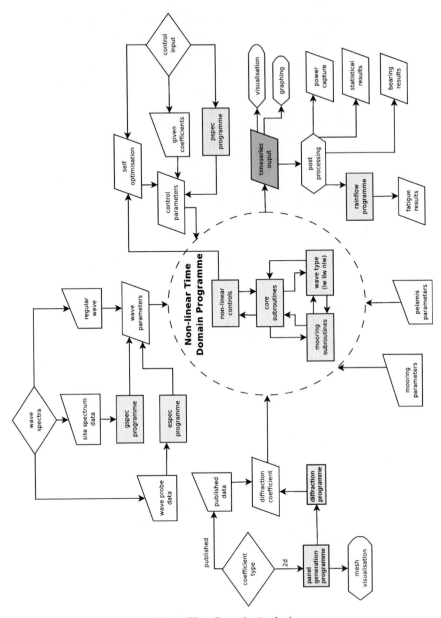

Fig. 5.22. Flow Chart for Non-Linear Time Domain Analysis

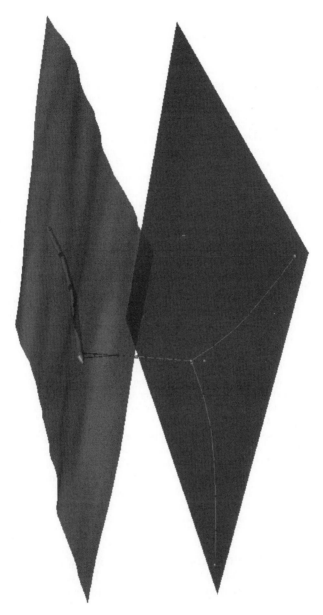

Fig. 5.23. Screenshot of the 3d visualisation interface of the PEL suite, developed by PWP

Table 5.2 Auxiliary Programs

Program	Description
Dpg **Viewmed** **Dpm**	A 3-D linear frequency domain wave diffraction program developed at the University of Strathclyde provides a high accuracy model of the interaction of infinitesimal amplitude waves with a rigid body. The **dpg** program generates a flat panel representation which may be viewed with the **viewmed** program. The panel representation is then used by the **dpm** program to evaluate wave diffraction, radiation, and hydrodynamic interaction coefficients for the rigid sections of Pelamis, as required by the PEL suite of programs. 2-D hydrodynamic coefficients may also be approximated by selecting a long enough cylinder, for use in the strip theory PEL programmes.
Gspec	Facility for generating wave spectra files for use by PEL time domain programs or the pspec program. The spectra can be directly read in as spectral components or generated from specified parameters using a prescribed spectral form (ie PM-spectra). The measured spectra may then be transformed to a different water depth under the assumption of no power loss, or by applying a power filter to account for dissipation due to sea-bed friction. The frequency components of the spectra can be re-sampled to provide a set of frequencies from which the time domain realisations may be efficiently evaluated.
Espec	Decomposes experimental wave records to provide corresponding wave spectra data files for PEL suite, allowing comparison between numerical and experimental results for mixed seas. Spectral components are evaluated to correspond to the three wave models used in the PEL suite: linear, second-order or Lagrangian.
Pspec	Iteratively runs the frequency domain model to select optimal controls to achieve response within specified constraints. Output powers are weighted with respect to site-specific wave data to give annual average power prediction.
Rainflow	Carries out rainflow analysis of time-series for evaluating fatigue damage fraction for various structural components and features.
Visualisation	Provides 3-D animation of time-domain simulation of waves, Pelamis, and mooring. The user can zoom and change viewpoint etc during animation.
Graph	General graphing program for time-domain simulation. Outputs time-series of any variable e.g. joint angles & moments, hydraulic system pressures, absorbed and generated powers, mooring forces.

Fatigue analysis was also carried out on the mooring components using the rainflow facility available within Orcaflex applied to line tension results from a set of eighteen wave spectrum simulations chosen to represent the EMEC Orkney site. Tests have also shown that primary mooring line and tether annual damage agree reasonably with those produced from model testing.

In order to verify the performance of Orcaflex in modelling the Pelamis, the system has been modelled to replicate the 20th scale tests performed in Nantes. Nine test cases were used for correlation – six regular waves, a short term spectrum wave group and a longer steep spectrum. Further checks were also performed to assess the performance of different types of wave modelling etc, with some limitations being identified.

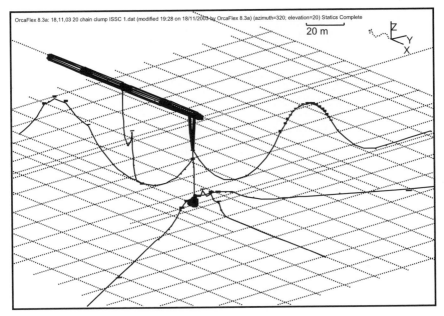

OrcaFlex 8.3a: 18,11,03 20 chain clump ISSC 1.dat (modified 19:28 on 18/11/2003 by OrcaFlex 8.3a) (azimuth=320; elevation=20) Statics Complete

20 m

Fig. 5.24. Screenshot from the Orcaflex simulation package showing the full-scale Pelamis prototype mooring set-up

5.4.2 Experimental Modelling

Physical models have been constructed at the 80th, 50th, 35th, 21st, 20th and 7th scales. They have tended to grow in scale as the modelling budget increased, and as the demands for more detailed data and greater functionality became evident.

The most elaborately controlled and instrumented model so far has been the 20th. Recently, a 21st scale model similar in detail has been tested (optimisation of the next generation of Pelamis machines). The scale is appropriate for power testing in readily available wave tanks, and for survival testing in the large wave basin at Ecole Centrale de Nantes. Figure 5.25 shows a motorised joint axis and the strain-gauged spider that connects it to the orthogonal axis. Moment, angle and velocity are measured in all 6 joint axes. In addition, pressure is measured around the foremost cylinder, mooring and tether line tensions, tether line angle and roll moment, and wave height down the length of the model. The model joints are controlled by a microcontroller that, via the motors, can apply arbitrary spring and damping, and model the full-scale ram characteristics, including the stepwise application of moment. Figure 5.26 shows the front two cylinder sections of the model, connected by a joint, covered with a neoprene rubber fairing. The electrical cables exit the joint to the right.

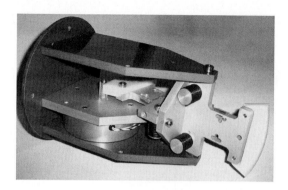

Fig. 5.25. Motorised joint axis and the strain-gauged spider

Fig. 5.26. Detail of the 20th scale model

The 20th model has been used to test joint control strategies under power and survival regimes, mooring configurations and failure modes. It has provided measurements of response amplitude operators and power performance, envelopes of joint moments and angles, and mooring loads and structural pressures in extreme waves. Figure 5.27 shows the comparison between the output of the numerical simulation and the experimental results obtained for the capture width.

The 7th scale Pelamis was a technology demonstration model, large enough to include representative systems in but small enough to handle and transport without large cost and to fit into large tanks such as l'Ecole Centrale de Nantes. It employed a hydraulic power take-off system functionally similar to the full-scale system and served as a platform for developing the control hardware and software used in the full-scale machine.

The PTO system was developed independently prior to construction of the 7th and full-scale Pelamis with the use of laboratory test rigs. The 7th scale PTO test rig, initially actuated by hand, was later adapted for actuation by a ball-screw operating under closed-loop control to perform the role of the waves. Pre-prototype hydraulic circuit and component test assemblies were designed and constructed for ad hoc experimentation. The 7th scale test rig was used for a set of tests designed to demonstrate the operation of the Pelamis PTO, test implementations of basic control algorithms, and to verify the mathematical model developed for computer simulation.

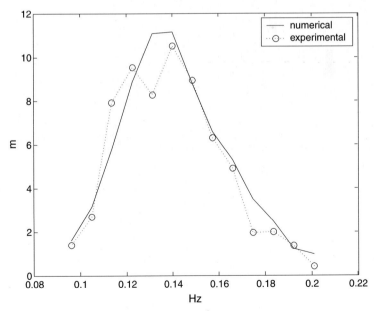

Fig. 5.27. Comparison of PEL simulation and 20th scale model experimental measurements of capture width

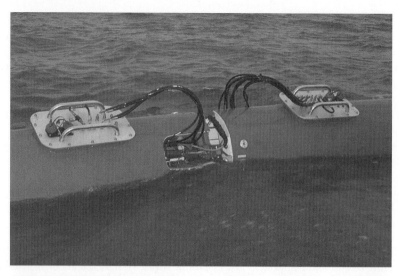

Fig. 5.28. Photographs of the 7th scale model joint during a sea-trial and the whole machine in the narrow towing tank at l'Ecole Centrale de Nantes

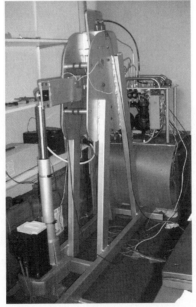

Fig. 5.29. Top: a fully assembled 7th scale power pack positioned upside down from the orientation in which it is installed. The rams are connected to the manifold via flexible hoses fitted to ports extending through the hatch. Bottom: the 7th scale test rig, fitted with a ballscrew actuator under position control, being used to test a 7th scale power pack.

The 7th scale rig provided an essential platform to develop an understanding of the various practical issues surrounding the operation of the PTO such as valve timing, compressibility, delays, and measurement. The lessons learnt and techniques developed during these small-scale experiments fed directly into the full-scale design. Experimental results from the test rig were also be used to verify the numerical models included in the PEL simulation suite.

A full-scale joint test rig was constructed in the winter of 2002 and used for further verification and adaptation of the full-scale PTO, and for extensive operational and cycle testing for assessment of components. Figure 5.30 shows the full-scale rig, where the actuation structure can be immediately spotted (see right-hand side of the photograph). Such land demonstration of the full-scale power take-off mechanism should be encouraged and supported via governmental grants. Even though the capital cost is high, the risk of skipping such stage (and eventually the cost) and embarking in the construction of a full-scale prototype is much higher.

Figures 5.31 and 5.32 show cycling pressures in the eight ram chambers (push and pull) of an axis of the 7th and full-scale joint rigs, respectively. The experimentally measured pressures are shown along with those resulting from running the same position signal and control signals through the PTO simulation. The agreement is extremely close.

Fig. 5.30. The full-scale joint rig and associated control system: see the power take-off rams within the rig and the outer rams which are used for load simulation

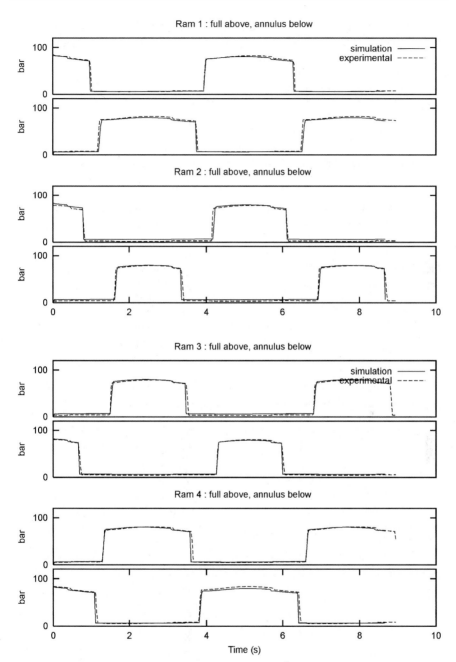

Fig. 5.31. Pressures for each ram chamber, for one axis: 7th scale rig

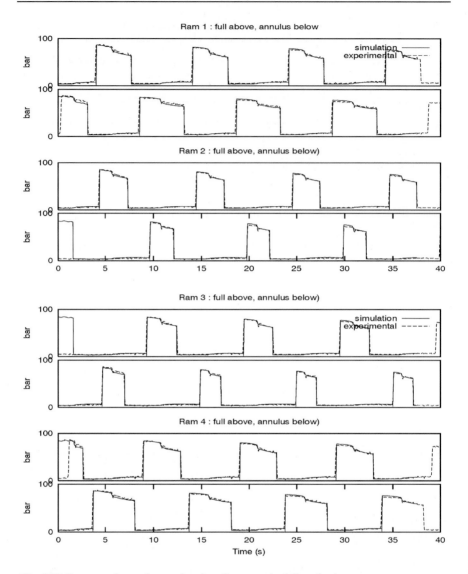

Fig. 5.32. Pressures for each ram chamber, for one axis: full-scale rig

5.4.3 Outcomes

The PEL suite has been of enormous use in every aspect of the Pelamis development programme. It will continue to be used in the future to further optimise the design of the machine and its control algorithms.

PEL also plays a vital role in project development where the likely yield of specific sites must be assessed and the machine configuration adapted. The power table, used to describe the power absorption characteristics of a specific machine configuration with respect to wave height and period of spectra, is derived using PEL. It is envisaged that this method of quantifying machine performance with respect to specific sites will become the standard for assessing the viability of projects, similar to the power curve currently used in the wind industry.

The verification of the prototype Pelamis by WS Atkins, which provided enough confidence for the machine to be granted insurance at a commercial rate, was made possible by exhaustive analysis using the PEL suite. This included:

- Fatigue analysis of stress in key components
- Simulation of partially flooded conditions
- Complete simulation of control and power take-off system including all pertinent effects
- Analysis of behaviour under systems failures

The development and testing of control algorithms is largely dependent on the PEL simulation. The virtual machine provides a platform that is not only com-

H_{sig} (m) \\ Power period (T_{pow}, s)	5.0	5.5	6.0	6.5	7.0	7.5	8.0	8.5	9.0	9.5	10.0	10.5	11.0	11.5	12.0	12.5	13.0
0.5	idle	idle	idle	idle	idle	idle	idle	idle	idle	idle	idle	idle	idle	idle	idle	idle	idle
1.0	idle	22	29	34	37	38	38	37	35	32	29	26	23	21	idle	idle	idle
1.5	32	50	65	76	83	86	86	83	78	72	65	59	53	47	42	37	33
2.0	57	88	115	136	148	153	152	147	138	127	116	104	93	83	74	66	59
2.5	89	138	180	212	231	238	238	230	216	199	181	163	146	130	116	103	92
3.0	129	198	260	305	332	340	332	315	292	266	240	219	210	188	167	149	132
3.5	-	270	354	415	438	440	424	404	377	362	326	292	260	230	215	202	180
4.0	-	-	462	502	540	546	530	499	475	429	384	366	339	301	267	237	213
4.5	-	-	544	635	642	648	628	590	562	528	473	432	382	356	338	300	266
5.0	-	-	-	739	726	731	707	687	670	607	557	521	472	417	369	348	328
5.5	-	-	-	-	750	750	750	750	737	667	658	586	530	496	446	395	355
6.0	-	-	-	-	-	750	750	750	750	750	711	633	619	558	512	470	415
6.5	-	-	-	-	-	-	750	750	750	750	750	743	658	621	579	512	481
7.0	-	-	-	-	-	-	-	750	750	750	750	750	750	676	613	584	525
7.5	-	-	-	-	-	-	-	-	750	750	750	750	750	750	686	622	593
8.0	-	-	-	-	-	-	-	-	750	750	750	750	750	750	750	690	825

Significant wave height (H_{sig}, m) (row axis)

Fig. 5.33. The power table for a given Pelamis configuration as derived using the frequency domain element of the PEL simulation suite

pletely risk free but repeatable, and tests can be run iteratively for optimisation purposes. Furthermore, changes to the PTO systems, moorings, and the machine configuration can be made in conjunction with related control algorithm changes. A holistic approach to the development of the Pelamis is made possible with minimum cost and/or risk.

A remaining weakness of the PEL suite is that results are inaccurate for the largest waves. Orcaflex provides us with less detailed simulation capability in large waves. While numerical work continues in this area, physical model testing remains vital for the survival regime.

5.5 Discussion

In this chapter the importance of numerical and experimental modelling when developing a wave energy converter was emphasised. Each section could be expanded into a single chapter, but given the scope of this book the main objective is to provide a starting point for those who are novices to the wave energy field.

In 5.1, frequency domain modelling is presented at the expense its time domain equivalent. This is due to the number of applications using both approaches and to the ease-of-use that frequency domain tools provide. In addition, the number of numerical models which directly implement solutions of the Navier-Stokes equations are still a rarity, but their use could increase in the near future. A good starting point for those interested in the field is given in the review conducted by McCabe (2004).

Section 5.2 gives insight to both tank and wavemaker design. The different options that allow the accurate simulation of both shallow and deep water waves are detailed. The influences of geometrical parameters like tank width and paddle size were also studied, and considerations regarding the absorption of reflected waves were presented.

Section 5.3 is intrinsically linked with 5.2 as it provided generic guidelines for the use of experimental facilities. Continuous validation of numerical simulation by means of experiments in wave tanks is essential and it is fundamental that such notion is clear to both developers and researchers.

Finally, section 5.4 demonstrated the integration of numerical and experimental modelling programmes in a commercial environment, by exemplifying the progress made when developing the Pelamis WEC. Several stages that eventually lead to a full-scale prototype were described.

References

Alves M (2002) Incident Wave Identification. MARETEC/AWS Internal Report 2/2002. Lisbon, Portugal

Bai K, Yeung R (1976) Numerical Solutions of Free-Surface Flow Problems. Proc 11th Sym Naval Hydrodyn. London, UK, pp 609–641

Baltazar J, Falcao de Campos JA, Bosschers J (2005) A Study on the Modelling of Marine Propeller Tip Flows using BEM. Congreso de Métodos Numéricos en Ingeniería. Granada, Spain

Biesel F, Suquet F (1951) Les apparails generateurs de houle en laboratoire. La Houille Blanche, 6, 2, 4, and 5. Laboratory Wave Generating Apparatus English version Project report 39. St Anthony Falls Hydraulic Laboratory, Minnesota University

Boccotti P, Filianoti P, Fiamma V, Arena F (2007) Caisson breakwaters embodying an OWC with a small opening – Part II: a small scale field experiment. Ocean Eng 34(5–6):820–841

Brito-Melo A, Sarmento A, Clément A, Delhommeau G (1998) Hydrodynamic Analysis of Geometrical Design Parameters of Oscillating Water Columns. Proc 3rd Eur Wave Energy Conf, Vol. 1. Patras, Greece, pp 23–30

Brito-Melo A, Hofmann T, Sarmento A, Clément A, Delhommeau G (2000a) Numerical Modelling of OWC-shoreline Devices Including the Effect of the Surronding Coastline and Non-Flat Bottom, Proc 10th Int Offshore Polar Eng Conf, Vol. 1. Seattle, USA, pp 743–748

Brito-Melo A, Sarmento A (2000b) Numerical Study of the Performance of a OWC Wave Power Plant in a Semi-Infinite Breakwater. Proc 4th Eur Wave Energy Conf. Aalborg, Denmark, pp 283–289

Cruz J, Payne G (2006) Preliminary numerical studies on a modified version of the Edinburgh duck using WAMIT. Proc World Maritime Technol Conf. MAREC stream, Paper 027. London, United Kingdom

Cruz J, Salter S (2006) Numerical and Experimental Modelling of a Modified Version of the Edinburgh Duck Wave Energy Device. J Eng Maritme Env – Proc IMechE Part M 220(3):129–148

Delauré Y, Lewis A (2000a) An Assessment of 3D Boundary Element Methods for Response Prediction of Generic OWCs. Proc 10th Int Offshore Polar Eng Conf, Vol. 1. Seattle, USA, pp 387–393

Delauré Y, Lewis A (2000b) A comparison of OWC response predicton by a Boundary Element Method with scaled model results. Proc 4th Eur Wave Energy Conf. Aalborg, Denmark, pp 275–282

Delauré Y, Lewis A (2001) A 3D Parametric Study of a Rectangular Bottom-Mounted OWC Power Plant. Proc 11th Int Offshore Polar Eng Conf, Vol. 1. Stavanger, Norway, pp 584–554

Delauré Y, Lewis A (2003) 3D hydrodynamic modelling of fixed oscillating water column by a boundary element methods. Ocean Eng 30(3):309–330

Eatock TR, Jeffreys E (1985) Variability of Hydrodynamic Load Predictions for a Tension Leg Platform. Ocean Eng 13(5):449–490

Eça L, Hoekstra M (2000) An Evaluation of Verification Procedures for Computational fluid Dynamics. IST-MARIN Report D72-7

Evans DV (1976) A theory for wave-power absorption by oscillating bodies. J Fluid Mech 77(1):1–25

Evans DV, Jeffrey DC, Salter SH, Taylor JR (1979) Submerged cylinder wave energy device: theory and experiment. Appl Ocean Re 1(1):3–12

Evans DV (1981) Power from Water Waves. Annu Rev Fluid Mech 13:157–187

Evans D, Linton CM (1993) Hydrodynamics of wave-energy devices. Annex Report B1: Device Fundamentals/Hydrodynamics, Contract JOU2-0003-DK. Commission of the European Communities

Falnes J (2002) Ocean Waves and Oscillating Systems. Cambridge Univ Press

Gilbert G, Thompson DM, Brewer AJ (1971) Design curves for regular and random wave generators. J Hydraulic Res 9(2):163–196

Haskind M (1957) The exciting forces and wetting of ships in waves. Izv Akad Nauk SSSR. Otd Tekh Nauk 7:65–79

Havelock T (1942) The damping of the heaving and pitching motion of a ship. Philos Mag 33(7):666–673

Hess J (1990) Panel methods in computational fluid dynamics. Annual Review of Fluid Mechanics, Vol 22, pp 255–274

Hess J, Smith A (1964) Calculation of nonlifting potential flow about arbitrary three-dimensional bodies. Journal of Ship Research, Vol. 8, pp 22–44

Hughes SA (1993) Physical Models and Laboratory Techniques in Coastal Engineering. World Scientific

Jeffrey DC, Keller GJ, Mollison D, Richmond DJ, Salter SH, Taylor JM, Young IA (1978) Study of mechanisms for extracting power from sea waves. Fourth Year Report of the Edinburgh Wave Power Project. The University of Edinburgh

Katory M (1976) On the motion analysis of large asymmetric bodies among sea waves: an application to a wave power generator. Naval Architecture, pp 158–159

Lamb H (1932) Hydrodynamics, 6th edn. Cambridge University Press

Le Méhauté B (1976) An Introduction to Hydrodynamics & Water Waves. Springer-Verlag

Lee C-H, Maniar H, Newman JN, Zhu X (1996a) Computations of Wave Loads Using a B-Spline Panel Method. Proc 21st Sym Naval Hydrodynam. Trondheim, Norway, pp 75–92

Lee C-H, Newman JN, Nielsen F (1996b) Wave Interactions with Oscillating Water Column. Proc 6th Int Offshore Polar Eng Conf, Vol. 1. Los Angeles, USA, pp 82–90

Lee C-H, Farina L, Newman J (1998) A Geometry-Independent Higher-Order Panel Method and its Application to Wave-Body Interactions. Proc 3rd Eng Math Appl Conf. Adelaide, Australia, pp 303–306

Lin C-P (1999) Experimental studies of the hydrodynamic characteristics of a sloped wave energy device. PhD Thesis, The University of Edinburgh

Linton CM (1991) Radiation and diffraction of water waves by a submerged sphere in finite depth. Ocean Eng 18(1/2):61–74

Maniar H (1995) A three-dimensional higher order panel method based on B-splines. PhD Thesis, Massachusetts Institute of Technology

Martins E, Ramos FS, Carrilho L, Justino P, Gato L, Trigo L, Neumann F (2005) CEODOURO: Overall Design of an OWC in the new Oporto Breakwater. Proc 6th Eur Wave Energy Conf. Glasgow, UK, pp 273–280

McCabe AP (2004) An Appraisal of a Range of Fluid Modelling Software. Supergen Marine Workpackage 2 (T2.3.4)

Mei CC (1976) Power Extraction from Water Waves. J Ship Res 20:63–66

Mei CC (1989) The Applied Dynamics of Ocean Surface Waves. Adv Ser Ocean Eng, Vol. 1. World Scientific [revised edition in 2005]

Mynett AE, Serman DD, Mei CC (1979) Characteristics of Salter's cam for extractgin energy from ocean waves. Appl Ocean Res 1(1):13–20

Nebel P (1992) Optimal Control of a Duck. Report of the Edinburgh Wave Power Project. Edinburgh, UK

Newman JN (1976) The interaction of stationary vessels with regular waves. Proc 11th Sym Naval Hydrodynam. London, UK, pp 491–501

Newman JN (1977) Marine Hydrodynamics. MIT Press

Newman JN (1985) Algorithms for the free-surface Green's function. J Eng Math 19:57–67

Newman JN (1992) Panel methods in marine hydrodynamics. Proc 11th Australasian Fluid Mech Conf, Keynote Paper K-2. Hobart, Australia

Newman JN, Lee CH (1992) Sensitivity of Wave Loads to the Discretization of Bodies. Proc 6th Behav Offshore Struct (BOSS) Int Conf, Vol. 1. London, UK, pp 50–63

Newman JN, Lee CH (2002) Boundary-Element Methods in Offshore Structure Analysis. J Offshore Mech Artic Engin 124.81–89

Ogilvir TF (1963) First- and second-order forces on a submerged cylinder submerged under a free surface. J Fluid Mech 16:451–472

Ouslett, Datta (1986) A survey of wave absorbers. J Hydraulic Res 24:265–279

Payne G (2002) Preliminary numerical simulations of the Sloped IPS Buoy. Proc MAREC Conf. Univ Newcastle upon Tyne, UK

Payne GS (2006) Numerical modelling of a sloped wave energy device. PhD Thesis, The University of Edinburgh

Pinkster JA (1997) Computations for Archimedes Wave Swing. Report No. 1122-O. Delft University Technology, Delft, The Netherlands

Pizer D (1992) Numerical Predictions of the Performance of a Solo Duck. Report of the Edinburgh Wave Power Project, Edinburgh, UK

Pizer D (1993) The Numerical Prediction of the Performance of a Solo Duck. Proc Eur Wave Energy Sym. Edinburgh, UK, pp 129–137

Pizer D (1994) Numerical Models. Report of the Edinburgh Wave Power Project, Edinburgh, UK

Pizer D, Retzler C, Henderson R, Cowieson F, Shaw M, Dickens B, Hart R (2005) PELAMIS WEC – Recent Advances in the Numerical and Experimental Modelling Programme. Proc 6th Eur Wave Energy Conf. Glasgow, UK, pp 373–378

Prado MGS, Neumann F, Damen MEC, Gardner F (2005) AWS Results of Pilot Plant Testing 2004. Proc 6th Eur Wave Energy Conf. Glasgow, UK, pp 401–408

Retzler C, Pizer D, Henderson R, Ahlqvist J, Cowieson F, Shaw M (2003) PELAMIS: Advances in the Numerical and Experimental Modelling Programme. Proc 5th Eur Wave Energy Conf. Cork, Ireland, pp 59–66

Roache PJ (1997) Quantification of Uncertainty in Computational Fluid Dynamics. Annu Rev Fluid Mech 29:123–160

Roache PJ (1998) Verification and Validation of Computational Science and Engineering. Hermosa Publishers

Roache PJ (2003) Error Bars for CFD. 41th Aerospace Sci Meet. Reno, USA

Romate JE (1988) Local Error analysis in 3-D Panel Methods. J Eng Math 22.123–142

Romate JE (1989) The Numerical Simulation of Nonlinear Gravity Waves in Three Dimensions using a Higher Order Panel Method. PhD Thesis, Universiteit Twente

Salter SH (1974) Wave Power. Nature 249:720–724

Salter SH (1978) Wide Tank User Guide. Fourth Year Rep, Vol 3(3). Edinburgh Wave Power Project

Salter SH (1981) Absorbing wave-maker and wide tanks. Proc Directional Wave Spectra Applicat. Am Soc Civil Eng, pp 185–202

Salter SH (2001) Proposals for a combined wave and current tank with independent 360° capability. Proc MAREC 2001. IMarEST, Newcastle, UK, pp 75–86

Sarpkaya T, Isaacson I (1981) Mechanics of Wave Forces on Offshore Structures. Von Nostrand Reinhold Company

Skyner D (1987) Solo Duck Linear Analysis. Report of the Edinburgh Wave Power Project. Edinburgh, UK

Sykes R, Lewis A, Thomas G (2007) A Physical and Numerical Study of a Fixed Cylindrical OWC of Finite Wall Thickness. Proc 7th Eur Wave Tidal Energy Conf. Porto, Portugal

Standing MG (1980) Use of Potential Flow theory in Evaluating Wave Forces on Offshore Structures. Power from Sea Waves. In: Count B (ed) Proc Conf Inst Math Appl. Academic Press, London, pp 175–212

Sumer BM, Fredsøe J (1997) Hydrodynamics around Cylindrical Structures. Adv Ser Ocean Eng 12. World Scientific

Taylor J, Rea M, Rogers DJ (2003) The Edinburgh Curved Tank. Proc 5th Eur Wave Energy Conf. Cork, Ireland, pp 307–314

Van Daalen E (1993) Numerical and Theoretical Studies of Water Waves and Floating Bodies. PhD Thesis, University of Twente, The Netherlands

Vugts JH (1968) The Hydrodynamic Coefficients for Swaying, Heaving and Rolling Cylinders in a Free Surface, Report No. 112 S. Netherlands Ship Research Centre TNO

Yeung RW (1982) Numerical methods in free-surface flows. Annu Rev Fluid Mech 14:395–442

Yemm R, Pizer D, Retzler C (2000) The WPT-375 – a near-shore wave energy converter submitted to Scottish Renewables Obligation 3. Proc 3rd Eur Wave Energy Conf, Vol. 2. Patras, Greece, pp 243–249

6 Power Take-Off Systems

In this chapter the main mechanisms that can be implemented to convert wave into mechanical and/or electrical energy are discussed. Such mechanisms are often called power take-off (PTO) or power conversion systems (the first term is adopted throughout the book). The review is directly linked with the most commonly used options and to those which are linked with the technologies described in Chapter 7. Firstly air turbines, used in Oscillating Water Columns, are focused (6.1), while in 6.2 the principles of linear generators (direct drive) are addressed. Section 6.3 is devoted to hydraulic power take-off systems and details regarding an alternative to electricity production (desalination) are given in 6.4.

6.1 Air Turbine Design for OWCs

Richard Curran and Matthew Folley

Queens University Belfast
Belfast
Northern Ireland, UK

6.1.1 Introduction

This section considers two of the most popular types of air turbine that are used in Oscillating Water Column (OWC) wave energy conversion systems: the Wells turbine and the Impulse turbine, although the Denniss-Auld turbine is also briefly introduced. Since the conception of the Wells turbine (Wells, 1976) at Queen's University Belfast (QUB) there has been considerable effort devoted worldwide to the development of the basic design. Similarly, there has been a considerable amount of development effort directed towards the self-rectifying impulse turbine that was first suggested by Kim et al. (1988). The development of the Wells turbine initially included investigations into the geometric variables, blade profile, and number of rotor planes (Raghunathan, 1995), and also the use of guide vanes (Setoguchi et al., 1988). Two further design enhancements have been the pitching of a monoplane turbine's blades and the counter-rotation of a biplane's rotors (Raghunathan and Beattie, 1996). Full-scale plant systems (Falcão et al., 1994) have utilised more advanced turbine configurations, encouraged by the successful pioneering of prototype plants such as the $75\,kW$ QUB Islay plant that used basic monoplane and biplane turbine configurations with standard symmetrical NACA profile blades (Whittaker et al., 1997a, 1997b). The development work for Impulse turbines also included investigation of the basic design parameters but much of the work then went on to focus on the design of the guide vanes and whether these would be fixed or pitching, or even self-pitching (Setoguchi et al, 2001). The Impulse turbine has been installed in several plants in Asia and there is still much interest in it as an alternative to the more widely used Wells turbine (used in plants in the UK, Portugal, India and Japan); finally the Denniss-Auld turbine has been used in the Port Kembla Australian plant (Curran et al., 2000; Finnigan and Auld, 2003; Finnigan and Alcorn, 2003).

The aim of this section is to give an overview of the key theoretical design issues concerned with the use of air turbines for wave energy conversion, including the basic performance of several types of turbine design, and then to present some operational aspects. The review also considers the integration of the turbine with the OWC type systems that typically use either a Wells or an Impulse turbine. Finally, the discussion will draw some general conclusions and highlight future developments.

6.1.2 Integrated systems: Engineering Design Requirements

An air turbine forms part of an integral system that consists of a capture device, which also includes an electrical generator. In addition, the performance of the capture device will be affected by its physical environment and may form part of an array. The turbine must be connected to the capture device by some sort of ducting arrangement, and the generator must be coupled to the turbine with minimal interference to the air flow exiting the system. One can consider the primary system integration entailing the optimal arrangement and sizing of the capture device, the turbine and the generator. The key generator design parameter to be considered in terms of the impact on the turbine's performance generator is the speed, which operationally will be determined by the required torque/speed characteristic and the generator power rating. Consequently, this section will tend to focus on the interaction between the turbine and the performance of the capture device through the Oscillating Water Column (OWC) principle. The subsequent inclusion of the generator in this design methodology is not thought to change the global design optima of the complete system as it can be sized and configured to have the necessary characteristics that have been determined by the turbine-OWC system. However, in reality there will be a small deviation off the optima due to aspects such as both fixed and varying generator losses. However, the electrical system will have a large influence if a very active control strategy is adopted where the speed is allowed to vary considerably to better match the provision of the pneumatic power at the air velocities that are preferred by the turbine.

This section addresses the optimisation of the core interdependence between the turbine and the OWC in order to maximise the potential for energy conversion. There are two matching principles that need to be addressed, namely that:

1. The turbine should provide a damping level (which effectively restricts the air-flow exiting the system) that maximises the conversion of wave energy to the kinetic energy that determines the motion of the OWC and which is ultimately converted to pneumatic energy by the excitation of the air immediately above the OWC.
2. The turbine should be able to maximise the conversion of pneumatic energy to mechanical energy (and subsequently electrical energy) over the range of flow rates produced as the air exits the plenum chamber immediately above the OWC. A high value of turbine applied damping reduces flow rate and vice versa.

OWC Interface

The performance of an OWC can be modelled by the well known equation describing the damped motion of a body oscillating with a single degree of freedom due to a time varying force:

$$F(t) = m\frac{d^2 y}{dt^2} + B\frac{dy}{dt} + Ky,$$ (6.1)

where F is the applied force at time t, m is the mass of the body, B is the damping, y is the displacement and K is the spring restoring constant due to the buoyancy of the OWC. The spring restoring constant K is given by: $K = A_C \rho_W g$, where A_C is the surface area of the OWC, ρ_W is the density of water and g is the gravitational acceleration. A direct analogy can be drawn with a mechanical mass-spring-damper system; however two additional terms must be included in order to account for the waves generated by the body as it oscillates. The mass of the column must include the added mass, which is frequency dependent, as well as the entrained mass. It is thus convenient to consider the mass of the column as its effective mass, M_E, which is the sum of the entrained and added masses. Secondly, the damping should be considered to consist of two components: the applied damping, B_A, which extracts energy from the system and is provided by the turbine, and the secondary damping, B_2. The secondary damping itself consists of two further components: the radiation damping due to the generation of waves by the column, and the loss damping due to energy losses which increase substantially with incident wave power.

The established solution (Mei, 1976) to the equation of motion for sinusoidal excitation is:

$$W_{OWC} = \frac{\frac{1}{2} B_A \omega_{OWC}^2 |F|^2}{(K - M_E \omega_{OWC}^2)^2 + (B_A + B_2)^2 \omega_{OWC}^2},$$ (6.2)

where W_{OWC} is the average power output and ω_{OWC} is the angular frequency of oscillation. The level of turbine applied damping which maximises the power output of the OWC is then given by:

$$B_{A_{OPT}} = \sqrt{B_2^2 + \left[\frac{K - M_E \omega_{OWC}^2}{\omega_{OWC}} \right]^2} . \qquad (6.3)$$

Figures 6.1 and 6.2 present the results of a series of hydraulic model tests carried out in order to determine the influence of applied damping on an OWC's power output and bandwidth response respectively. Figure 6.1 shows the results for four different values of significant wave height. At low levels of applied damping it can be seen that the power output is very sensitive to changes in the applied damping. Furthermore, the power output is maximised at a certain level of applied damping which is known as the optimum applied damping. Subsequently, the power output decreases as the applied damping is further increased although the performance sensitivity is much less than for under-damping. Under-damping thus greatly reduces the efficiency of the column to an extent which over-damping does not, thus providing a further incentive to over-damp the system design.

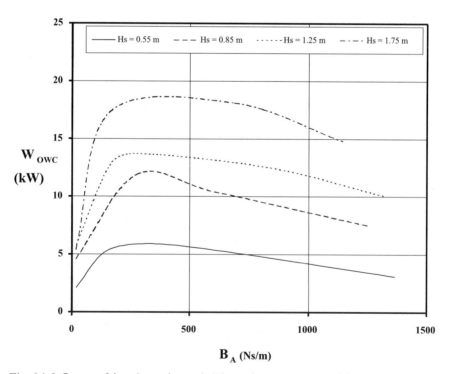

Fig. 6.1. Influence of damping and wave height on the power output of the OWC

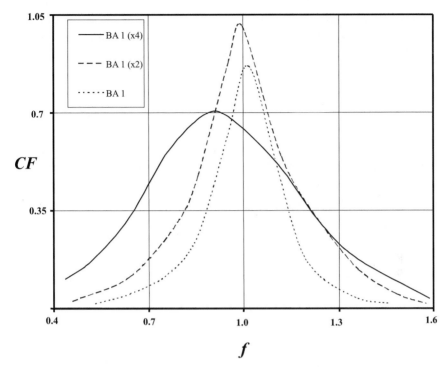

Fig. 6.2. Influence of damping and frequency ratio on the capture factor of the OWC

Figure 6.2 shows the performance bandwidth response for three different damping levels; at an initial value of B_A1 and for double (*x2*) and quadruple (*x4*) the initial level. The capture factor, *CF*, in Fig. 6.2 is defined as the ratio of the capture width of the device to the column width, is plotted against the frequency ratio: $f = \dfrac{\omega_{OWC}}{\omega_O}$, where ω_O is the resonant frequency. The capture factor is the amount of power 'captured' by the OWC relative to the value incident on the width of coastline spanned by the device (thus it is often called the relative capture width; see Chapter 3). It can be seen that the performance must be optimised across a range of frequencies in order to maximise the output for any particular set of wave conditions. It can be concluded that the optimal applied damping is thus a function of the wave period, incident wave power, and tide level (Stewart, 1993). Furthermore, no single level of damping will optimise the performance of the column in every individual sea although there will be a global optimum which will maximise the output over a range of mixed frequency seas, i.e. over a prolonged period.

Turbine Performance

Assuming conservation of mass flow, the damping, B_A, applied to the OWC by the turbine can be expressed as:

$$B_A = \Delta P_0 \frac{A_C}{V_C} = \Delta P_0 \frac{A_C^2}{A_A V_A}, \qquad (6.4)$$

where A_C and A_A are the cross-sectional areas at the water column surface and turbine duct respectively, and V_C and V_A are the respective air velocities. Eq. (6.4) assumes that the airflow is incompressible as the turbine Mach numbers have been restricted through design to values of less than 0.5, particularly for the highest velocity vectors experienced at the blade outer tip position. The same equation can be expressed in terms of the turbine damping ratio: $B_R = \dfrac{P*}{\phi}$, where $P*$ and ϕ are the pressure and flow coefficients, respectively (both non-dimensional). This leads to Eq. (6.5) and with further manipulation to Eq. (6.6):

$$B_A = 4\rho_W \left(\frac{A_C^2}{A_A} \right) U_t \left(\frac{P^*}{\phi} \right); \qquad (6.5)$$

$$B_A = \tfrac{1}{2} \pi \rho_W \omega D_t^3 A_R^2 (1 - h^2) \left(\frac{P^*}{\phi} \right), \qquad (6.6)$$

where the tip diameter is given by: $D_t = \dfrac{2U_t}{\omega}$, the column-to-duct area ratio is: $A_R = \dfrac{A_C}{A_A}$, the turbine duct area is: $A_A = \tfrac{1}{4} \pi D_t^2 (1 - h^2)$, and the hub-to-tip ratio is: $h = \dfrac{D_h}{D_t}$, where subscripts h and t denote the hub and tip radii respectively. Figure 6.3 plots Eves empirically based characteristic of B_R against the number of rotor planes N_p and each plane's solidity S, shown as a dashed line (Eves, 1986); an additional solid curve is plotted following relationship:

$$\frac{P^*}{\phi} = 0.525 N_p \tan\left(\frac{\pi}{2} S \right), \qquad (6.7)$$

where $S = \dfrac{A_B}{A_A}$, A_B is the total blade area in plan view per plane, and correlation factor $= 0.525$. However, the empirical results are taken from small-scale results and therefore designers would be expected to adjust the correlation coefficient upward, in order to take into account the scale effects due to the variation in Reynolds Number Re. The Reynolds Number is based on the chord length c and is defined by: $Re = \dfrac{Vc}{\upsilon}$, where V is the relative airflow velocity and $\upsilon = 1.5x10^{-5} \ m^2/s$ is the kinematic viscosity of air.

The original coefficient was estimated from steady state model tests (Eves, 1986) that were carried out at a mid-span (average) Reynolds Number of $Re_{mid-span} = 1.5x10^5$ while other authors correlated their higher Reynolds Number work, as shown in Fig. 6.3. The new model results (Curran and Gato, 1997) shown were performed at $Re_{mid-span} = 5.5x10^5$ and are denoted by the clear diamonds. They also included data from the Islay prototype turbine acting in random oscillating flow denoted by the solid circles in Fig. 6.3, including two sets of conditions at $Re_{mid-span} = 1.2x10^6$ and $Re_{mid-span} = 2x10^6$, referred to as 0012 Islay and 0015 Islay respectively, in reference to the NACA 0012 and NACA 0015 blade profiles used. Consequently, the authors adjusted the correlation factor from 0.342 to 0.525, as shown by the dashed and solid curve fits in Fig. 6.3 respectively. They have assumed the mathematical form of the original Eves (1986) correlation work, as there are insufficient larger scale data to warrant change from a trend so readily evident at small scale. It is significant that Fig. 6.3 shows that the damping ratio was higher than that seen at small-scale, due to the effects of increased life realised at higher Reynolds Numbers. Consequently, air turbines that are designed with Eves correlation would over-damp the system and although the OWC is less sensitive to this, it would affect the optimal range of flow conditions for the intended wave conditions.

Eqs. (6.6) and (6.7) can be substituted to give Eq. (6.8), leading to Eq. (6.9) (noting that r_t and A_R are interrelated):

$$B_A = 0.2625\omega D_t^3 A_R^2 N_p (1-h^2)\tan\left(\frac{\pi}{2}S\right); \qquad (6.8)$$

$$B_A = f(\omega, r_t, A_R, N_p, h, S). \qquad (6.9)$$

Single aerofoil theory suggests that a Mach Number of 0.5 ($M = \dfrac{V}{a}$, where V is the relative air velocity and a is the speed of sound) is a reasonable free stream Mach number at which localised sonic conditions may be considered to be influential. Additionally, airflow cannot be assumed to be incompressible at higher values of M and so this value is a constraint at the blade tip, which consequently determines the maximum turbine diameter. Alternatively, the operational Re of the turbine must be high enough to ensure that the full potential of the blades' aerodynamic performance is realised.

Finally, given that the pneumatic power generated by the OWC is expressed by:

$$W_{pneumatic} = \Delta P_o Q = B_A V_A A_R^{-1}. \qquad (6.10)$$

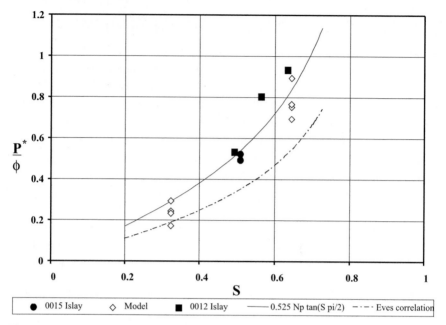

Fig. 6.3. Correlation of large scale results for prediction of the damping ratio

Eqs. (6.4), (6.8) and (6.10) can be used to show that the total power output of the turbine is:

$$W_{converted} = \eta W_{pneumatic} = \eta \left[0.265 \omega D_t^3 A_R N_p (1-h^2) \tan\left(\frac{\pi}{2} S\right) \right] V_A, \quad (6.11)$$

where η and V_A are instantaneous values of turbine efficiency and axial air velocity which are determined respectively, by the turbine efficiency characteristic and the flow distribution generated by the OWC.

System Optimisation

It is necessary to design a system with a level of applied damping which represents the global optimum in order to achieve the maximum accumulated value of energy converted from incident wave energy, with reference to Eq. (6.11) and Eq. (6.3). The latter tends to be the constraint that determines the lower threshold for the applied damping, as evident in Fig. 6.1. However, there is a considerable range above this value where the performance of the OWC is relatively insensitive to the level of applied damping. Figure 6.1 also shows that there is a level of damping above $B_{A_{OPT}}$ which can be beneficial in mixed seas. Therefore, a range of damping levels above the theoretical optimal value $B_{A_{OPT}}$ should be investigated in order to maximise the potential conversion determined ultimately by Eq. (6.11) for the turbine.

There are a number of additional energy losses incurred at the turbine-generator stage on a plant which need to be considered. The turbine converts pneumatic power into mechanical power that is reduced by further downstream losses prior to electrical power export, where the total instantaneous converted power is given by:

$$W_{converted} = W_{elect} + W_L + W_{inertial} , \qquad (6.12)$$

where W_{elect} is the exported electrical power, W_L is the total power losses and the inertial power $W_{inertial}$ is given by $W_{inertial} = \dfrac{dE_{inertial}}{dt}$, where $E_{inertial} = \dfrac{1}{2} I \omega^2$ and I is the combined inertia of the turbine-generator unit. Apart from the mass of the turbine-generator unit the amount of inertial energy $E_{inertial}$ stored instantaneously will also depend on the resistance of the electrical generator and the degree of power smoothing required to meet the electrical supply specifications. However, with regard to Eq. (6.12), $\dfrac{dE_{inertial}}{dt} \to 0$ when optimising over a sufficiently lengthy period of time.

The losses W_L can be further divided into the hub windage losses $W_{Lwindage}$, and the generator losses $W_{Lgenerator}$:

$$W_L = W_{Lwindage} + W_{Lgenerator} + W_{Bearings} . \qquad (6.13)$$

The mechanical bearing loss $W_{Bearings}$ is very small but can be added for completeness. The hub windage loss can be estimated by using the well known equation for the skin friction drag. For a unit width section of thin plate, the drag force produced by an air flow over the surface is given by: $F_D = \frac{1}{2} \rho V^2 l C_f$, where C_f is the friction coefficient and l is the length of section. Therefore, for a turbine hub, the drag losses can be calculated incrementally in annular rings of width δr, where the total windage loss will be equal to the integral between the hub's centre, $r = 0$, and hub's outer radius, $r = r_h$. The length is given by the circumference of each annular strip while the radial increment δr is equivalent to the plate width. Therefore, the estimate of the hub windage (where $W_{Lwindage} = T\omega$ and the torque T is $T = F_D r$) is given by integrating with respect to radius:

$$W_{Lwindage} = N_p \rho_W \omega^3 \pi C_f \int_{r=0}^{r_h} r^4 dr , \qquad (6.14)$$

where $W_{Lwindage}$ is the total power loss due to hub windage (blade drag uses chord length c rather than circumference $2\pi r$ and it is multiplied by the blade number) between the shaft and the hub's outer radius, on one side of the rotor.

The generator losses $W_{Lgenerator}$ are primarily incurred due to the machine's iron loss and copper loss. The iron losses are fixed and are determined by the power rating of the particular machine being used and can be approximated to 2.5 % of the generator's power rating. The generator's copper loss varies according to the square of the stator current, which in-turn is a function of the power output and load voltage. Typically, the copper loss increases to a maximum of approximately 1 % of the generator rating at full rated load, and consequently could be approximated to an average value of 0.5 % of the rating, corresponding to the generator operating on average at half the power rating. Cumulatively, the generator losses can be approximated to:

$$W_{L\,generator} = 0.03R, \tag{6.15}$$

where R is the generator rating.

The total amount of energy E_{total} which has been converted, during the time period t_1 to t_2, is given by: $E_{total} = \left[\sum_{t_1}^{t_2} \{E_{converted} - E_{losses}\} \right]$. Therefore, it can be seen that the equation which expresses the value of power being exported instantaneously W_{elect} must be optimised over the range of flow velocities at the plant in order to maximise E_{total}:

$$W_{elect} = \left[\eta B_A V_A A_R^{-1} \right] - [W_L]. \tag{6.16}$$

It can be conclude that for a turbine operating at a particular speed, the key design drivers are the instantaneous turbine efficiency, the axial flow velocity (i.e. energy levels) and the optimal applied damping. This establishes turbine applied damping B_A as one of the primary design parameters that actually maximises the potential system's power conversion. For the other two drivers, the turbine efficiency characteristic is pre-determined by the initial turbine configuration and the air velocity is governed by the incident wave power regime and the OWC's efficiency characteristic. Eq. (6.16) represents an objective function based on performance which could be used in the optimisation procedure although a minimal of cost per unit of energy output would be the ultimate objective.

System Improvement

Wave energy conversion systems would benefit greatly in terms of overall performance with control strategies and other techniques that react to the system's conditions in order to maximise the power output, according to Eq. (6.16). A simple limiting technique is to use a blow-off valve that can be opened in order to release energy from the system at a specified limiting value of pneumatic power. Effectively, this reduces the flow velocity to the range over which the turbine most efficiently converts power. Alternatively, a variable speed generator could be used so that the rotational speed can be increased during the larger sea states, and vice-versa in the smaller sea states, over an incremental range to be specified (Justino and Falcão, 1998; Sarmento et al., 1990). The principle utilised in the control strategy is that the applied damping increases with the increase in speed, according to Eq. (6.8), which leads to a reduction in the flow rate, which results in the turbine remaining at a more efficient angle of attack. However, it should be noted that these conversion gains at the turbine-generator stage may be offset by a negative impact on the performance of the OWC, as described by Figs. 6.1 and 2.2, due to the strong performance coupling previously described.

6.1.3 Air Turbine Design Configurations

The Wells Turbine

Many of the plants built to date around the world have tended to be Oscillating Water Columns with a Wells turbine-generator power take-off. The various OWC concepts and configurations (Count, 1980; Salter, 1988; Curran, 2002) couple in a similar manner to the Wells turbine with regard to the matching strategy relative to the hydraulic performance. The Wells turbine utilises symmetrical blades located peripherally on a rotor at 90 degrees stagger to the airflow so that an alternating airflow drives the rotor predominately in a single direction of rotation (Raghunathan et al., 1985; Inoue et al 1987; Gato and Falcão, 1989). The axial air velocity at the turbine combined with the rotational velocity determines the relative velocity and angle of attack onto the turbine blades, typically of a symmetrical NACA profile (Jacobs and Sherman, 1937). Aerodynamic lift and drag forces are generated perpendicular and parallel respectively, to the relative velocity and can be resolved into resultant axial and tangential forces in the direction of flow and rotation respectively. Relevant to the aerodynamic performance of the turbine blades has been the considerable increase in the use of Computational Fluid Dynamics (CFD) tools to investigate the physics of the blade profile interaction with the air flow. This has included work into the optimisation of the blade profile (Gato and Henriques, 1994), investigation of the 3-D turbine aerodynamics (Thakker et al, 1994; Dhanasekaran and Govardhan, 2005) and detailed investigation of the design (Watterson and Raghunathan, 1997).

A Wells turbine rotor (see Fig. 6.4) consists of several symmetrical aerofoil blades set around a hub at an angle of 90 degrees with their chord lines normal to the axis of rotation. For absolute air flow velocity, V_a, which is axial at the inlet and tangential rotor velocity, U_t, at a radius, r, from the axis of rotation, the relative velocity, W, is at an angle of incidence, α, to the blade chord. This will generate a lift force, L, and a drag force, D, normal and parallel respectively to the relative velocity, W. These forces can be resolved into tangential and axial directions to the rotor given by:

$$F_T = L\sin\alpha - D\cos\alpha, \qquad (6.17)$$

and

$$F_X = L\cos\alpha + D\sin\alpha. \qquad (6.18)$$

For a symmetrical aerofoil in an oscillating airflow, the magnitude of F_T and F_X, will vary during a cycle but the direction of F_T will remain the same, being independent of reciprocating flow. However, at the very lowest flow rates F_T will be negative due to the aerodynamic drag on the blades and also at the very highest flow rates where lift has been lost due to boundary layer separation. For a turbine which is well matched to the sea distribution, F_T will result in the generation of positive torque and power for the majority of the wave cycle. Furthermore, the stalling action of the aerofoils will act as a limiter during extreme wave conditions. The normal component, F_X will result in an axial thrust force which has to be borne by suitable bearings. The expressions for F_T and F_X are given in dimensionless coefficients C_T and C_X, where:

$$C_T = C_L \sin\alpha - C_D \cos\alpha ,$$

(6.19)

and

$$C_X = C_L \cos\alpha + C_D \sin\alpha .$$

(6.20)

The lift and drag are subject to 3-D flow (Horlock, 1956) and it has been shown that the blade performance varies considerably from the corresponding single aerofoil data (Curran et al., 1998) due to interference effects between turbine blades (Raghunathan and Beattie, 1996). In general drag dominates over lift at the lowest angles of attack, after which lift prevails to a maximum value at the stalling incidence, at which point the boundary layer separates from the blade surfaces leading to loss of lift and a rapid increase in the dominance of drag on performance (Abbot and Von Doenhoff, 1959). The forces which determine the driving force are always resolved in the same direction of rotation, due to the blade symmetry, but the resultant axial force alternates during the wave cycle and needs to be borne by axial thrust bearings.

It is evident from Fig. 6.5 that the operational region of the Wells turbine is poor at low angles of incidence, nominally 0–18 degrees, i.e. $\approx 0\text{-}\alpha_{stall}$. The performance characteristic in this region is therefore of utmost importance in terms of turbine design. The main performance parameters are efficiency, pressure drop

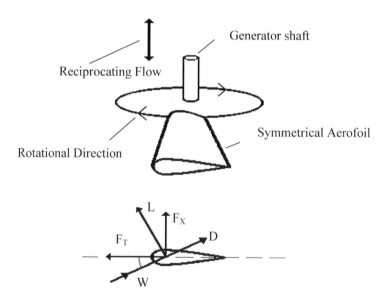

Fig. 6.4. Principle of operation of the Wells turbine

and torque. The aerodynamic characteristics of an aerofoil blade and therefore the Wells turbine are strongly dependent on the incidence of the airflow onto the blade. This can be observed from the experimental results shown in Figs. 6.5 and 6.6 for monoplane and biplane Wells turbines.

The frequency of bi-directional and random air flow in a wave energy device is very low, e.g. nominally 0.1 Hz, and therefore the air flow through the device is normally assumed to be quasi-steady. The values of efficiency and pressure shown are regarded as instantaneous values with respect to flow ratio or air flow incidence. The results show typical variation of η and $\Delta p_0{}^*$ with ϕ. For a given rotor speed, p^* was proportional to the pressure drop across the rotor, Δp, and ϕ was proportional to the flow rate, Q. For a Wells turbine rotor, a linear relation-

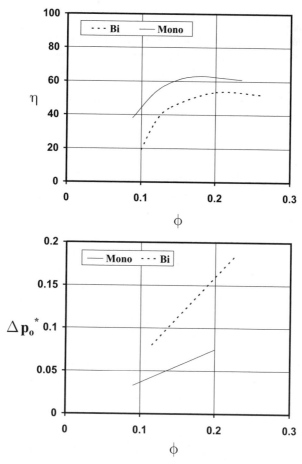

Figs. 6.5 and 6.6. Non-dimensional efficiency and pressure drop relative to the flow coefficient for the monoplane and biplane Wells turbines, respectively

ship exists between pressure drop and flow rate, except for very low flow rate values. This is true for a turbine either with low or high solidity and for both a monoplane or biplane turbine. A linear pressure drop-flow rate characteristic is well suited for matching the turbine with a wave energy device which uses the principle of the oscillating water-air column. A constant value of damping produced by the turbine can be chosen to optimally damp the OWC according to the chamber geometry and the incident wave conditions. Also, the variations in flow are minimised if one compares the ratio to that of an orifice damper which shows an exponential relation.

The aerodynamic efficiency, η, increases with ϕ up to a certain value after which it decreases. The reason for this is that at higher values of flow incidence the boundary layer on the aerofoil blades tends to separate and subsequently fully separate, there-by increasing the drag and reducing the lift. The phenomenon of flow separation, known as turbine stall, can be severe and could in extreme circumstances cause the driving torque to become negative. This is rather more gradual on thick aerofoils and so results in a more gradual drop in η with ϕ. Therefore, a Wells turbine with thick aerofoils will be able to operate over a wider range of flow rates without a significant drop in performance. At very small values of ϕ the efficiencies are negative. In oscillating airflow it can be shown that the power absorption at small incidence is a small proportion of the total power output if the turbine is well matched to the pneumatic power distribution. The time averaged cyclic efficiency has been shown to be typically 5–10 % lower than the peak efficiency of a cycle. However, there will be periods when it no longer becomes viable to operate the system as the power imported to overcome the system losses out-weighs the converted power.

The pressure-flow ratio or damping relation is related to the axial force generated, where for lower angles of attack the axial component is largely determined by the linear lift component (Curran et al., 1998). As mentioned earlier, the pneumatic energy level that any turbine accommodates varies with configuration, including solidity, size and speed, while further performance deviations will also be evident in the stall region. One turbine configuration may show an increased damping level in the post-stall region, such as for the guide-vane design, while another may show a decrease, as evident with the variable-pitch design, while that of the counter-rotating configuration remains relatively constant. It is beneficial to at least maintain the pressure-flow ratio into the stall region as this tends to restrict the flow rate to values that are closer to the turbine's efficient performance range. Some of these considerations are addressed further along this section, which compares the long-term productivity of a number of advanced Wells turbine configurations.

Advanced Wells Turbine Configurations

The following sections introduce a number of variations to the basic Wells turbine configuration that can be termed advanced configurations. The productivity of these configurations needs to be predicted in order to facilitate practical comparison. Such comparison has hitherto been difficult to undertake as nothing short of a full productivity analysis fails to fully address the productivity expectations of the turbine types. Classically, wave energy turbines have been compared in terms of their efficiency over a flow range (Curran and Gato, 1997). However, the pressure drop and resultant pneumatic energy range over which the turbines operate may vary greatly for like flow ranges. Furthermore, the efficient conversion of power determines economic feasibility and consequently is a more practical basis for comparison. It was assumed that the turbines operate in the same typical pneumatic power distribution but that in each case, the turbine's sizing can be optimised to maximise that configuration's output. A comparison of the results for torque, pressure drop and efficiency is shown in Figs. 6.7, 6.8 and 6.9 respectively.

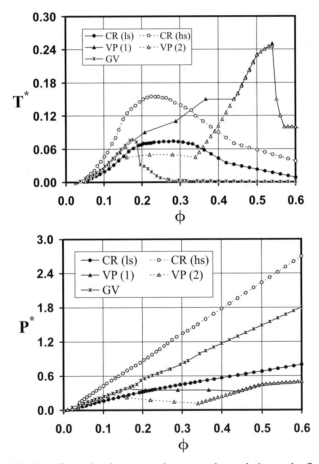

Fig. 6.7 and 6.8. Non-dimensional torque and pressure drop relative to the flow coefficient for the low (ls) and high solidity (hs) Counter-rotating turbine (CR), variable pitch turbine (VP) with two control strategies, and guide vane Wells turbine (GV)

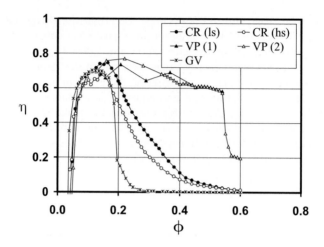

Fig. 6.9. Efficiency relative to flow coefficient for the low (ls) and high solidity (hs) Counter-rotating turbine (CR), variable pitch turbine (VP) with two control strategies, and guide vane Wells turbine (GV)

Variable-Pitch Monoplane

In principle, a variable-pitch turbine limits the angle of attack by rotating the blade about its radial axis so that the turbine can maintain an efficient angle of attack, requiring a control strategy for pitch-control that can also be implemented at the lowest flow rates (Gato and Falcão, 1989). However, the pitching process also reduces the effective solidity of the turbine and consequently the concept is disadvantaged by the need to pitch ever further to meet the increased flow induced by the pitching action, a direct consequence of the reduced damping ratio (Curran and Gato, 1997). Due to the variation in the damping ratio, it is not clear to what degree the turbine's instantaneous efficiency or torque output should be the driving variable that determines the required pitch angle. The turbine blades can be pitched when the torque produced by the turbine begins to fall substantially at stall and thereafter the pitching rate can be made to match that of the angle of attack, i.e. equivalent-angle control. Alternatively, the blades can be pitched in order to always maximise efficiency rather than just maintain torque at stall.

The equivalent angle method is denoted by *VP1* in Figs. 6.7 through 6.9, while the maximal efficiency method is denoted by *VP2*. It was found that the maximal efficiency method pitched at a substantially lower flow rate of $\phi = 0.13$ than that of the equivalent-angle method where $\phi = 0.165$. Furthermore, the pitch rate was greater where the rate of *VP1* was equal to the increase in the angle of incidence over that of stall. It should be noted however, that the intention of the study was not to find the exact optimal control mechanism but to determine which control method was more likely to maximise productivity. Additionally, the performance at the more extreme stall flow rates had to be estimated as the experimental data did not extend over the full range necessary for analysis. A comparison of the results shown in Figs. 6.7 and 6.8 shows that the equivalent angle method did maintain the torque and pressure levels respectively, while Fig. 6.9 shows that the efficiency was actually lower than for the maximum efficiency method. It is clear from the results that the variable-pitch turbine operates with high efficiency over a significantly greater flow range in contrast with the other configurations being investigated (introduced in the following sections), although the pressure drop remains low after the pitch control is initiated.

Fixed Guide-Vane Monoplane

The main aim of a fixed guide vane configuration is to maximise the turbine's conversion efficiency over a limited but optimal flow range, while forfeiting conversion at more extreme flow rates. Consequently, this configuration has a smaller operational flow range outside of which the performance is poor due to flow separation off the guide vanes at those angles of attack not accommodated in the aerodynamic design of the vanes. The guide vanes are used to optimally direct the airflow to and off the blades for a given profile, thereby also ensuring axial airflow from the turbine and minimal air-swirl losses for the exit stage. Moveable guide vanes are not considered but offer the possibility of optimising the upstream and downstream vane angles, whereas fixed vanes must be deployed in a symmetrical and compromised orientation. The rotor of any air turbine imparts a swirl component to the airflow due to the viscous interaction of the blade surface on the air particles. Consequently, in a similar level of complexity to the variable-pitch turbine, the vanes could be controlled by a pitching mechanism although this additional capability is generally deemed to be not worth the extra complexity and cost, in both non-recurring and maintenance terms. The sharp decrease in torque output at stall due to the limiting vane angle is very evident in Fig. 6.7 but it can also be seen from Fig. 6.8 that there is a slight increase in the damping ratio.

Counter-Rotating Biplane

The main turbine performance issue being addressed in the counter-rotating configuration is the recovery of air swirl exiting the turbine. It was noted above that the rotor of a Wells turbine imparts a swirl component to the airflow relative to the work done. However, an alternative to utilising guide vanes is to rectify the swirling airflow with the opposing action of a counter-rotating downstream rotor. This alternate concept also has an added attraction in benefiting from the favourable interaction evident between the rotors during turbine stall, resulting in a more gradual loss of efficiency. Another major feature of multi-plane machines is the wide range of damping ratios and pneumatic power levels that can be accommodated. Both high and low solidity counter-rotating configurations were tested and are included in Figs. 6.7 through 6.9. It is readily evident that a change in solidity results in a significant variation in the performance and that the efficiency decreases as solidity is increased. Notwithstanding, the performance characteristics are regular with a reasonable range in efficiency relative to flow rate.

Impact on Productivity

The predicted annual distribution of pneumatic power is presented in Fig. 6.10 in terms of occurrence and actual contribution to power (converted power, W). The latter is given in terms of contribution to pneumatic power rating where the sum is the annual rating while multiplying across by time yields the contribution to energy. It is evident that the most frequent power band to occur, $400\,kW$, is actually lower than the average figure of $500\,kW$ and that the highest power band is almost four times the average. Figure 6.11 presents an example of the relation of pneumatic power to converted power for the low solidity counter-rotating turbine while Fig. 6.12 presents the conversion results for all configurations.

Figure 6.11 shows that the lower distribution of converted power mirrors the shape of the pneumatic power distribution as the turbine is well designed to operate efficiently over the range of pneumatic power levels. The distribution of converted power would become more skewed if the diameter was inappropriate and provided unsuitable flow rates at the pneumatic power bands that contribute most energy, i.e. through mismatching. Figure 6.12 is surprising as there is little to distinguish one turbine from another, especially considering the variation that was so prominent in Fig. 6.9.

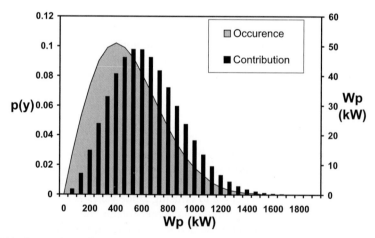

Fig. 6.10. Comparison of pneumatic power occurrence and actual contribution

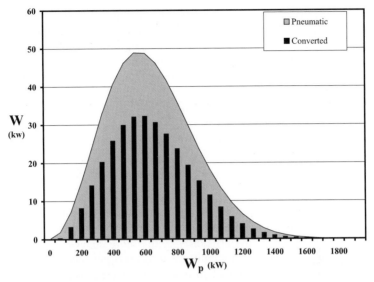

Fig. 6.11. Comparison of pneumatic and converted power for the low solidity counter rotating turbine

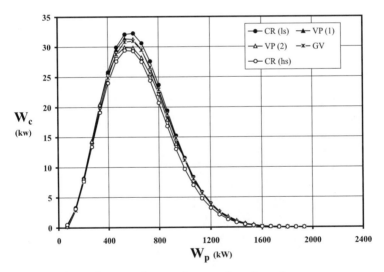

Fig. 6.12. Comparison of converted power for several configurations

It can be seen that the high solidity counter-rotating turbine was least efficient over the majority of the power bands and that the other turbines only really differed in mid-range. The low solidity counter-rotating turbine achieved the highest efficiency of 64 % while the equivalent-angle pitching turbine was next with 62 %. However, one must consider the pneumatic power range within which the turbines operate in order to understand why the variable pitch turbines were not superior, given the considerably wider range of efficient flow rates suggested by Fig. 6.9. Therefore, the torque and efficiency of the turbines can be considered relative to a coefficient that incorporates pneumatic power, as suggested through Eq. (6.21):

$$\Pi_p = \frac{W_p}{\pi \rho_A \omega^3 R_t^5 \left(1-h^2\right)} .$$

(6.21)

The torque and efficiency from Figs. 6.7 and 6.9 was re-plotted and is presented in Figs. 6.13 and 6.14 respectively, where the first is divided to facilitate more detailed inspection of the efficiency at the more congested power points. Comparing Figs. 6.10 and 6.13 highlights the direct relation between torque output and pneumatic power input, and that all turbines generate similar levels when operating at lower power coefficients, although the variable pitch turbines have considerably superior range. Also, it is interesting to note that the higher efficiencies of *VP (2)* now cover a much smaller operational range when pitching due to the reduced pressure drop evident from Fig. 6.8, resulting in the pneumatic power remaining relatively constant.

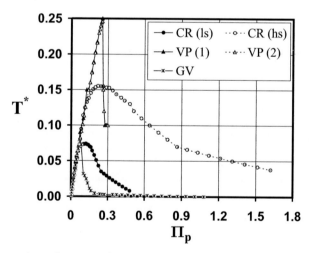

Fig. 6.13. Comparison of torque and pneumatic power conversion

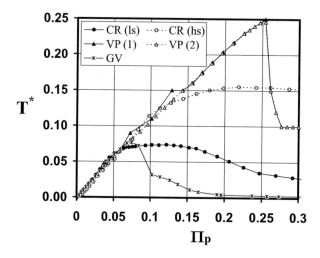

Fig. 6.13. a Comparison of torque and pneumatic power conversion (zoom on the horizontal axis)

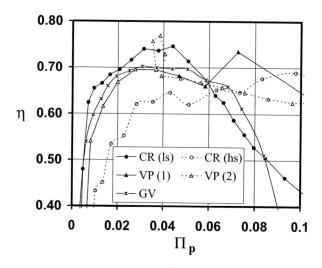

Fig. 6.14. Comparison of efficiency and pneumatic power conversion

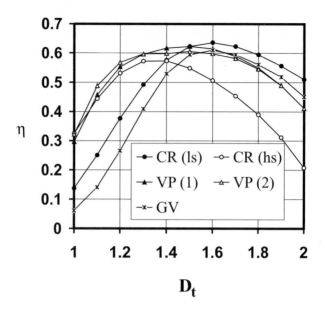

Fig. 6.15. Effect of diameter change on overall efficiency

Figure 6.15 presents the variation in efficiency with tip diameter in order to show the degree of sensitivity in the determination of the optimal diameter that maximises productivity. However, the figure is also representative of the sensitivity to power fluctuations to a change in either diameter or input pneumatic power, which will have the same ultimate effect in causing the air velocity and angle of attack to change, as evident in Eq. (6.13). It can be seen that as expected the variable pitch turbines is the least sensitivity while in all other cases slightly higher tip diameters would present a more robust design, as the characteristics are more steeply sloped towards the lower values.

Impulse Turbines

The use of a self-rectifying impulse turbine for wave energy conversion was first suggested in 1988 (Kim et al., 1988). The axis of rotation of the self-rectifying impulse turbine is aligned with the principle direction of air flow within a duct, with guide vanes used to deflect the flow onto the rotor as shown in Fig. 6.16. To maximise the power capture the setting angle of the upstream and down stream guide-vanes needs to be different, which due to the oscillatory nature of the flow means that the guide vanes would ideally move to provide the optimum setting angle depending on the instantaneous direction of flow. This has been achieved by using pivoting guide vanes that use the aerodynamic moment induced by the movement of the air to change the setting angle both upstream and downstream of the rotor, denominated "self-pitch-controlled" guide vanes. Potential problems

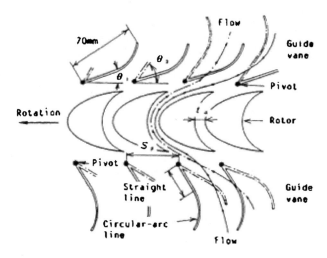

Fig. 6.16. Self-rectifying impulse turbine (Setoguchi et al., 2001)

with maintenance of such a complex mechanism led to the mechanically simpler, though fundamentally less efficient solution, where the guide vanes do not move, leading to the denomination of "fixed" guide vanes (Maeda, 1999).

In addition to the classification of "self-pitch-controlled" and "fixed" guide vane configurations, it is possible to sub-divide the "self-pitch-controlled" guide vanes into "mono-vane" and "splitter" types. In "mono-vane" type guide vanes the whole vane moves around the pivot, whilst for "splitter" type guide vanes only the section of guide vane closest to the rotor moves, engaging with stationary curved stators; for "fixed" guide vanes plate and aerofoil type guide vanes have been investigated (Setoguchi et al., 2001). A further modification to the design of the guide vanes has also been made by linking the two sets of guide vanes so that they move together. This was done because it was found that at times the downstream guide vanes did not move appropriately in response to the aerodynamic couple experienced (Setoguchi et al., 1996). In addition to these differences in the design of the guide vanes, it is also possible to vary the number, setting angles and dimensions of the guide vanes to provide an infinite number of possible guide vane arrangements. Although these different guide vane configurations have some influence on performance as discussed in Setoguchi et al. (2001), it has been found that the critical parameters are the upstream and downstream setting angles and the guide vane length to pitch ratio. For "self-pitch-controlled" guide vanes the optimum upstream setting angle is 15 to 17.5 degrees, the optimum downstream setting angle is 55 to 72.5 degrees and the guide vane pitch to length ratio should be less than 0.65. For "fixed" guide vanes the optimum setting angle is 30 degrees.

Prior to discussing the specific performance of self-rectifying impulse turbines, it is useful to outline their general characteristics relative to the Wells turbine, against which they are typically compared. Unfortunately this simple comparison has been complicated by the use of different turbine configurations by different research teams so that comparisons using readily available data cannot easily be made. In particular there are large differences between different research teams on the cyclic efficiency of the Wells turbine. However, the following differing characteristics are generally accepted:

- The optimum flow coefficient of the impulse turbine is significantly larger than that for the Wells turbine;
- The impulse turbine is self-starting, whilst the Wells turbine has poor starting characteristics;
- The impulse turbine has no stall condition thereby maintaining relatively high efficiency at high flow coefficients, unlike the Wells turbine that has a clear stall point;
- The pressure – flow relationship of the impulse turbine is non-linear, unlike the Wells turbine that has an almost linear relationship.

The larger optimum flow coefficient of the impulse turbine means that for the same power the turbine is smaller and spins at a slower angular velocity than a similarly rated Wells turbine. Coupled to the absence of stall, this means that typically the impulse turbine is quieter than the equivalent Wells turbine. In addition, the slower rotational velocity means that windage losses of the impulse turbine are likely to be smaller, whilst the potential for energy storage using a directly coupled flywheel is reduced. While these factors are unlikely to have a large effect on the power output the turbine, they may have a large influence on design, which means that certain circumstances may dictate the choice of the turbine largely irrespective of their differences in performance. For example, the lower noise generation of an impulse turbine may make it more suitable for locations with a mixed use, such as breakwaters, where the noise generated could be a nuisance to people working nearby. Conversely, if steadier power generation is required then the Wells turbine may be preferred due to its larger potential for energy storage, which could be used to smooth the power output.

The pressure – flow relationship of a turbine influences the damping applied to the motion of the water column and thereby influences the amount of energy extracted by the oscillating water column plant. At any instant there is an optimum amount of damping that should be applied to the water column to maximise the power capture. In small amplitude monochromatic waves the optimum damping condition equates to a linear relationship between the water column velocity and damping force. This is achieved with a linear pressure – flow relationship, implying that the Wells turbine has more suitable characteristics for wave energy conversion. However, the optimum relationship differs for the oscillating water column when it is subjected to more realistic wave conditions, although determining what constitutes the optimum relationship cannot easily be defined, varying with the particular wave characteristics and oscillating water column design. It remains unclear if in realistic conditions whether the pressure – flow characteristics of the Wells turbine remains superior, or whether the impulse turbine characteristics are typically more suitable.

Figure 6.17 illustrates the typical characteristics for an impulse turbine. For this particular configuration it can be seen that the peak efficiency of the turbine is approximately 60 %, with a broad bandwidth of high efficiency due to the absence of stall in an impulse turbine. In oscillating and realistic flow conditions, this broad bandwidth translates into a relatively high efficiency of 48 % for an impulse turbine with self pitch controlled guide vanes and 35 % for an impulse turbine with fixed guide vanes (Setoguchi et al., 2001).

In addition to the experimental work that has been described above, computational fluid dynamic studies have also been performed to investigate the performance of impulse turbines (Thakker and Hourigan, 2005). CFD studies have enabled the design of impulse turbines to be focused on two important effects: the tip clearance losses (Thakker and Dhanasekaran, 2003) and the guide vane losses (Thakker and Dhanasekaran, 2005). As would be expected the efficiency of the turbine decreases as the tip clearance increases due to increased leakage around the rotor. However, the CFD models indicate that if the tip clearance is less than 1 % of the rotor diameter then the losses are negligible, whilst for tip clearances greater than 4 % the effect on efficiency is negligible, though by this point the peak efficiency has reduced by approximately 25 %. CFD studies on the effect of guide vanes have so far concentrated on the fixed guide vane configuration. These studies have shown that whilst the upstream guide vane accounts for a minimal amount of pressure drop, the downstream guide vane can account for a large proportion of the pressure drop leading to a reduction in efficiency.

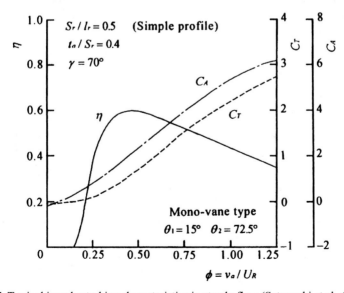

Fig. 6.17. Typical impulse turbine characteristics in steady flow (Setoguchi et al., 2001)

The Denniss-Auld Turbine

The Denniss-Auld turbine was developed at the University of Sydney, Australia in collaboration with the Australian company Energetech who used the turbine in the design of their first plant at Port Kembla (Curran et al., 2000; Finnigan and Auld, 2003; Finnigan and Alcorn, 2003). This turbine is aerodynamic in principle with aerofoil blades that pitch in order to provide an optimal angle of attack, in a similar manner to the variable pitch Wells turbine. However, the blades are located on the periphery of the rotor hub in a neutral position that is parallel to the axial direction of the flow rather than tangential to the direction of rotation as for the Wells and Impulse turbines. Also, the blade profile is symmetrical about the maximum thickness line at half chord as well as being symmetrical about the chord line itself. Consequently, the blade is rotated about its centre of gravity and pitches into the direction of rotation to optimise the angle of attack. The aerofoil shape was determined as that which gave maximum lift at zero incidence and was achieved by merging the two forward sections with NACA 65-418 profiles. This is a laminar flow profile with a maximum camber of 6% chord and a maximum thickness of 18% chord.

The performance of the turbine is presented in Fig. 6.18 in terms of the efficiency and pressure drop relative to the flow coefficient (Curran et al., 2000). It can be seen from the efficiency characteristic that the turbine performs at higher flow rates than the Wells and Impulse turbines but otherwise there is a lot of similarity in terms of running losses at the lowest flow rates, a peak efficiency (that is more similar to that of a Wells turbine), and then decreasing efficiency levels at the highest flow rates when losses tend to dominate.

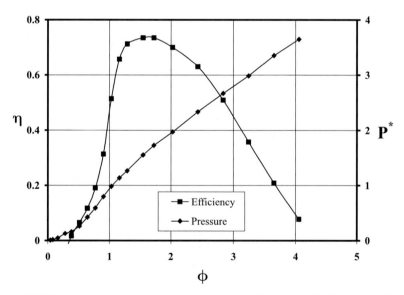

Fig. 6.18. Efficiency and pressure drop against flow coefficient for the Denniss-Auld turbine

6.1.4 Operational Findings for the Wells Turbine

Conclusions and designs derived from numerical and laboratory-based modelling of air turbines can only be partially valid. When a turbine is mounted in an oscillating water column, the real circumstances of its installation will influence its performance in potentially unexpected ways. Invariably, unlike laboratory based turbine testing facilities, the design of the turbine and duct are constrained and modified by constructional and cost factors that are not present with model turbines. From their nature these operational issues are difficult to predict; however an example of the effect of these operational issues is provided by the analysis of the performance of the contra-rotating Wells turbine installed in the LIMPET oscillating water column (Boake et al., 2002).

The turbine duct for LIMPET was fully instrumented so that the pressure drop across any of the valves or turbine rotors could be measured; however, more importantly the duct allowed the insertion of a hot-wire anemometer both upstream and downstream of the contra-rotating turbine, which enabled the flow profile across the duct to be measured and analysed (Folley et al., 2002). The analysis of the flow profiles, together with the information on the pressure drop across the turbine identified two operational issues that were not identified during the laboratory-scale testing. These two operational issues were: the effect of an uneven flow profile during the intake stroke and the reduction in instantaneous turbine torque due to unsteady flow conditions (Folley et al., 2006).

During laboratory testing much care is taken to ensure that there is minimal variation in the flow across the turbine duct. A mesh is typically used to minimise flow turbulence and the wind-tunnel duct taper angles chosen to avoid flow separation in a rapidly diverging duct. Unfortunately, these ′perfect′ conditions are more difficult to achieve on a real plant, with the consequence that the flow conditions for installed turbines are often far from ′perfect′, though results from the $75\,kW$ oscillating water column prototype installed on Islay indicated that Wells turbines were relatively robust when the flow is turbulent due to the presence of control valves within the duct (Whittaker et al. 1997a; 1997b) with a minimal reduction in performance identified. However, high levels of noise from the LIMPET turbine meant that the acoustic attenuation chamber (silencer) the end of the turbine duct had to be re-designed, which whilst generating a minimal additional pressure drop created a large mal-distribution of the intake flow. This is illustrated in Fig. 6.19 (the circumferential positions were approximately evenly spaced with positions 4/5 being at the top of the duct), with the flow at the bottom of the duct being over twice the flow at the top of the duct, indicating a dramatic mal-distribution. The poor flow profile means that the amount of circulation around any particular blade of the turbine is increasing and decreasing every full rotation of the turbine. This high frequency fluctuation in circulation tends to inhibit the generation of circulation, thus reducing the pressure drop across the turbine when compared to a turbine with an ideal flow distribution.

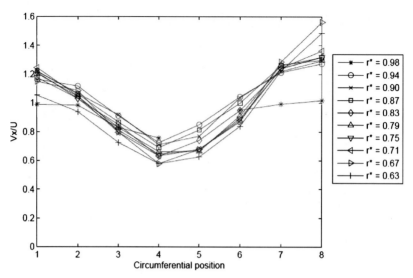

Fig. 6.19. Circumferential flow distribution during the intake stroke (from Folley et al., 2006)

However, this effect is only small as illustrated by only an 8 % reduction in the damping coefficient during the intake stroke of the LIMPET turbine. The turbine torque characteristic is also affected by the reduction in circulation, with a 25 % reduction in the torque coefficient at stall when compared to the exhaust stroke, which does not have an azimuthal variation in flow profile. Figure 6.20 shows the difference between the intake stroke (negative flow coefficients) and the exhaust stroke (positive flow coefficients) on the turbine torque coefficient. It is also interesting to note that during the intake stroke the downstream rotor has a higher peak torque, possibly indicating that the upstream rotor has helped to redistribute the flow and thereby reduce the fluctuations in flow experienced by the downstream rotor blades.

Figure 6.20, in addition to illustrating the difference between the intake and exhaust strokes, also illustrates that the torque coefficient of the LIMPET turbine is lower than that derived from the constant flow model testing. This is further illustrated in Fig. 6.21, which shows the variations in turbine efficiency with flow coefficient. It is clear that something other than the mal-distribution of flow has a dramatic effect on the turbine efficiency.

A number of studies into unsteady flow conditions (Setoguchi et al., 1988; Alcorn et al., 1998; Muman et al. 2004) have identified a hysteresis in the pressure-flow characteristic of Wells turbines due to differences in the wake behind accelerating and decelerating flow processes. However, the contra-rotating Wells turbine demonstrates a more significant difference in performance between constant and unsteady flow conditions.

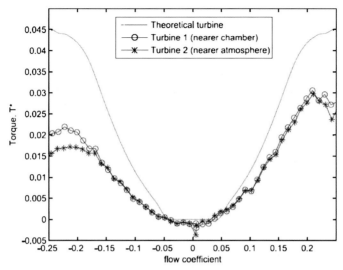

Fig. 6.20. Torque coefficients for both LIMPET rotors during intake and exhaust (from Folley et al., 2006)

It is convenient to define three aspects of unsteady flow: the accelerating and decelerating flow processes already identified in previous studies, the reversal of the flow through the turbine and the characteristic time constant of the turbine for stable flow conditions to be achieved. In comparing the LIMPET contra-rotating Wells turbine with other Wells turbines, the only significant difference that has been noted in model testing is the extended time it takes for the contra-rotating turbine to stabilise following changes in flow. In mono and bi-plane Wells turbines the turbines would adapt to the new flow conditions without any apparent delay; however the contra-rotating turbine could take 20–30 seconds to stabilise. It is suggested that this is the primary reason for the reduced efficiency of the LIMPET turbine in comparison to predictions based on constant flow testing.

Although the instability of the turbine during constant flow testing could have been used as an indicator that a more complex scenario existed, the lack of test facilities that can provide carefully controlled oscillating flow conditions limits the opportunity for investigation in this area. As new refinements to turbines are developed, doubts must exist on their actual benefit if only constant velocity unidirectional testing is used. In addition, these results must call into question the extensive use of quasi-steady state numerical modelling of turbines, since such models would not identify these essentially transient and possibly chaotic characteristics (although to date quasi-steady state modelling for other forms of Wells turbines has produced adequate predictions). It would thus seem critical that appropriate test facilities are available and that turbines should, as far as practicable, be tested in realistic conditions, which includes the configuration and layout of the turbine duct.

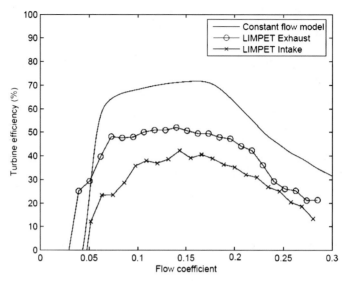

Fig. 6.21. Comparison of LIMPET and constant flow model turbine efficiencies (from Folley et al., 2006)

6.1.5 Discussion and Conclusions

It is evident that there are a number of alternative air turbines that can be successfully utilised in wave energy conversion systems. The two most common types are the Wells turbine and the Impulse turbine. Relative to the Impulse turbine, the Wells turbine operates over a smaller and slower range of air flow rates but typically operates at higher rotational velocities and higher peak efficiencies. The Denniss-Auld turbine was briefly considered and it was shown that it operated over the widest flow range while achieving relatively high efficiency, albeit still at a development stage with a pitching mechanism.

It was highlighted that any of the turbines needs to be properly integrated into the energy conversion chain, with particular reference to the OWC. The turbine needs to provide a suitable level of damping to the OWC in order to maximise its hydrodynamic performance. However, the level of damping provided by the turbine also determines the range of flow rates that will be generated from the pneumatic power provided by the OWC. Therefore, the diameter, speed and solidity of the turbines in particular need to be carefully selected in order to provide the majority of the pneumatic power at flow rates that correspond to a high efficiency. It was also highlighted that these parameters would be set at significantly different levels for either a Wells or Impulse turbine, for the same OWC rating, due to the different performance characteristics mentioned previously. In this respect it is advantageous that the Impulse turbine would have a smaller diameter and lower rotational speed, although the standard monoplane or biplane Wells turbines are mechanically simple and do not require guide vanes, whether fixed or moving.

A number of more advanced Wells turbines were considered in some detail, including the variable pitch monoplane, fixed guide vane monoplane, and counter-rotating biplane. It was concluded that of these three design types, the low-solidity counter-rotating turbine provided the best performance in terms of efficiency and range. However, while still being less complex than a standard configuration Wells turbine or a Impulse turbine with guide vanes, there is a requirement for a second generator which may have significant impact on the overall economic viability, relative to the cost of the converted electrical energy.

There has been a considerable amount of development work carried out for both the Wells and Impulse turbine. This has included experimental studies, numerical modelling and Computational Fluid Dynamic (CFD) modelling. The CFD modelling has not been addressed in detail as there has been little validation possible for rotating blades, in terms of flow velocities and detailed pressure measurements for example. However, the CFD work has contributed to the understanding of the aerodynamic performance and has been used in the optimisation of blade profile in particular, and in investigating the flow through the turbine from the OWC chamber. This CFD work may contribute to new developments in the future.

Another aspect touched on briefly is the economic analysis of the design for the various alternative turbine types. As more plants are now being commissioned, the manufacture and operational issues are becoming more prevalent in terms of the financial effectiveness of the designs. This may lead to new developments that are driven by different requirements, potentially exploiting new materials and focusing on the life cycle analysis and optimisation of the various turbine types and technologies being exploited. It is likely that researchers will continue to come up with new developments, such as boundary layer control (Raghunathan, 1995), and these innovations will have to be assessed in terms of the overall relevance of to the impact on overall power conversion and cost effectiveness. Completely new designs may also be considered, such as the Darius-type design.

There has also been a considerable amount of work over the years in the use of speed control, to control the operational range of the turbine relative to the OWC performance. This may well offer the greatest potential increase in the overall performance of the turbines when assessed on a longer term annual basis. Naturally, this focus may lead to new innovations in the turbine design that help in the improved exploitation of speed control.

It can be concluded that air turbines are still an important component of a significant number of wave energy devices being developed globally. Although many new concepts for wave energy conversion are being proposed, it is highly likely that commercial exploitation will strongly feature the use of air turbines. Consequently, researchers need to also focus on operational performance as well as developing new fundamental work. This may lead to new improved future innovations that are driven by the need to analyse the role of the turbine in the context of the OWC scenario (and economics), rather than from a purely aerodynamic basis.

6.2 Direct Drive – Linear Generators

Oskar Danielsson, Karin Thorburn, Mats Leijon

Uppsala University
Uppsala
Sweden

It may prove advantageous to utilise linear machines for applications that involve linear or reciprocal motion as the mechanical interface often can be reduced compared with systems based on rotating machines. Such configuration enables a direct drive approach and the schematic for a linear generator system with direct drive conversion is shown in Fig. 6.22. The idea of direct drive is appealing since it enables simple systems with few intermediate conversion steps and reduced mechanical complexity, but the demands on the generator increase.

One of the first concepts where linear generators were suggested as power take-off, filed as a US patent in 1980 (Neuenschwander, 1985), was later changed to a system with rotating generators. Direct drive linear generators where at that time ruled out as heavy and inefficient. Progress, mainly in permanent magnet materials but also in power electronics, has dramatically changed the conditions for linear generators and they have lately appeared as one of the most commonly used power take-off in point absorber wave energy converters (Mueller, 2002).

Using a directly coupled linear generator as power take-off in the wave energy plant reduces the structure and mechanic components to a minimum, but poses new requirements on the generator. Those requirements are to a great extent determined by the nature of ocean waves. The generator also needs to be connected to the grid, which must be taken into account in the design specification. It is important to consider the whole chain of conversion when designing linear generators for wave energy applications. The demands on directly coupled linear generators in wave energy applications are extraordinary regarding torque, overload capability, and compatibility with converter systems.

This sections explains the fundamental principles of linear generators in direct drive wave energy conversion systems and gives an overview of the basic types of linear machines suitable for wave energy conversion. The section ends with a case study of an offshore trial of a full-scale linear generator concept.

6.2.1 Principles of Direct Drive Wave Energy Conversion With Linear Generators

Direct drive in wave energy conversion is similar in principal to direct drive in wind energy, where the moving part of the generator is directly coupled to the energy absorbing part - the wind turbine. The basic principles of a linear generator are illustrated in Fig. 6.23. It consists of a moving part, the translator, on which

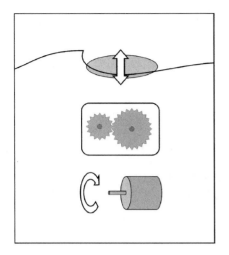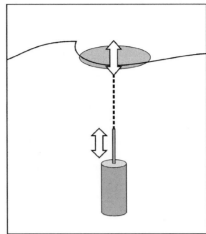

Fig. 6.22. Basic idea of direct drive wave energy conversion with linear generator. To convert the slow reciprocal motion of the ocean wave into a fast rotating motion, required by conventional high speed rotating generators, some kind of intermediate conversion step is necessary. A slow speed linear generator can be coupled directly to a point absorber and no intermediate system is needed

magnets are mounted with alternating polarity. The translator moves linearly next to a stationary stator that contains windings of conductors, the armature windings. Between the translator and the stator is a physical gap, the air gap. A voltage is induced in the windings as the magnetic field changes due to the translator motion, all in accordance with Faraday's law.

A number of direct drive WEC concepts that uses linear generators have been suggested over the years. Almost all of these concepts involve a point absorber and a reference system, where the relative linear motion between the two to drives the generator. Figure 6.24 illustrates four main concepts. The first three concepts all comprise floating point absorbers but have different reference frames, namely, the sea floor (a), a floating structure (b), and a damper plate (c). The last concept (d) uses a submerged gas filled vessel to absorb the power of the wave, here illustrated with the sea floor as reference. Other concepts have also been suggested, e.g. with the use of a mass as reference; these are however not illustrated here for simplicity.

Two different WEC systems will be used as examples. Both systems have been tested offshore and use linear generators as power take-off, but differ widely in several aspects. The first system is of type (a) in Fig. 6.24, and is representative for a small and slimmed system where control and auxiliary systems is reduced to a minimum (Leijon et al., 2006). The system has been tested offshore on the Swedish west coast in a low energy wave climate (Waters et al., 2007). It consists of a point absorber buoy with a diameter of $3\,m$, and a design significant wave height of $1.6\,m$ and a nominal power rating of $10\,kW$. This concept will be further

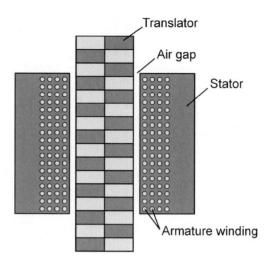

Fig. 6.23. Basic principles of a linear generator

described in the case study at the end of this section. The other concept is based on principle (d) in Fig. 6.24. It is known as the Archimedes Wave Swing (AWS) and has been tested offshore outside the coast of Portugal (Prado et al., 2006). This represents a larger WEC with higher complexity in terms of more auxiliary systems. The absorber has a diameter of 9.5 m and a rated power of 1 MW (see also Chapter 7).

Damping

The point absorber and the reference system form an oscillating system where the waves are the driving force and the generator acts as a damper. The absorption level of the device strongly depends on the damping of the generator. Optimal damping is determined by a number of factors where, above all, are the size and weight of the absorber; the speed is also essential (Falnes and Budal, 1987). Different approaches are used to increase the absorption of energy by the WEC and these can differ between a system that uses active control or tuning of resonance frequency to optimise the energy absorption and a passive system far from resonance. Generally, the optimal damping is lower for a resonating system than for a system which is not in resonance. There can also be a need for additional damping in harsh climates to reduce the translator speed and stroke length as well as loads on end stops. In the AWS this problem is addressed by adding hydrodynamic dampers, which can provide the extra damping needed (Polinder et al., 2004). In Table 6.1 typical damping, given as force per speed, is given for the two example systems.

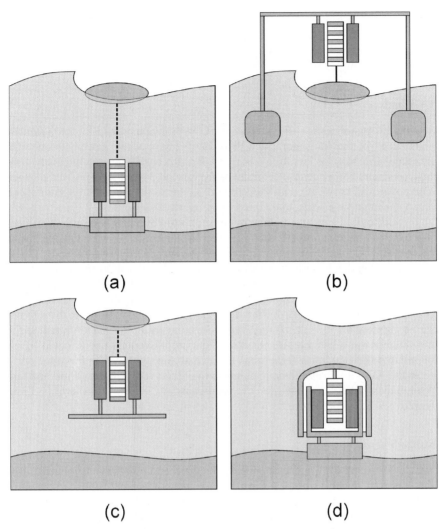

Fig. 6.24. Different concepts of direct drive WECs using linear generators. (*a*) Floating point absorber with the sea floor as reference system, (*b*) floating point absorber with floating structure as reference, (*c*) floating point absorber with submerged damper plate as reference, and (*d*) submerged gas filled vessel with sea floor as reference

Reaction force

One of the most obvious differences to conventional generators is the slow speed. The slow speed is a direct consequence of the direct drive concept. The speed of the moving part of the generator, the translator, will to a first approximation, be determined by the vertical speed of the sea surface, which is in the order of 1–2 *m/s*. This is 15–50 times slower than conventional synchronous rotational generators and the reaction force, consequently, needs to be 15–50 times larger to give the same output power. To achieve the required damping force, a direct drive generator will need to be significantly larger than a high speed generator. The air gap areas for the two examples WECs are given in Table 6.1.

Overloads

The output power from a direct drive WEC will be determined by the instantaneous speed of the translator since there is no intermediate power storage. The translator will more or less follow the waves which results in a continuously varying speed, and consequently, a continuously varying power. As a result the generated power will vary both on a short timescale, due to the varying translator speed, and on a longer time scale, due to the changing wave conditions.

It is difficult to estimate the maximum short term overload resulting from steep wave fronts impinging on the device. Experimental results show that sudden peaks reaching eight times the average power may be encountered (Leijon et al., 2006). Short-term overloads pose demands on voltage stability and on rectifier and converters. The thermal load from a short term overload has limited effect on the machine, whereas the long term overload can cause heating problems. The long term overload will be determined by the most extreme wave condition at the site and the absorption factor of the plant. Even though the absorption factor is reduced as the wave power increases, the maximum long term overload power could be expected to be several times higher than the annual average power of the generator.

In the AWS the hydrodynamic dampers limit the overloads, both short and long term, which reduces the need for overload capacity. Typical short term and long term overloads are given for the two examples in Table 6.1.

Stroke length

The stroke length of the generator should be matched to the wave heights and is in the range of meters. Compared with other linear machine applications this is a relatively long pitch length. The induction in the machine will be reduced if the stator and translator have equal length. In order to maintain a constant induction air gap area for a longer part of the stroke length, either the stator or the translator must be longer than the other. The Ohmic losses will increase as the conductor length increases if the stator is made longer, whereas a longer translator only cost more initially. It is therefore more common to have a translator longer than the stator. The total stroke lengths for the two WEC systems are given in Table 6.1.

Grid connection

The translator motion of a linear generator is in most cases reciprocal and the speed varies continuously as the translator accelerates and decelerates near the turning points. This will give a varying frequency of the induced voltage and, furthermore, the order of the phases in a multiphase machine will be interchanged

Table 6.1. Basic prerequisites for linear generators in Uppsala University project and Archimedes Wave Swing wave energy converters

Parameter	Uppsala University project, 3 *m* point absorber	Archimedes Wave Swing
Nominal speed	0.7 *m/s*	2.2 *m/s*
Nominal power	10 *kW*	1 *MW*
Nominal reaction force	14.3 *kN*	454 *kN*
Long term overload	2–5 times nominal load	2 times nominal load
Short term overload	~ 10 times nominal load	2 times nominal load
Stroke length	1.5 *m*	7 *m*
Air gap area	2.1 *m²*	20 *m²*
Load	Passive rectifier	Current source inverter (passive)

when the translator changes direction. A direct coupling of a linear generator to a utility grid is thus practically impossible. This problem is addressed in Danielsson (2006) for wave energy applications where a number of connection solutions are presented. In practice, the current from a linear generator needs to be rectified before it can be converted into a 50 or 60 *Hz* AC with power semiconductor components.

6.2.2 Linear Generators

The internal structure of a linear machine has, in most cases, its origin from a rotating machine. The geometry of a linear generator can be produced from a rotating generator by an imaginary cutting and unrolling of the rotating machine. This is illustrated in Fig. 6.25, which shows the cross-section of a rotating generator, an arc shaped generator, and finally a linear generator. The linear geometry also opens up for tubular machines, which have a circular cross-section along the axis of motion. A tubular machine can be produced by rotating the cross-section geometry of a flat linear machine along an axis parallel to the direction of motion. This is illustrated in Fig. 6.26. Tubular machines forms a new type of geometry not encountered in the rotating machine fauna.

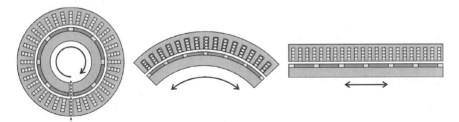

Fig. 6.25. Rotary becomes linear

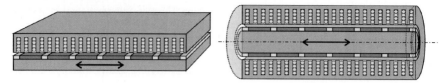

Fig. 6.26. Flat and tubular linear generator

There are however some marked differences in both the functioning and design of linear generators:

- Direct connection to grid is not possible and a rectification step is necessary.
- The supporting structure of a linear generator differs from that of a rotating machine. In contrary to a rotating machine, a linear machine cannot use one axis to fix the translator but will need several points of fixation to maintain the geometry of the air gap.
- The magnetic circuit of a linear machine differs in one significant aspect from a rotating machine - the linear machine will have open magnetic circuits in both ends, whereas the magnetic circuit of a rotating machine forms a closed circle. Unlike a rotating machine, the active area of the air gap can vary as the translator moves in and out of the stator. Furthermore, these ends have influence on the magnetic flux and the magnetic flux pattern of linear machine is thus different from a rotating machine (Danielsson, 2006).

Basic generator theory

When the permanent magnet on the translator moves in relation to the stator an electromotive force (*emf*) is induced in the armature windings. This *emf* will, if the armature winding is coupled to a load, drive a current in the armature winding. This current, in turn, creates a magnetic flux that interacts with the flux of the permanent magnet and results in a force on the translator. The mechanical energy, mediated by the translator, can in this way be converted to electrical energy, which is consumed in the load. This is the basic principle of a permanent magnet generator.

A generator can be described electrically by a simple lumped circuit diagram. Here a simplified theory will be presented given the scope of this review. The lumped circuit diagram of one phase of a synchronous generator is illustrated in Fig. 6.27. One phase of the generator is modelled by an electromotive force *emf*, a resistance, a reactance, and a load. The *emf*, E, is the voltage induced by the permanent magnet flux wave. This voltage is also called the *no-load voltage* since it is the voltage that will be measured over the phase terminals if no load is connected. Inside the generator there is a resistive voltage drop, due to the winding resistance, R_g, and an inductive voltage drop modelled by the synchronous reactance, X_s. The load, Z_l, could be either purely resistive or have a reactive component.

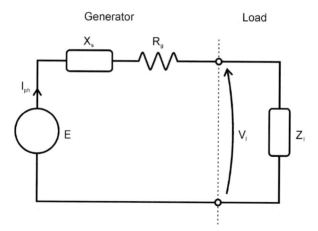

Generator   Load

Fig. 6.27. Lumped electric circuit diagram of one phase of a synchronous generator

Induced emf

The permanent magnets on the translator are mounted with alternating polarity, which will create a magnetic flux wave with alternating direction. This flux is the *permanent magnet induced flux* of the machine and the flux wave will follow the translator as it moves. Figure 6.28 shows a schematic illustration of the permanent magnet induced flux and the armature windings of one phase. The flux encircled by one coil, shaded in the figure, will be a function of the translator position and will change as the translator moves. From the Faraday's law of induction, $E = -\dfrac{d\psi}{dt}N$, we can calculate the induced *emf*:

$$E = \omega \psi_{pm} N \, , \tag{6.22}$$

where ω is the angular frequency, ψ_{pm} is the permanent magnet induced flux per pole, and N is the total number of coil turns. The angular frequency is given by the translator speed v and the distance between the poles, the *pole pitch* w_p, as follows:

$$\omega = 2\pi \frac{v}{w_p} \, . \tag{6.23}$$

A small pole pitch and high speed thus gives a high angular frequency. The magnet induced flux per pole is determined by the magnetomotive force provided by the permanent magnets and the magnetic reluctance of the magnetic circuit. In practice all direct drive machines utilise high energy permanent magnets of Neo-dymium-Iron-Boron type, which can produce high magnetomotive force for relatively small magnet heights. The magnetic circuit consists to a large extent of fer-romagnetic steel, which reduces the magnetic reluctance of the circuit. The magnetic reluctance depends on the geometry of the magnetic circuit, and especially the cross section of narrowest parts in the flux paths, the bottle necks, and

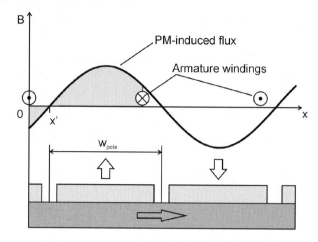

Fig. 6.28. Permanent magnet flux wave in air gap; the integration of the flux over one coil width is shaded

the magnetic properties of the steel. This will be discussed further for the different generator topologies. The total number of coil turns in a generator is equal to the number of poles, N_{pole} multiplied with the number of turns in each pole, N_{turn}: $N = N_{pole}N_{turn}$.

Synchronous reactance

The armature winding consists of a set nestled coils, which will have associated inductances. The *synchronous inductance, L_s,* is a combination of the self inductance of one phase together with the mutual, phase shifted inductances of the other phases. The *synchronous reactance, X_s,* models the voltage drop caused by the self-induced armature flux and is given by multiplying synchronous inductance with the electric frequency:

$$X_s = \omega L_s .$$

(6.24)

The electric frequency varies with the translator speed and the synchronous reactance will thus be a varying parameter.

The synchronous inductance can be divided into two parts, one which is associated with the flux that is coupled with the translator, the *main inductance L_m,* and one which is associated with the leakage fluxes, the *leakage inductance L_l;* the result is $L_s = L_m + L_l$. The main inductance plays an import role in the generator since it represents the magnetic coupling between the armature winding and the permanent magnets of the translator. The inductance of one coil is proportional to the square of the number of coil turns. The main inductance of the generator can thus be expressed as:

$$L_m \infty N_{coils} N_{turns}^2 .$$

(6.25)

The load

The current from a directly driven linear generator is by necessity rectified before it is converted and delivered to a grid. The rectification can be passive or active. A simple diode rectifying bridge represents a passive rectifier and is characterised by having a power factor equal to one, *i.e.* the load voltage and current will be in phase. A passive rectification can, in the lumped circuit, be represented by a purely resistive load[1] and $Z_l = R_l$. In machines with high synchronous reactance the available active power can be considerably increased if the load has a power factor different from one (Xiang et al., 2002). This can be provided by using an active rectifier. In an active rectifier the power factor can be controlled and the load voltage can be made to lag or lead the current. There are also possibilities to do phase compensation by using capacitor banks (Chen et al., 1998). Both active rectification and phase compensation with capacitor banks increase complexity and add components to the system. The load will then be described by the general expression: $Z_l = R_l + jX_l$.

Output power

Knowing the components of the lumped circuit the active power in the load can be calculated:

$$P_l = E^2 \left(\frac{R_l}{\left(R_l + R_g\right)^2 + \left(X_s + X_l\right)^2} \right).$$ (6.26)

The active output power of a generator depends on the induced *emf*, the synchronous reactance, the armature winding resistance and the load. The induced *emf* varies with the translator speed and the output power will thus vary as the translator changes speed during a wave period. By changing the load the output power for a given translator speed can also change. By varying the load, the damping of the generator the generator can be varied. This can be used as control strategy to control the power absorbed by the WEC. Especially in harsh climates it can be necessary to over-damp the WEC, e.g. to reduce forces and stroke length.

Flux ratio

As mentioned above, generators with relatively high synchronous reactance need to be phase compensated in order to increase the power from the generator. The need for phase compensation is traditionally described in terms of *power factor*. However, in these applications, where the generator often is connected to a load via a passive rectifier, the power factor is constant and equal to 1. The need for

---

[1] A rectification bridge is a non-linear component, which cannot be modeled accurately with linear theory. A resistive load, however, resembles a passive rectification bridge by having a power factor equal to one, i.e. the voltage and the current is in phase.

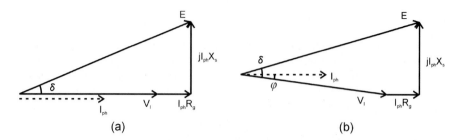

Fig. 6.29. Phasor diagram of synchronous generator, (*a*) purely resistive load and (*b*) phase compensated load

phase compensation can also be described by the *flux ratio* between the permanent magnet induced flux and the armature current induced flux. The concept was introduced by Harris et al. (1997) and was used to describe the problem of low power factor in TFPM (transverse flux permanent magnets) motors. The flux ratio can be produced from a phasor diagram of the voltage components in the generator. Two phasor diagrams are illustrated in Fig. 6.29, where (a) represents a purely resistive load and (b) a phase compensated load. The voltage components, $I_{ph}X_s$ and E, are proportional to the armature winding induced flux, ψ_{arm}, and the permanent magnet induced flux, ψ_{pm}, respectively. The quota between the two,

$$\frac{I_{ph}X_s}{E} \infty \frac{\psi_{arm}}{\psi_{pm}},$$ (6.27)

can be given as a figure of merit for the need of phase compensation. A low quota means low need for phase compensation and a high quota indicates a large need for phase compensation. Another, often used term, is the *load angle*. The load angle is the angle between the *emf*, E, and the load voltage, V_l, and is indicated with δ in the figure. The load angle is often used as a measure of how hard a machine is loaded.

6.2.3 Linear Machines Topologies

There exist today a large number of different linear machines types most of them derived from rotating machines (Boldea and Nasar, 1999; McLean, 1988). Not all linear machines are suitable for wave energy conversion and only a handful fulfils the basic prerequisite stated in Table 6.1. Induction machines and field wound synchronous machines can at an early stage be ruled out since they cannot compete with the permanent magnetised machines at the slow speed at hand (Polinder et al., 2005). Today there are mainly three main classes of machines that have been investigated for wave energy applications, longitudinal flux permanent magnet generators also known as synchronous permanent magnet generators, variable reluctance permanent magnet generators with transverse flux permanent magnet generators as a special case, and air-cored tubular permanent magnet generators.

Longitudinal flux permanent magnet generator (LFPM)

The name refers to the flux path in the yoke which is in the longitudinal direction. This machine is also known as a *synchronous* permanent magnet generator, which reflects that the armature winding flux and the permanent magnet flux are moving synchronously in the air gap. This is however the case for all generators described here and the name can be somewhat misleading. LFPM generators have been investigated with finite element methods (Danielsson et al., 2006) and analytical methods (Polinder et al., 2004) for wave energy applications and tested offshore in both the AWS project and the Uppsala University project.

A cross-section of the magnetic circuit of a longitudinal flux permanent magnet generator is illustrated in Fig. 6.30. The main magnetic flux is illustrated with dashed lines and the direction of the flux is indicated with arrows. The magnetic flux from one magnet crosses the air gap and is conducted by the stator teeth through the stator coils. In the stator yoke the flux is divided into two paths, which return through the stator teeth, via the air gap and through the adjacent magnets. The steel plate in the translator connects the magnetic flux at the back of the magnets. The LFPM has an inherently small synchronous reactance and the stator construction is simple and robust. The geometry of LFPM, however, limits the stator teeth width and cross-section area of the conductors for a given pole pitch. Increasing the tooth width to increase the magnetic flux in the stator or increasing the conductor cross-section demands a larger pole pitch and the angular frequency of the flux is thus reduced. This sets a limit for the induced *emf* per pole and consequently the power per air gap area. The basic features of a LFPM are summarised in Table 6.2.

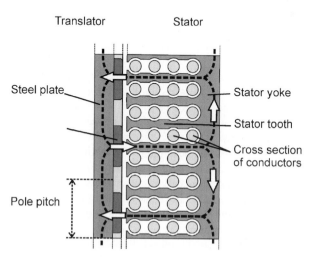

Fig. 6.30. Cross-section of a LFPM generator, magnetic flux path is illustrated with a dashed line and the direction of the flux with arrows

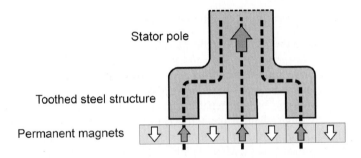

Fig. 6.31. Basic principles of a variable reluctance machine

Variable reluctance permanent magnet generator (VRPM)

Variable reluctance machines form a whole set of different generator geometries. The limiting coupling between the cross section area of the magnetic circuit and the pole pitch, found in synchronous machines, is avoided here by using alternate return path through a toothed steel structure. The basic principles are illustrated in Fig. 6.31. This enables large magnetic fluxes combined with a very short pole pitch but the synchronous reactance of these machines tends to be very high. Different types of VRPM have been suggested for use in direct drive WECs (Baker, 2003).

Transverse flux permanent magnet generators (TFPM)

The TFPM was introduced by Weh et al. (1988) and has been investigated for wave energy applications by Mueller (2002), and more generally, by Harris et al. (1997). The geometry of TFPM machine is illustrated in Fig. 6.32, which shows a double sided TFPM with the translator with permanent magnets in the middle and a set of C-cores on the top and bottom providing closed magnetic flux paths around the two coils. As the translator moves the permanent magnets will couple their flux alternately with the coil above and with the coil below. The illustration to the right (*b*) show the flux paths of two parallel magnets, with the flux path of the adjacent magnets illustrated in shaded nuance. This special arrangement avoids the conflict between high cross-section areas of both yoke and armature conductor combined with small pole pitch. Accordingly, both PM-induced flux and armature current flux can be high while pole pitch remains small and these kinds of machines can produce a very high force per pole. The drawback is the relatively high synchronous inductance and that they are intended to work at relatively high armature current level. Phase compensation or some kind of power converter with high power rating is needed to fully appreciate the generator performance. Another drawback is the relatively complex stator composition, which is made of several independent C-cores of laminated steel. The main characteristics of a TFPM are summarised in Table 6.2.

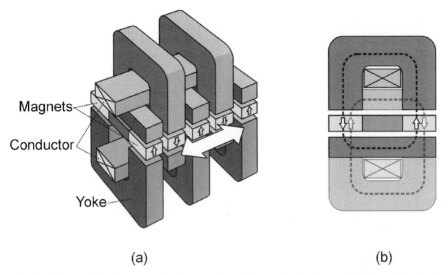

(a)   (b)

Fig. 6.32. Transverse flux permanent magnet machine. (*a*) 3D-view of double sided TFPM with the translator with permanent magnet in the middle and two sets of C-cores, one on above and one below the magnets (*b*) 2D-wiev along the axis of motion, flux path of two parallel magnets are illustrated together with the flux path o the adjacent magnet pair (shaded)

Tubular air cored permanent magnet generator (TAPM)

Lately, an air cored tubular permanent magnet generator has been suggested for wave energy applications (Baker and Mueller, 2004). The principles are illustrated in Fig. 6.33, which shows a 3D-view (*a*) and a cross-section (*b*) of a TAPM. Here the magnets are magnetised in the axial direction and placed with altering direction of the flux. Flux concentrators are placed between the magnets and a varying flux wave is created outside the translator. Other configurations are also possible, *i.e.* surface mounted permanent magnets. By removing the steel in the stator the normal forces are virtually eliminated. This is a large constructional advantage

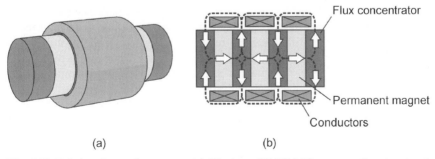

(a)   (b)

Fig. 6.33. Tubular air cored generator. (*a*) 3D-view of TAPM (*b*) cross-section showing the flux paths

Table 6.2. Basic features of different linear generators

	Power per air gap area[*] (kW/m^2)	Flux ratio IX/E	Pros	Cons
LFPM	~25	0.1–0.5	Low synchronous reactance. Simple and robust stator.	Low power rating per air gap area.
TFPL	~50	1.6–2.6	High power per air gap area.	High synchronous reactance. Complex stator.
TAPM	>12.8[‡]	<0.3	No normal forces in the air gap.	Very low power per air gap area.

[*]The power is given for a speed of 1 m/s
[†]Synchronous reactance per unit is the synchronous reactance divided by the load resistance
[‡]Experimental values from Baker (2000)

since normal forces in other permanent magnet linear generators are in the order of 200 kN/m^2 (Nilsson et al., 2006). Significant support structures are needed to overcome these forces and maintain a constant air gap width. The drawback is that the magnetic reluctance in the magnetic circuit is increased considerably, since the distance the flux travels in air now is in the range of the pole pitch and not, as in steel stator machines, just over the air gap. The flux of an air cored machine is thus much smaller and the power per air gap area is thus considerably lower. The main characteristics of a TAPM are summarised in Table 6.2.

6.2.4 Case Study: 3 m Point Absorber

Uppsala University in Sweden has developed and tested a WEC on the Swedish west coast in the spring of 2006. The basic principles of the concept are illustrated in Fig. 6.34. The WEC consists of a floating buoy, coupled via a rope directly to the translator of the linear generator. The generator is placed the seabed and the tension in the rope is maintained by springs connected underneath the translator. A end stop at the top of the generator limits the stroke length of the translator. The concept is unique in several aspects and protected by a number of patents.

The basic idea behind this concept is simplicity. Auxiliary systems, such as dampers and active rectifiers are avoided. As stated in the introduction, a reduction in complexity often increases the requirements posed on the generator. The short time overload on the generator is, as can be seen in Table 6.1, expected to be up to 10 times larger than the nominal load. Excess capacity in damping will also be needed since no additional damping is provided.

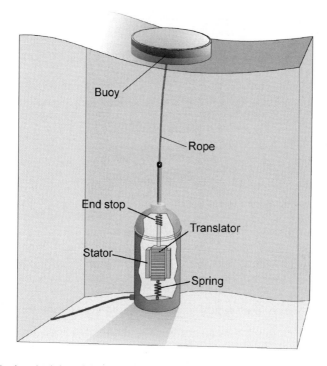

Fig. 6.34. Basic principles of the Uppsala University WEC concept

Generator design

The generator is of the LFPM type with surface mounted Neodymium-Iron-Boron permanent magnets. Standard cables with circular cross-section are used as armature windings. In contrast to other generators, where the design to a larger extent is determined by the thermal loading, the heat generation is a minor problem for this generator since the generated power per volume is so small. Furthermore, the WEC will be surrounded by sea water with a temperature in the range of 4–8°C. The main issues in this generator are the *size*, the *efficiency*, and the *load angle*. The goal of the design study is to construct a small generator with high efficiency and low load angle at nominal load (Danielsson et al., 2006). In contrast to other wave energy converters where the generator only is a minor component, this generator will actually carry a large part of the total construction cost of the wave energy converter. A reduction in generator size will be directly reflected in the cost of the plant. Furthermore, the efficiency has a direct influence on the energy delivered to the grid and will thus affect the annual revenue of the plant. A nominal wave with wave height 1.6 m and a period of 4.6 s was chosen, based on the moderate wave climate on the Swedish west coast. The nominal load of the generator was set to 10 kW at a speed of 0.7 m/s.

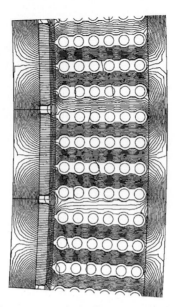

Fig. 6.35. Cross-section of a segment of the linear generator showing the magnetic flux lines at load

A combined field and circuit equation model is utilised to model the electromagnetic behaviour of the generator. The finite element equations and the circuit equations are solved simultaneously. Such models permit transient simulations, where the translator speed is continuously varying. This tool is used in a design study over five hundred simulations from the basis of a first laboratory prototype generator (Leijon et al., 2005). The simulated magnetic flux during load is illustrated in Fig. 6.35. The main features of the generator are summarised in Table 6.3.

Table 6.3. Main features of generator at nominal load

Nominal power	$10\,kW$
Main voltage	$200\,V$ (r.m.s)
Armature current r.m.s.	$28.9\,A$
Fundamental frequency	$7.0\,Hz$
Synchronous reactance	$0.44\,\Omega$
Load angle	$6.6°$
Pole width	$50\,mm$
Air gap width	$3\,mm$
Total air gap area	$2.08\,m^2$
Efficiency	86%
Hysteresis losses	$0.53\,kW$
Eddy current losses	$0.04\,kW$
Resistive losses	$1.0\,kW$

Experiment

Based on the design study, a first prototype generator was constructed. An illustration of the prototype generator is shown in Fig. 6.36. The prototype generator was half the length of the designed generator and the air gap area was $1.04\,m^2$. The generator has four sides with the square shaped translator in the middle and four separate stator packages on the outside facing the translator sides. The translator and stator are attracted to each other by magnetic forces. By using a multiple side translator, these forces can be reduced, since the forces of the opposing sides of the translator act in different directions. However, the forces increase rapidly with decreasing air gap widths and a small horizontal displacement of the translator will give a resulting force on the translator, acting in the direction of the smallest air gap. A support structure is necessary to maintain a constant air gap width and to counteract the significant attractive forces that are developed between the translator sides and the stator packages. The normal forces proved to be a large challenge and the smallest realizable air gap width was $5\,mm$, which is $2\,mm$ larger than the original design.

A hoist system with a $75\,kW$ induction motor was connected to the test generator. The translator was pulled at a constant speed of $0.86\,m/s$ and the phases where connected to a purely resistive load. The resulting voltages are presented in Fig. 6.37 together with the simulated voltage. The amplitude of the voltage is changing although the speed is constant. This is a result of the changing active air gap area as the translator moves in and out of the stator.

Full-scale offshore testing

Full-scale on-site testing of the wave energy converter adds a number of new challenges, but provided information and experiences impossible to be obtained in a laboratory (Waters et al., 2007). For the first time the full dynamics of the wave energy converter concept could be studied. The plant is illustrated in Fig. 6.38. The generator is placed inside a watertight steel structure, which is mounted on a concrete foundation. The linear motion is translated by a piston through a water tight piston sealing. The rope is guided at the top of the structure by four rolls, which ensure that the piston force only has a vertical component. The buoy is cylindrical and has a diameter of $3\,m$ and a height of $0.8\,m$ and is connected to the piston with a Vectran rope. The plant is placed at a depth of $25\,m$ and the structure is pressurized continuously during the submersion to a final pressure of $2.5\,bar$. A sub-sea cable connects the test site with a land station. The on-shore measurement station is equipped with heat sink variable resistive loads.

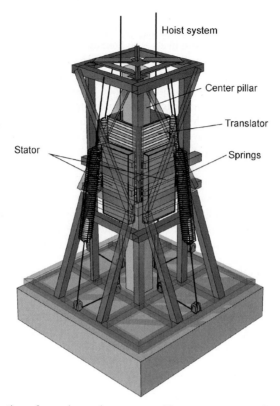

Fig. 6.36. Illustration of experimental generator with support structure is made transparent over the active parts of the generator

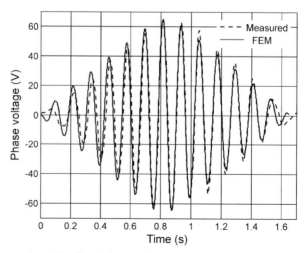

Fig. 6.37. Measured and simulated phase voltage from laboratory prototype

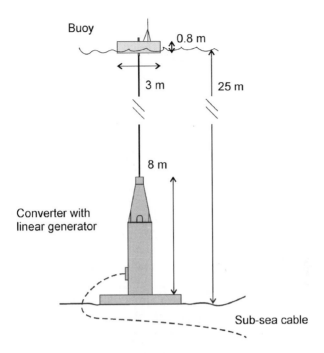

Fig. 6.38. Illustration of full-scale offshore wave energy converter

The magnetic circuit of the generator has the same design as the laboratory generator except for some small modifications. The length of the stator was increased and the air gap area was $2.04\,m^2$ and the stroke length is $1.8\,m$. The support structure was modified into a significantly stiffer structure and the designed air gap width of $3\,mm$ was achieved without problems. An illustration of the generator is shown in Fig. 6.39.

The complete WEC was launched in March 2006 outside Lysekil on the Swedish west coast. The voltages over the load were continuously measured and all important electrical entities in the system, such as power, resistive losses in generator and cable, no-load voltage, load angle, etc. can be estimated by using the lumped circuit model. Moreover, both the position and speed of the translator can be determined from the phase voltages, knowing that two successive zero-crossings of one phase correspond to a movement of the translator by one pole pitch. A sample of the main voltages is illustrated in Fig. 6.40. A typical pattern is achieved where the voltages varies both in frequency and in amplitude as the translator follows the motion of the waves.

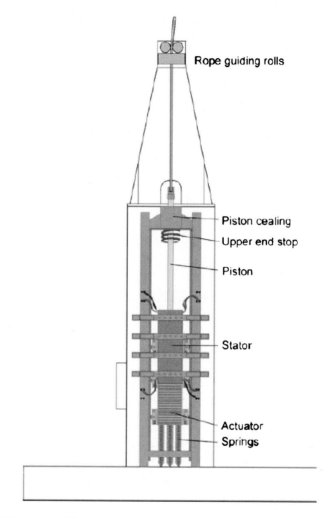

Fig. 6.39. Illustration of linear generator employed in the offshore WEC

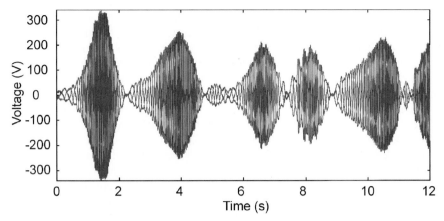

Fig. 6.40. Three phase voltages measured over the load resistances from the offshore WEC

6.3 Hydraulics

Jamie Taylor

University of Edinburgh & Artemis Intelligent Power Ltd
Edinburgh
Scotland, UK

"It is easy to make a device that will respond vigorously to the action of sea waves. Indeed it is quite hard to make one that will not. However the conversion of the slow, random, reversing energy flows with very high extreme values into phase-locked synchronous electricity with power quality acceptable to a utility network is very much harder." (Salter et al. 2002)

6.3.1 Introduction

If it was possible to make a perfect wave energy converter, it might be compared with a graceful and clever dancer who only ever gets to work with a mad orchestra. As the band swings without warning from jig to waltz, to punk, to minuet, to raga, to jig, to bebop, to reel – sometimes all of these at once – the dancer instantly finds each new tempo without losing a single step. Fascinated by the equivalent wave energy conversion problem, engineers have suggested many ways to improve the amount of energy that their devices might capture from changing seas. Salter et al. (2002) consider the range of such *control strategies* that have been proposed, ranging from fixed amplitude and variable force, through pre-emptive over-damping, to full complex-conjugate control (Nebel, 1992) and stiffness modulation. Each control strategy defines a relationship between the motions induced on the system by wave action and the corresponding reaction forces that should be provided by the power take-off system. The simplest and best known idea is to make force proportional to the instantaneous velocity. The ratio of these is then referred to as the *damping coefficient* and damping can be compared to the occasional domestic sensation of stirring a jar of syrup.

Whatever the strategy, a wave energy converter must have power take-off machinery that can be ordered to do things according to the changing values of digital and analogue commands – switch on, switch off, increase, decrease, connect, disconnect. A simple power take-off may have very few things that can be changed, but for the best performance and to deliver the cheapest energy it will be better if it can provide many such control-variable 'hooks' for the commanding computer to attach to. High-pressure oil-hydraulic systems are often particularly well suited to this sort of task. However, machines of sufficient size for larger wave energy devices are generally not available commercially. Spurred on by this need, a great deal of effort has gone into designing pumps and motors that are particularly suitable for wave energy devices.

Most of the topics discussed in this section are treated in greater depth in Salter et al. (2002). A deeper understanding of the behaviour of fluids can be gained from a comprehensive textbook on fluid mechanics such as Douglas et al. (2005). Three handbooks that may be useful in the design of high-pressure oil-hydraulic systems are Hunt and Vaughan (1996), Majumdar (2001) and Chapple (2002). The latter is published in conjunction with the British Fluid Power Association. A specialist book that explores the detailed design of high-pressure oil-hydraulic machines has recently become available in English, Ivantysyn and Ivantysynova (2000). For readers in the UK, the monthly trade magazine Hydraulics & Pneumatics is available on free subscription and is a valuable source of information regarding commercially available equipment and practices.

6.3.2 Spline Pumps

As a first example, Fig. 6.41 shows Stephen Salter's (1974) initial hydraulic solution to the conversion problem referred to at the start of this section. Each of the forty or so wave absorbing vanes of this classic wave energy converter rotates about the centre of the common supporting cylindrical member. The paraxial ridges on the cylindrical member and the inward facing ridges on the vane comprise double-acting spline-pumps. With non-return valves rectifying the flow, deoxygenated and decarbonated water is forced through the one metre diameter manifold pipes at a pressure of 400 pounds per square inch (*psi*) to a water-turbine and generator which are shared by all of the vanes. The water returns to the spline-pumps through the low-pressure manifold which is provided by the hollow centre of the cylindrical member. The system is a closed-circuit, high-pressure water, hydraulic transmission that incorporates an integral custom designed pump.

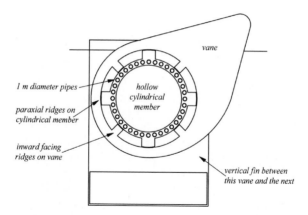

Fig. 6.41. Vertical section through early 'vane' showing spline pump (Salter, 1974)

6.3.3 Increasing Complexity

Five years later, the power take-off of the duck, as it was now called, had changed beyond recognition. Figure 6.42 shows how prolonged exposure to the energy conversion problem had led to a startling new arrangement. The treated water of the 1974 design had been replaced by mineral oil and its pressure had been increased by a factor of more than ten. The entire power take-off system was now enclosed within a closed and evacuated steel canister. This was located within the nose of the duck which could now go through much bigger angles because there weren't any spline-pump ridges to crash into one another. As it was rocked by waves of ever-changing periods and amplitudes, constant-frequency grid-quality electric current would be transmitted down the cable from the power take-off tube to the network. This grand design, which exploited the precession of gyros as a means to synthesise large inertial reaction, seemed to provide all of the good things that wave engineers dreamt of: complete environmental protection of the power conversion machinery from the sea; a generous amount of energy storage (in the spinning gyros); super-clean hydraulic oil (centrifuged by the gyros); high-class synchronous generation (thanks to a decoupled transmission); relative ease of control by computers (a new kind of valve); and high energy conversion efficiency (extensive use of hydrostatic bearings).

The elaborate arrangements of the gyro duck (Salter, 1980; 1982) reflect both the difficulty of designing robust and efficient wave energy converters and the historic fact that the target market for early UK wave energy designers was perceived as being for 2 *GW* installations. Accordingly, every part could be designed for the job. After 1979 the power take-off for the duck continued to evolve. The upper inset image in Fig. 6.42 shows a string of ducks at sea. The hollow cylindrical member of the spline-pump duck was now replaced by an articulated spine and it had

Fig. 6.42. A 1979 artist's impression (by L. Jones) showing duck with complete gyro power-take-off system contained within a sealed canister. The close-up view shows two axial-piston motors driving a gyro flywheel and two ring-cam pumps capturingenergy from the precession of the gyro-frame. Note that the bearing arrangement shown here between duck and spine is purely symbolic

been experimentally discovered that the power generated by the duck could be increased if the compliance of the spine joints could be actively controlled by high-pressure hydraulic rams, Fig. 6.43. The same rams could also provide damping forces that could contribute to the net generated power of the whole system (Salter, 1993).

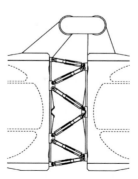

Fig. 6.43. Side-view of the joint between two spines, coupled together by a 'coronet' of 12 high-pressure hydraulic cylinders to give six degrees of controlled freedom (Salter, 1993)

What is still of particular interest is that the 1979 duck contained two new high-pressure oil-hydraulic machines. They can be seen in the cut-away view on the right of Fig. 6.42. One of these machines was a large diameter low-speed ring-cam pump that captured wave energy from the duck via the precession of the gyro frames and the other was an axial-piston variable swash-plate motor. Two of the latter are seen driving each gyro flywheel and another was assigned to drive the generator. Still more of the variable axial-piston machines would subsequently be assigned to control the spine joint rams.

The reason for going to so much trouble in the design of the power take-off was that the designers believed that, if they were going to have to pay very large sums of money for the front-end of the wave energy converter, its construction yards, moorings, underwater cables, sub-stations and maintenance fleet and then somehow persuade mariners to stay out of its way once at sea, they were obliged to do their best to get as much energy as possible from it. The generator of any wave power plant will spend most of its time operating at a fraction of its name-plate rating. To get the best possible energy return, the power take-off machines must therefore have the highest possible part-load efficiencies. They should also be easy to control with computers. The hydraulic components that were commercially available in 1979 were generally not designed with these requirements as priorities.

Of course nobody has got round to building a $2\,GW$ wave energy converter system, and so the approach of most current wave energy designers is necessarily different to that of the team who designed the gyro duck (the University of Edinburgh, John Laing Ltd & the Scottish Offshore Partnership). Most systems that have been built have been prototypes. In place of the public money that it was hoped would finance the designs of the 1970s, much of the money has come from private investors who have to reduce their risks by reducing the amount of technical innovation. Hence, the hydraulics used in the wave energy systems that have gone to sea are generally based on commercially available components and machinery.

The effort that went into designing the gyro duck later led to a new kind of hydraulic machine. Artemis Intelligent Power Ltd, a University of Edinburgh spin-out company that was set up by Win Rampen and Stephen Salter in 1994 to continue the work that had been started for wave energy and to progressively bring it to market, came up with the phrase Digital Displacement^{TM} to describe this development. The features of their machines will be described later. In the meantime it should be emphasised that it takes a long time before new designs of large hydraulic pumps and motors can be launched on the market with the necessary warranties for tens or hundreds of thousands of hours of operation. This is why it has always been important, but not necessarily easy, to continue the development of the kinds of specialised power conversion equipment that is appropriate to wave energy in parallel with the development of the hydrodynamic and naval architectural aspects.

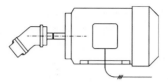

Fig. 6.44. Relative sizes of a typical fixed-displacement hydraulic pump (on left) and an electric motor (based on a 1,500 *rpm*, 110 *kW* system)

6.3.4 Advantages of Oil-Hydraulics

Hydraulic engineers often seem to have a hard time compared with their electrical friends who don't have to spend hours filling their wires with electrons or removing air bubbles from every corner of their circuits. Electricians don't worry about slipping on dirty electrons that have leaked onto the floor. Hydraulic hoses break too often, hydraulic connectors leak too easily, hydraulic oil ruins clothes and needs constant attention. It must be filtered, de-watered, de-aired, and either cooled or heated depending on its mood. The super-sensitivity of its viscosity to temperature frustrates machine design. Electricity is in a class of its own when it comes to transmitting energy over almost any distance but hydraulic circuits of more than a few metres in length can seem maddeningly lossy.

However, remarkably small and light-weight hydraulic machines can handle enormous forces and hydraulics has no rival when trying to capture power from large objects that are being pushed around relatively slowly by enormous wave forces. The net forces created by a typical hydraulic pressure of 350*bar* (5,000*psi*) are around fifty times greater than those from the magnetic circuits of the best electrical machines. This is why an electrical machine is best employed when it can use high velocity to compensate for its relatively low torque. It's also why a hydraulic pump that is driven by an electric motor of equivalent power usually looks very small in comparison (Fig. 6.44). Where necessary, hydraulics can also provide static forces for indefinite lengths of time with little expenditure of energy.

6.3.5 Hydraulic Circuits

In the late 1990s, Richard Yemm and his company Pelamis Wave Power Ltd started to develop their Pelamis device (Yemm et al., 2000). The prototype which was launched in 2004 is undoubtedly the most advanced floating wave energy converter that has gone to sea. Although he took some ideas from the duck's actively-controlled spine, Yemm decided to use only commercially available components in the power take-off system in order to boost investor confidence. Hydrostatic transmissions that are built for applications such as heavy winches and excavators generally use variable displacement pumps to deliver continuously variable pressure and flow. Their overall efficiencies can be well below 60 % when operating away from their full-load ratings. The Pelamis designers tried to find a way to use conventional components without incurring such inefficiencies.

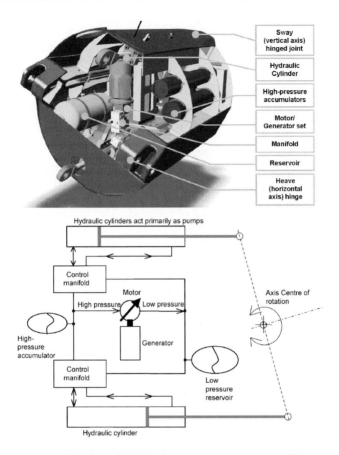

Fig. 6.45. A power module from the Pelamis wave energy converter. Top: cut-away view from solid-model. Bottom: Simplified hydraulic circuit of one active joint (Henderson, 2006)

Ross Henderson (2006), who has been closely involved since the start of its development, describes the Pelamis power take-off, as consisting of "sets of hydraulic cylinders that pump fluid, via control manifolds, into high-pressure accumulators for short-term energy storage. Hydraulic motors use the smooth supply of high-pressure fluid from the accumulators to drive grid-connected electric generators." Figure 6.45, from his paper, shows a cut-away view and a simplified hydraulic circuit of one of the Pelamis power modules.

The accumulators that provide the crucial decoupling between the hydraulic cylinders and the hydraulic motors are devices that store hydraulic energy through the compression of a gas within a pressure vessel. The gas, usually nitrogen, is separated from the fluid within a bladder or by a free-piston. Rising fluid pressure compresses the gas and admits more fluid to the vessel. Falling pressure expands the gas and expels fluid back into the circuit. At the risk of great thermodynamic simplification, the accumulator behaves like a spring. If it is of adequate volume, and with a suitable gas pre-charge pressure, it can provide enough energy storage to decouple the *primary* wave side of the power take-off from the *secondary* generator side.

Largely to avoid the danger of *cavitation* (the creation and collapse of damaging vapour bubbles due to low pressure transients), it is normal practice to operate hydraulic machines with a small *boost* pressure, up to 5 *bar*, at the intake port. The low-pressure accumulator shown in the hydraulic circuit schematic of Fig. 6.45 helps to maintain this boost pressure.

In the Pelamis power take-off, different combinations of chambers within the hydraulic cylinders, are used to pump oil directly to the accumulator. The chambers are switched in and out by electronically controlled valves during each wave cycle (within the "control manifolds" of Fig. 6.45). The reaction torque about each Pelamis heave or sway hinge can thus be varied through a range of values depending on the accumulator pressure and the number of cylinder chambers that are connected, (see Fig. 6.46). The pressure in the accumulator depends on the difference between the rate at which energy is supplied to it from the primary circuit and the rate at which it is taken from it by the secondary generator-driving circuit. With three heave joints, three sway joints, and six generators in the prototype machine, this allows the Pelamis control engineers to try out a number of different control strategies.

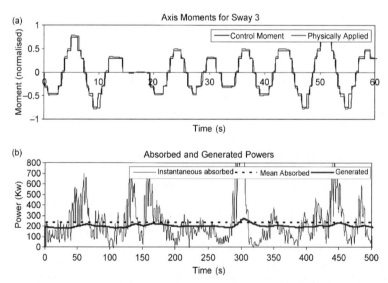

Fig. 6.46. Traces from a computer simulation of a Pelamis sway joint (Henderson, 2006). The steps in the reaction moment provided by the separately switched ram compartments are clearly seen in the upper image. The lower image compares the rapidly changing instantaneous power absorbed in the primary circuit with the slowly varying power of the secondary generation circuit

With reference to the emerging generation of Digital Displacement^{TM} machines, Henderson notes the development of "a novel digital hydraulic pump/motor, originally intended for a wave energy application, capable of offering a continuously variable transmission of hydraulic power at much higher efficiencies than conventional hydrostatic transmissions". He wisely concludes that "while the technology is nearing commercial application, it is still at the prototype stage and unsuitable for immediate deployment in a WEC". He considers the losses in the Pelamis primary transmission, from oil compressibility, bearings and seal friction in the cylinders and flow losses in pipes and valves and estimates that with good design these can be kept below 20 % over a wide range of operating conditions. He also points out that the primary transmission can absorb instantaneous power levels that are ten times higher than the average power that the secondary transmission is designed for (see Fig. 6.46).

6.3.6 Linear Pumps

Because of their familiarity in impressively powerful everyday machines such as diggers and cranes, linear hydraulic cylinders or rams as used in the Pelamis power take-off are an obvious choice for use in wave-driven pumps. Salter et al. (2002) report the availability of rams up $24\,m$ in length with diameters up to $1\,m$. The Pelamis rams are protected from the sea by flexible rubber bellows. The Ceramax plasma-sprayed coating process used by the Dutch company Rexroth Hydraudyne for the rods of some hydraulic cylinders may allow them to be used directly in sea-water without further protection. The maximum velocities that rams can deal with are governed by the material of the hydraulic seal and the temperature rise across it due to shear loss and friction. With adequate cooling, velocities up to $30\,m/s$ seem to be possible.

6.3.7 Rotary Pumps

For some wave energy devices such as the duck, rotary hydraulic pumps may be more appropriate in the primary stage of the power take-off. However, it may not be possible to find a pump that can develop sufficient torque at the very low angular velocities produced by waves. The Swedish company Hägglunds claim that their Marathon, which has a long record of service in exposed marine conditions, is by far the largest hydraulic motor in the world. It is of interest here because hydraulic motors can usually be used in reverse as pumps with little modification (the reverse is generally not true). Table 6.4 includes the maximum ratings of the biggest machine in the Marathon range.

Machines such as the Marathon have fixed displacements. They are designed to be used in systems where the flow or the pressure can be varied by external means.

Table 6.4. Some details of the Hägglunds MB4000 (from Hägglunds, 2006)

Displacement	251.3	*litres/rev*
Specific torque	4,000	*N/bar*[1]
Rated speed	8	*rpm*
Maximum speed	12	*rpm*
Maximum pressure	350	*bar*
Max. intermittent output power	1.58	*MW*
Case diameter	1,460	*mm*
Length excluding shaft	2,095	*mm*
Shaft diameter	460	*mm*
Weight	10.75	*Ton*

[1] $1 \, bar = 10^5 \, Pa \, (Pa = N/m^2)$

6.3.8 Ring-cam Pumps

A rotary-machine topology that is often used for slow, high power machines such as the Marathon is sketched in Fig. 6.47. A central cylindrical block is driven by the shaft and incorporates within it a number of cylinders within which pistons slide freely. Rollers are pressed by the cylindrically shaped pad at the foot of each piston against an undulating multi-lobed cam that forms the outer part of the pump case. As in the design of all hydraulic machines, a detailed understanding of hydrostatic and hydrodynamic processes is required to ensure that all moving surfaces that are in close proximity are always separated by a film of oil so as to avoid metal to metal contact.

Figure 6.47 does not show the valves and galleries that connect the cylinders alternatively to the low-pressure and the high-pressure ports of the machine. Neither does it show compression springs that might be fitted between pistons and cylinders to help hold the rollers always in contact with the ring-cam. Some form of radial guides would usually be fitted to support the circumferential forces on the rollers to ensure that only axial forces are transmitted between pistons and cylinders. The machine of Fig. 6.47 has a twelve lobed cam and thirteen pumping-

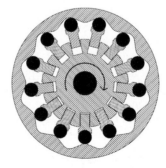

Fig. 6.47. Diagram of a ring-cam pump. In this case, with 12 cam lobes and 13 cylinders, there would be 156 piston strokes per revolution

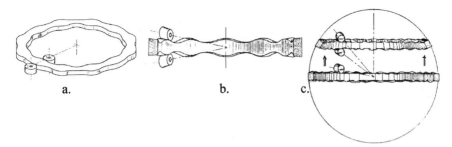

Fig. 6.48. Transformation of ring-cam topology for gyro duck (Salter, 1980) a) duplex, lobes on outside as well as inside of cam; b) twisted cam with conical rollers; c) cam moved from the equator to the tropics

groups. Every pumping-group is worked by every lobe, and so during a single revolution there are a total of 156 pumping operations. In a low angular-velocity drive, the multiple lobes of the ring-cam provide a means of stepping up the frequency and combined volume of pumping operations, to produce a machine that has a high power-density. Salter et al. (2002) suggest that the value-for-money of a ring cam machine becomes steadily better as it becomes larger and more powerful because, whilst the cost depends on the sum of the costs of the pumping modules and the lobes, the value depends on their product.

Figure 6.48 shows the progressive transformation of a simplex ring-cam into the open duplex arrangement of Fig. 6.42 that was designed to capture power from the precession of large gyro-frames. The first stage of transformation, shown in the left-hand image, is to add a second external series of cam lobes so that pairs of opposing pumping-modules can be placed around the ring. Most of the mechanical stresses are then transmitted locally through the thickness of the cam and directly between module pairs rather than around the circumference of the whole ring. In the central image of Fig. 6.48, the cam surfaces are twisted through ninety-degrees and the rollers become conical to allow a more compact placement of pumping-modules. Finally, to free up room for the gyro flywheel drive motors, the ring-cam migrates, as shown in the right-hand image, from the equator position of the gyro frame to the tropic position.

6.3.9 Hydraulic Motors

Whatever the type of pump that is used in the primary circuit, the secondary circuit will almost certainly require a hydraulic motor that can drive a synchronous or an induction generator at speeds around 1500 or 3000 *rpm*. Of the commercially available machines, axial-piston bent-axis types, made by companies such as Parker or Bosch Rexroth, are the most obvious choice for generator drive.

Axial machines can be converted to variable displacement by the introduction of a swash-plate. Salter et al. (1988) reported the design by Robert Clerk (1908-1993) and the construction by Matthew Rea of a refined version of such a machine. Clerk's original specification was for flywheel storage applications, and he was particularly concerned to obtain the lowest possible losses when idling, along with a capacity for high torque and high speed. These attributes were also appropriate for the high-speed motors inside the power canister of the gyro duck.

Referring to the simplified sectional view of Fig. 6.49, the motor operates as follows. The cylinder block 'A' rotates past the ports in the port-head 'B'. The shaft 'C' is coupled to the cylinder block by drive lugs 'D'. Ball-ended rods 'E' connect the pistons 'F' to the drive-plate 'G' which is forced axially against an angled swash-carrier 'H'. The angle of the swash-carrier can be changed by a piston which is not shown, and the friction between it and the drive-plate is reduced by the use of hydrostatic support pads. The reaction between the swash-carrier and the drive-plate induces output-torque which is passed to the shaft through Clerk's tri-link mechanism which is shown half-way along the shaft. The machine's displacement per revolution clearly depends upon the angle of tilt of the swash-carrier because this directly controls the length of the pistons strokes.

The use of spherical bearings in the tri-link, push-rods and main-shaft greatly reduced the accuracy of alignment that was required in the construction and assembly of the motor and made it tolerant of dimensional changes under high stress. Clerk also used hydrostatic bearings throughout the machine to reduce loss and extend life. Hydrostatic bearings require great attention to oil cleanliness, but if this can achieved the complete separation that they provide between running surfaces, makes it possible to consider the use of working fluids such as water-emulsions, that have no natural lubricity.

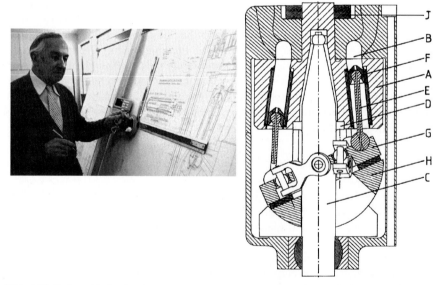

Fig. 6.49. Robert Clerk, shown in 1985 working on his high-speed axial-piston variable displacement machine of which a simplified section is shown on the right. (Photo: University of Edinburgh. Drawing: Salter et al., 1988)

Fig. 6.50. Matthew Rea, in 1985, with parts that he and Carn Gibson made for a 600 *kW* Clerk high-speed motor. On the left, the motor is shown assembled without its case. The angled swash-plate is clearly visible (Photos: University of Edinburgh)

Figure 6.50 shows a prototype of Clerk's high-speed motor that was built and tested at the University of Edinburgh.

6.3.10 Flow Commutation

Cylinders within rotary machines must be connected alternatively with the low-pressure and high-pressure ports that join them to the external system. The mechanisms that are used for this are comparable in function to the commutator of a *DC* brush motor. The port-plate arrangement used in the Clerk motor and typical of bent-axis and swash-plate machines in general, is shown in Fig. 6.51. Two kidney ports on a stationary block are connected respectively to the low-pressure and the high-pressure ports of the machine. A barrel with multiple ports rotates with the cylinders and this alternatively connects them to the low- and high-pressure ports. The design of such a flow-commutator requires great subtlety and experience and is possibly the hardest aspect of a rotary hydraulic machine to treat successfully.

6.3.11 Losses

The main processes of energy loss within rotating hydraulic machines are through churning, leakage, shear, compressibility and breathing.

The cases of most high-pressure oil-hydraulic machines remain fully flooded during operation. In high-speed machines the energy absorbed by the resultant

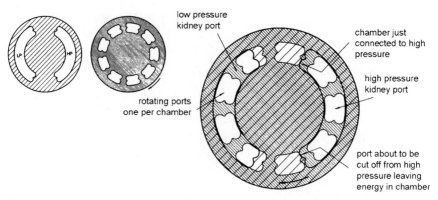

Fig. 6.51. Port-plates as used to provide flow-commutation in axial-piston machines. The stationary plate shown at top left is connected through to the low- and high-pressure ports. The rotating barrel at top centre has one port per cylinder. The two parts are shown on the right at a particular point in their relative rotation (modified from Salter et al., 2002)

churning can be several percent of the full power rating. The Clerk tri-link machine was designed to run 'dry' so as to eliminate these losses. Hydraulic fluid escapes through any gap that has a pressure difference across it. However some of these leakages may be intentional if they are designed as part of a hydrostatic-bearing to provide controlled separation between moving parts. Shear losses occur through the viscous forces generated between moving surfaces that are separated by a layer of oil. Leakage power-loss depends on the square of pressure and can be reduced by making the clearances finer, using higher viscosity fluid or by increasing the lengths of leakage paths. However these changes have the effect of increasing shear losses which depend on the square of speed. The machine designer has therefore to carefully balance the conflicting aims of reducing leakage and shear losses.

Compressibility losses arise because the working fluid has a bulk modulus that is not infinite, typical values for a hydraulic oil being around $1.6 \times 10^9 \, N/m^2$. Potential energy is absorbed by the compression of the oil as well as by elastic deformations of machine parts. If the compressed oil in a cylinder is suddenly depressurised, for instance as a result of poor timing of the commutation mechanism, the stored energy may be suddenly and wastefully dissipated with consequent noise, wear and energy loss. With mechanical flow-commutation and variable operating conditions, it is virtually impossible to avoid compressibility losses at some parts of the operating regime.

Breathing losses result from pressure drops caused by the rapid movement of fluid around bends and through changes of section in the generally complicated passages between cylinders and ports. These pressure drops are particularly critical on the low-pressure side because of the need to avoid cavitation and are compensated by boosting pressure by at least a few bar above ambient. This fixes the problem, but reduces the working pressure differential across the cylinders of the machine. Good breathing requires the sort of attention to detail shown in Fig. 6.52. It also needs space so that flow velocities can be kept as low as possible.

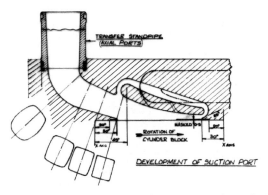

Fig. 6.52. Suction port detail for high-speed swash-plate motor, drawn by Robert Clerk

6.3.12 Active Valves

However well a conventional machine is designed there is a limit to how far its losses can be reduced, particularly when it is operating well below its rated capacity. One reason is that all of the cylinders experience the full pressure cycle at all times and so are always subject to leakage losses. Furthermore, flow-commutation by mechanisms such as port-plates or pintle-valves offers little timing flexibility, poor control of compressibility losses, and may compromise the space available for good valve breathing. An alternative approach where cylinders are only pressurised when they are required to contribute flow (in a pump) or provide torque (in a motor) is made possible by the kind of active low-pressure poppet-valve that was conceived of for the gyro duck low-speed ring-cam pump. An early drawing of this valve is shown in Fig. 6.53. Its further development is currently leading to a new generation of higher efficiency hydraulic machines that naturally interface to control computers.

To best capture energy from waves of varying heights and periods, some means had to be found in the gyro duck design of Fig. 6.42 to vary the reaction torque that the ring-cam pumps provided against gyro precession. It was also important to reduce losses at times of low incident power. The new active low-pressure intake valves provided the means to satisfy both requirements.

As the piston in each of the many pumping modules reached bottom-dead-centre it would have filled with oil from the low-pressure manifold and would be ready to start a power-stroke. At this moment in a conventional machine, the inlet-valve would close, the piston would start a compression stroke, the oil pressure within the cylinder would rise, the outlet-valve would open and a unit of high-pressure oil would be delivered to the high-pressure system. However, in the new design, when the piston was at bottom-dead-centre, a computer would decide whether or not another unit of high-pressure was required. If the answer was yes, then the operation would be as for the conventional machine. If the answer was no, a pulse of current would be sent to a coil in the new inlet-valve and the magnetic-bistable latch would hold the valve open so that the piston would merely return the un-pressurised oil to the low-pressure manifold.

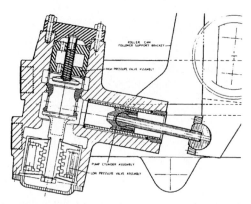

Fig. 6.53. Detail of pump module from the gyro duck ring-cam pump. The upper high-pressure outlet-valve is of conventional passive design. However, the low pressure intake-valve assembly at the bottom features a magnetic bistable latch and marks the beginning of Digital Displacement^TM hydraulics. From an unpublished Edinburgh-SCOPA-Laing report to the UK Department of Energy, 1979

By choosing an appropriate enabling-pattern for the intake-valves of the pumping modules, the displacement of the entire pump could be varied in discrete steps.

By only pressurising cylinders that are required to do work, the pressure-related losses of the pump would be greatly reduced and the part load efficiency would be increased. The physical layout made possible by the new active valve gave a new freedom to provide good breathing arrangements and impressive power density. The proposed 5 *m* diameter units would have a power rating of up to 30 *MW*.

6.3.13 Active Valve Machines

The design and construction of Robert Clerk's low-loss high-speed motor (Figures 6.49 and 6.50) was an impressive achievement. However its operation was subject to the limits imposed by mechanical commutation and so there was room for further improvement in its losses and its controllability. Active commutation can be applied to high-speed motors if the high-pressure valve as well as the low-pressure valve is of the active type. Such a machine is described by Salter and Rampen (1993) and various views of it have already been shown in an earlier section of this book (Fig. 2.15). The new machine used a radial rather than an axial configuration with the pistons working against a single-lobed crankshaft eccentric as shown in Fig. 6.54. The kinematics of this arrangement are near perfect.

In this version, the surface of the shaft eccentric forms part of a sphere. The piston big-end pads are designed to support themselves on this surface with a combination of hydrostatic and hydrodynamic lift. Springs provide additional

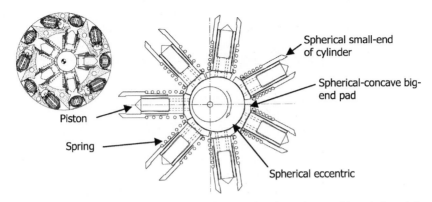

Fig. 6.54. Radial-piston geometry used for high-speed active-valve machines (adapted from Salter and Rampen, 1993)

acceleration force to hold the pads to the eccentric at high rotational speeds. The outer small-ends of the cylinders also fit to the body of the machine via spherical surfaces (see the inset in Fig. 6.54). This arrangement protects the pistons and cylinders from side loads and gives them the freedom to take up any dimensional changes in the machine under high load.

The radial arrangement of pumping-modules around the eccentric shaft places the piston pads at the centre of the machine where the linear velocity of the eccentric is at a minimum. This helps to reduce the shear losses. The valves are placed around the periphery of the machine where there is plenty of space to optimise the breathing arrangements. Figure 6.55 shows how several radial machines can be arranged along a common shaft to produce a more powerful machine, or one that can provide multiple services.

An application of a multiple-service machine to a wave energy power take-off is sketched in Fig. 6.56. In this case, the active-valve radial-machine provides a total of six services that are grouped in three pairs. The wave-excited part of the device is represented by the large piston at centre-left which moves relative to the bodies of the two single-acting hydraulic rams. The varying reaction forces provided by the rams are controlled by the active high- and low-pressure poppet-valves of the multi-eccentric machine. High force peaks and instantaneous overloads are passed directly to the gas accumulator which is managed via dedicated circuits on the multi-eccentric machine. The shaft of the machine drives a synchronous generator at 1500 *rpm* and can be compared with the summing junction of an operational amplifier. In this case, energy rather than currents are balanced along the length of the shaft. Depending on the control algorithm, power can flow in any direction in any of the services including the electrical machine. Any instantaneous energy deficits within the power take-off could then be supplied by the generator acting as a motor for part of the cycle. Ehsan et al. (1995) describe the time-domain simulation of such a system.

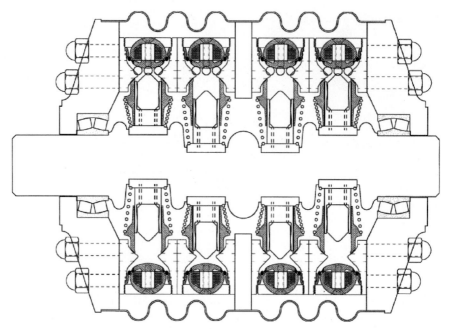

Fig. 6.55. A four-layer active-valve radial-piston machine. From Salter et al (2002)

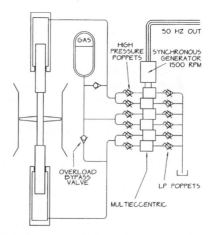

Fig. 6.56. Simplified schematic of a multiple-service, active-valve machine controlling two rams and an accumulator for use in wave energy devices such as the IPS Buoy or PS Frog (Salter et al., 2002)

6.3.14 Digital Displacement™

The phrase Digital Displacement™ was coined by Artemis Intelligent Power Ltd, a spin-off company from the University of Edinburgh, for the active-valve hydraulic machines that they are currently developing with industrial partners. Figure 6.57 shows components for the first such machine which was built by Artemis in the late 1990s.

This real 12 *cc* per revolution pump is small compared with the original vision of an active valve machine within the power canister of the gyro duck. Even so, it has a nominal power rating of around 20 *kW* at 3000 *rpm* and 350 *bar*. The demanding specification for hydraulic machines that would be suitable for power conversion in large wave energy devices provided the impetus for this new technology which has started to find application in automotive transmissions and mobile hydraulics. In the meantime the cylinder sizes and power capacities of such machines are being gradually increased to the stage where they can at last start to find application in wave energy devices (Payne et al., 2005).

Figure 6.58 shows the sequences of ones and zeros that are used to command the enabling or disabling of cylinders within Digital Displacement™ machines. The time-averaged result is a close linear fit between the flow demand signal and the flow into or out of the machine. This feature illustrates the suitability of such machines for certain open-loop applications including metering.

The final plots, in Fig. 6.59, compare the measured efficiency characteristics of a 250 *cc* bent-axis machine and predicted efficiency of a 192 *cc* Digital Displacement™ machine. The upper plots correspond to full displacement and the lower plots to 20 % displacement. The contours for the 192 *cc* Digital Displacement™ machine are based on a numerical model that was calibrated against data from tests of smaller prototype machines. The broad plateau of high efficiency operation across a wide range of speeds and pressures is typical of the improved per-

Fig. 6.57. The first Digital Displacement™ radial-piston pump built by Artemis Intelligent Power. Parts for the active low-pressure valve are in the left foreground. In the centre is a passive high-pressure check-valve. A 2 *cc* pumping module is at front right. The microcontroller is at back right. Photo: Artemis Intelligent Power Ltd

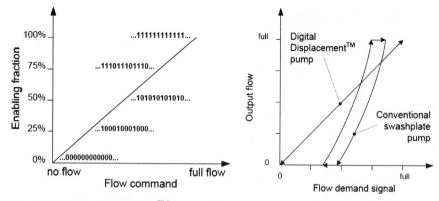

Fig. 6.58. Digital Displacement^{TM} pump control. On the left: binary valve control sequences are used to vary the enabling fraction between 0 % and 100 %. On the right: the resulting linear relationship of output flow to the demand signal. (Payne et al., 2005)

formance that has already been demonstrated by small Digital Displacement^{TM} machines. The systematic development of megawatt class Digital Displacement^{TM} machines is currently underway and these are likely to be of great benefit to wave energy power take-off systems.

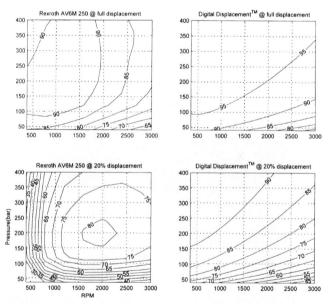

Fig. 6.59. Predicted efficiency contour plots of a Digital Displacement^{TM} pump (on the right) at full and 20 % displacement compared with an equivalent-sized commercially available machine (Payne et al., 2005)

6.4 Alternative Applications: Desalination

Matthew Folley,

Queens University Belfast, Northern Ireland, UK

João Cruz,

Garrad Hassan and Partners Ltd, England, UK

It would seem that anybody who has worked in wave energy has at sometime considered the potential for wave-powered desalination. However, few have progressed to investigate the challenges and prospects of wave-powered desalination in depth, so that within this subject there appears to be a lot of talk with very little substance. This section attempts to clarify the topic of wave-powered desalination, putting it in context with desalination technology in general and detailing the historical development of wave-powered desalination. Thus the section starts with an introduction into the context of desalination and a description of the different types of desalination technologies currently used. This is followed by a review describing the history of wave-powered desalination and details of the already approached technologies. The challenges that are faced by the use of wave power for desalination are also presented and the section is concluded with a discussion on the prospects for wave-powered desalination.

6.4.1 Introduction

The lack of available fresh water is considered by many to be one of the greatest challenges facing the world. Whilst water shortages affect countries from both the developed and developing worlds, it is in countries in the developing world where its effects are most pronounced. A lack of fresh water is directly associated with 80 % of all diseases in these countries and 30 % of all deaths worldwide. Desalination technologies, which convert salt water into fresh water, are used extensively to produce fresh water where there is a scarcity. The current worldwide desalination capacity is approximately $16,000,000 \, m^3/day$, providing fresh water for irrigation and human consumption in many arid parts of the world. It is estimated that this capacity is growing by approximately 1 % per annum; however this is lower than the estimated increase in demand for fresh water, which is growing at twice the rate of population growth; that is a rate of 3 % per annum. It is thus predicted that unless circumstances change the number of avoidable deaths due to a lack of fresh water will increase, with some authors predicting that future wars will be fought over this scarce commodity (Engelman et al., 2000; Barlow and Clark, 2002).

The obvious solution to the scarcity of fresh water is to increase the desalination capacity in the effected regions. However, desalination technologies are energy intensive, which places them in direct conflict with the other great challenge facing the world: the energy crisis. It is relatively easy to show that the minimum energy requirement for desalination is approximately $2.5 \, kJ/kg$ ($0.7 \, kWh/m^3$) of fresh water produced (physically it corresponds to the free energy change associated with the process of salt dissolution). But the majority of current desalination plants are based on distillation processes with a specific energy consumption of around 35 kJ/kg (10 kWh/m^3). Putting this is in context: standard agricultural practices require 0.1 m^3 of fresh water, equivalent to 3.5 kJ (1.0 kWh), to produce 1 kg of tomatoes. Although more modern technologies utilising reverse osmosis together with energy recovery from the reject brine flow have reduced this energy consumption to approximately 10.0 kJ/kg (3.0 kWh/m^3), the energy consumption for desalination remains significant.

In many regions the scarcity of both traditional, non-renewable, sources of energy and fresh water has understandably led to proposals of desalination plants powered by renewable energy, including wave power. Indeed, wave-powered desalination would seem to offer particularly good prospects because the two main requirements, salt water and energy, are both available in abundance. Whilst a few prototype wave-powered desalination plants have been developed, the development of desalination plants powered by wind and solar energy has been much more significant (Subiela et al., 2004; Bouguecha et al., 2005; García-Rodríguez, 2003; Raluy et al., 2004; Koklas, 2006). These have demonstrated that renewable energy powered desalination plants can be cost-effective and it is reasonable to expect that wave-powered desalination can be similarly successful.

Desalination plants powered by renewable energy have an additional advantage where the electrical distribution network is weak because they can help to increase the proportion of renewable energy utilised. In these locations the electricity is typically supplied by diesel generators using imported fuel making the cost of electricity from traditional power plants high, often 3–4 times higher than the cost where coal, gas or nuclear power stations supply power to a large distribution network. Ironically the exploitation of renewable energy to generate electricity is limited in these locations due to the weakness of the grid, exactly where it would be most profitable to locate renewable energy plants. For example, in Gran Canaria the peak power demand is 230 MW, however to maintain grid stability the maximum installed capacity of wind turbines is limited to 80 MW, although there is much more wind power (and wave power) available. By using renewable energy to power desalination plants directly, the problems with the weak grid can be circumvented and the total amount of renewable energy exploited increased. It is possible to think of the desalinated water as an alternative and cheap form of energy storage, with a capacity of 7–35 kJ/kg (2–10 kWh/m^3).

Finally, offshore wave-powered desalination plants have a number of environmental benefits when compared to traditional desalination plants. Firstly, the concentrated brine, which is a by-product of desalination, can be disposed of away from the coastline, reducing the environmental impact. Secondly, noise generated by the desalination plant would occur out of earshot from the local population and thirdly, a minimal amount of the valuable land close to the shoreline is required.

6.4.2 Desalination Technologies

Two basic processes can be used to separate salt from water: thermal processes and physical processes. Thermal process utilise a change in phase, exclusively liquid to gas in current technologies, to separate the salt from the water. Physical processes involve a physical barrier, a membrane, which allows the water across, but not the salts. In addition chemical processes can be used and involve changing the chemical composition to separate the salt by means such as precipitation. However, chemical processes are typically expensive and require a supply of chemicals and thus will not be discussed further.

Thermal Processes

Thermal distillation processes have been used for large-scale desalination for over 50 years and make-up the majority of commercial desalination plants. Earlier plants used a multi-stage flash (MSF) process which involves bulk heating of water that is subsequently passed through a number of stages. In each stage the pressure is reduced resulting in some of the water evaporating and condensing on heat exchangers that preheat the feed water. This technology is often coupled with a conventional thermal power station utilising its "waste heat". Multi-effect evaporation (MEE) is related to MSF, but rather than using the latent heat in the water vapour to pre-heat the feed water, it is used to provide additional heat for evaporation to the subsequent stage. MEE plants typically have a better thermal efficiency than MSF plants, however problems with scaling has limited their application. Finally vapour-compression processes involve reducing the pressure to cause the water to evaporate. The heat required to make the water evaporate comes either from a mechanical pump, and the process is then called mechanical vapour-compression (MVC), or from the expansion of steam and the process is called thermal vapour-compression (TVC). In both cases a heat exchanger is used to recover the energy from the product water and recycle it back into the process.

To determine the efficiency of thermal processes there are essentially two means of energy loss: the heat lost to the atmosphere through the walls of the plant and the temperature difference between the intake and brine/product waters. Thermal insulation is typically used to reduce the first loss, whilst heat exchangers are used to reduce the latter. Clearly, the more thermal insulation used and the larger the heat exchangers used the more efficient the process becomes. Each particular configuration lends itself to reduce these losses and it is typically found that MSF has the highest specific energy consumption of 100–$300\,kJ/kg$ (28–$83\,kWh/m^3$), MEE the second highest with 100–$200\,kJ/kg$ (28–$56\,kWh/m^3$) and VC the lowest with 15–$60\,kJ/kg$ (4–$17\,kWh/m^3$). The variations are due to different plant arrangements, plant size, operating temperatures and recovery-ratios (the ratio of product water to feed water).

Thermal processes are relatively maintenance-free due to the absence of moving parts. One problem that can occur is scaling. Scaling becomes progressively more problematic as the operating temperature and recovery-ratio increase and so can be avoided by limiting the operating temperature and recovery-ratio; however this typically increases the specific energy consumption of the process. Most commonly pre-treatment chemicals are added to the feed water to inhibit the precipitation of the salts. A potential alternative, which has not been commercially proven, is to use an anti-scaling coating, which stops the adhesion of the salts onto the plant surfaces, so that they continue to operate effectively.

A potential advantage of thermal distillation processes is that they produce a high quality output with a salt concentration typically less than $10 \, ppm$. However, for human consumption a salt concentration of less than $500 \, ppm$ is considered palatable, whilst the World Health Organisation limit for human consumption is $1000 \, ppm$. The maximum salt concentration acceptable for irrigation depends on the plant being grown, but is typically around $2000 \, ppm$.

Physical Processes

The only physical process used for seawater desalination is reverse osmosis (RO). Electrodialysis (ED), the other main physical process has thus far been found to be uneconomic for use with seawater. Reverse osmosis involves pressurising water so that it is forced through a membrane that allows the passage of the water but not the salts. The pressure required to force the water through the membrane increases with its salinity and so initially the use of RO membranes was limited to brackish waters with a salinity of less than $10,000 \, ppm$. For seawater the osmotic pressure is typically $25–28 \, bar$. Once the osmotic pressure has been overcome, the flow of water through the RO membrane is approximately proportional to the amount of excess pressure applied.

Together with the water a small amount of the dissolved salts leak through the RO membranes so that the product water typically has a salinity of around $100–500 \, ppm$. Because the leakage rate of salts depends primarily of the salinity gradient across the membrane and is independent of the feed pressure, higher product salinities occur with lower feed pressures and lower product salinities with higher feed pressures. For current membranes a feed pressure of $50 \, bar$ will produce a product with a salinity of approximately $150 \, ppm$.

Appropriate pre-treatment of the feed water is required to ensure that the RO process remains efficient and to ensure a reasonable longevity of the membranes; RO membranes typically need replacing every 2–5 years. The first stage in the pre-treatment is filtering the feed water to reduce the maximum particulate size to less than $0.1 \, \mu m$. This is normally achieved in two stages, a sand filter removing particles larger than $20 \, \mu m$, and then a cartridge filter, which removes the remaining particles larger than $0.1 \, \mu m$. However, if the geology at the location is appropriate then the feed water can be drawn from a "beach-well", utilising the natural porosity of the rock/sand/silt as a filter. The next stage is to sterilise the feed water, killing any organisms that may grow on the surfaces of the membranes. This may be achieved by using UV light, the addition of chlorine or other chemical treatment. Again, the use of a "beach-well" can eliminate the need for this stage, where organisms cannot survive in the rock sufficiently long to reach the feed water intake. The final pre-treatment involves treating the feed water to eliminate the problem of scaling, which increases with the recovery-ratio. This can be achieved by adding acid to the water to increase its pH, or chemical additives to stop the scale from forming. Depending on the chemicals added then some post-treatment may be required to remove the carbon dioxide produced.

The optimal pre-treatment processes required depend on the characteristics of the feed water and desalination plant. In some circumstances extensive pre-treatment is required, whilst in others it has been found that no pre-treatment has been required. However, if the correct pre-treatment has been applied then the reliability of a RO plant is similar to other desalination technologies, with a minimum level of maintenance.

Whilst membrane technology has continued to improve, the major advancement that has made RO suitable for desalination of sea-water is in energy recovery technology. Energy recovery involves recovering and recycling the energy contained in the discharge stream of un-desalinated water. Because the specific energy consumption of a RO plant increases with salinity the recovery-ratio has typically been limited to approximately 40%, so that a large amount of energy was lost in the 60% of pressurised sea-water that was not desalinated.

Early energy recovery technologies were simply reverse-running pumps or Pelton wheel turbines connected to the main high pressure pump of the plant to reduce the load on the drive motor. These were capable of recovering around 75% of the brine stream energy. More recently energy recovery technologies have been designed that transfer the pressure directly from the brine to the feed water, thereby eliminating the conversion of energy into shaft power, with resultant energy recoveries of over 95% (Andrews and Laker, 2001; Harris, 1999; Geisler et al., 1999; Stover, 2004). Consequently, the specific energy consumption for RO plants has recently plummeted with a number of plants claiming a specific energy consumption of less than $7.0\,kJ/kg$ ($2.0\,kWh/m^3$), though a more typical figure may be $10.0\,kJ/kg$ ($3.0\,kWh/m^3$).

An additional beneficial effect of the increased efficiency of energy recovery is that RO plants can operate economically at lower recovery-ratios, thereby avoiding problems associated with scaling and the need for this pre-treatment. By using UV light to sterilise the feed water the need for chemicals in the pre-treatment can be avoided and so the plant can more easily be operated in remote locations where the supply of chemicals could be problematic. This strategy has been used by ENERCON in an autonomous wind-powered desalination plant for remote locations (Paulsen and Hensel, 2005).

Costs of Desalination Technologies

Investigating the specific cost of desalinated water from the different technologies shows a vast range of costs for each technology, making it difficult to compare technologies effectively. Part of the reason for this is the lack of a standard methodology by which the costs can be calculated, but this is exasperated because a large proportion of the cost of a desalination plant is in its construction, therefore once it is built it is likely to be operated because the marginal cost of the desalinated water is acceptable, even though the specific cost based on the plant's full life-cycle cost may be high. Alternatively, it is possible to look at what is being built to indicate what the most profitable technology is currently, indicating that reverse osmosis is currently the preferred technology. However, it must be recognised that this is for standard plants powered by non-renewable energy sources.

The specific cost for each proposed plant will be minimised based on the plant construction and maintenance costs and the relative cost of energy in the specific circumstances; the assumptions used may no longer be valid for of wave-powered desalination plants.

The cost of desalination can be split into three basic cost centres: plant construction costs, operation & maintenance costs and energy costs. The contribution from each cost centre will differ for the different technologies and for different scenarios making it difficult to generalise. However, it is useful to understand the basic mechanics of the cost calculations so that the design optimisation of desalination plants can be understood. As an example, consider the case of a reverse osmosis plant. The capital cost of a medium-sized, $5,000 \, m^3/day$, RO plant is about £1000 per m^3/day, operation and maintenance costs are about £100 per m^3/day per year and energy costs are about $0.09 \, £/m^3$ (based on a specific energy consumption of $3 \, kWh/m^3$ and electricity cost of $0.03 \, £/ \, kWh$). If the plant life is 25 years and an 8 % rate of return is applied then the cost of water is $0.62 \, £/m^3$, with 41 % capital costs, 44 % operation & maintenance costs and 15 % energy costs; the plant has a capital cost of £ 5,000,000 and consumes $625 \, kW$ of power. Alternatively, where diesel generators are used the real cost of electricity is closer to $0.13 \, £/kWh$, so that if the same plant were used the cost of water would be $0.92 \, £/m^3$, with energy accounting for 42 % of the cost. However, in this scenario more RO membranes would be used to reduce the specific energy consumption, which increases the capital cost, but would reduce the specific cost of the water produced.

If only the energetic cost of the processes are considered, then comparison of the specific energy consumptions of the different desalination technologies indicates that vapour-compression and reverse osmosis technologies have a significantly lower specific energy consumption than the rest, with the energy recovery technologies developed in the last 5–10 years giving reverse osmosis a slight edge. However, it must again be emphasised that these energetic costs have been optimised for standard desalination plants; the relative costs for wave-powered desalinations plants may be different.

6.4.3 Current Status of Wave-Powered Desalination

Renewable energy schemes experienced a major boost during the 1970's oil crisis. Wave energy was no exception, and it began to be seriously studied around this time, although the first ideas from Girard and sons date back to the 18th century (McCormick, 1981). The main alternative to the traditional output (electricity) is typically assumed to be desalinated water. Wave energy devices are usually associated with direct conversion from wave to electrical energy. The global resource is of the same order of magnitude of the world's consumption of electrical energy (Isaacs and Seymour, 1973), which makes it one the most promising and attractive forms of renewable energy. Generic introductions to wave power technologies can be found in Salter (1989) and Clément et al. (2002). Several devices and schemes are nowadays on a pre-commercial stage, in a clear sign that wave energy is being seriously considered not only by the scientific community but also by the industrial world.

By adding the need of fresh water, the first attractive feature about wave powered desalination is to have both the energy resource and the raw material (seawater) in the same site. Davies (2006) presents both a review on wave-powered desalination schemes and an application exercise to some notably dry countries. African countries are assessed in detail, with the fresh water potential from wave powered desalination being quantified through a mathematical model. It is clearly pointed out that many other arid regions can be typical examples, like regions of the USA, Western Australia and Chile, and islands like the Canaries and the Maldives. The reduced number of such coupled schemes is addressed and the review is concluded with the notion that, due to the predicted increase of the problems surrounding fresh water supplies, the interest in such systems is likely to grow, either with further development of the current alternatives or with the emergence of completely new ones.

In this section, several concepts that reached the prototype stage are addressed. The technologies differ both in the desalination process used and in the way that it converts wave into mechanical energy.

The Delbuoy

The first device that used wave energy directly for the production of fresh water from seawater was the Delbuoy, a system studied and developed at the University of Delaware, USA. The concept introduced by Pleass (1974) is described in detail in Hicks et al. (1989), where the system's design, operating principle and sea trials results are presented. The main motivation was the same as today's main goal: such systems are useful to remote areas, with unreliable or even insufficient power sources or with expensive and polluting ones. Typically the optimal candidates to such technology are islands or arid costal regions exposed to large oceans.

The Delbuoy concept involves a buoy, which is subjected to the waves that pass by it, a linear pump and an anchoring system that interacts with single-pass reverse osmosis membranes to produce fresh water (Fig. 6.60). The basic operating principle relies on heave for the pumping motion, which produces the pressure required by the reserve osmosis module, as showed in Fig. 6.60.

Following nine years of research and development activities at the University of Delaware the system underwent sea trials, conducted by the Department of Marine Science of the University of Puerto Rico, in 1982. A total of seven full-scale prototypes (1 *m* radius buoy) were installed. The final configuration included six devices and the first commercial installations took place in 1989, in the Caribbean (Puerto Rico and Belize). This location could guarantee a device output from 1100 to 1900 litres of fresh water per day (Hicks, 2004).

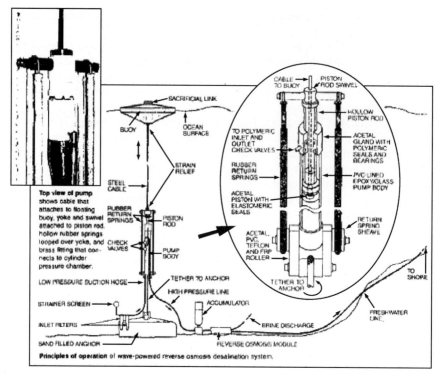

Fig. 6.60. The Delbuoy system (Hicks et al., 1989)

It is relevant to point out that the designers had in mind the fundamental issue of survivability. From Figs. 6.60 and 6.61 it is clear that most of the equipment was kept submerged, ensuring that under extreme conditions the system would not be compromised. A sacrificial linkage was therefore on the engineering list, and enabled the loss of the floater if the survivability of the remaining (expensive) components at the sea bed was at risk. It is possible to say that the solution was successful, as the system withstood two tropical storms at the test site.

An economic study of the system was also conducted, comparing the cost of the water produced with that from a conventional reverse osmosis scheme. It is legitimate to wonder why this system did not achieve a full commercial status. Dr. Hicks tries to answer some questions, include this one, in a communication to the Horizon International website (Hicks, 2004), detailing five reasons for this setback, which include "the loss of all of the equipment and infrastructure that was put in place to begin full commercialisation of the Delbuoy in St. Croix when hurricane Hugo devastated the island and the premature death of Dr. Pless, the inventor of the technology". Although its future application is still an unknown, its pioneer character will remain untouched.

a)   b)

                                        c)

Fig. 6.61. a) Desalination unit preparation on dock; b) Lowering the Delbuoy; c) Desalination unit being checked (Hicks, 2004)

The Vizhinjam OWC plant

In Sharmila et al. (2004) details are given about the commissioning of a rated 10,000 *l/day* reverse osmosis desalination plant coupled with the Vizhinjam demonstration Oscillating Water Column (OWC) plant in Kerala, India. One distinctive characteristic is that electricity produced from the waves is used for the desalination module, not the actual action of the waves that drives the desalination process. Thus this is an indirect process, similar to the schemes that involve other renewable energy sources, with one main advantage when compared to ones that use these other resources, which is the distance to the seawater. But the option is clearly not as appealing as the complete approaches offered by the Delbuoy and the Edinburgh Duck (see pp. 270), that are stand alone desalination units, independent from the equipment needed for the conversion of wave to electrical energy. The concept is generically described in Fig. 6.63, where it can be seen the clear distinctive nature between the desalination and electricity production stages, with all the advantages and disadvantages involved (e.g.: additional losses are introduced in the energy conversion chain but the use of batteries allows a steady input to the reverse osmosis module). The system underwent a series of tests and numerical modelling, and operates in the benefit of the local community. The same technological solution could be extended to the existing OWC plants spread throughout the world, or indeed any other wave energy converter that produces electricity.

Fig. 6.62. Wave energy plant at Vizhinjam (Sharmila et al., 2004)

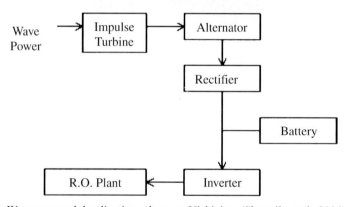

Fig. 6.63. Wave powered desalination scheme at Vizhinjam (Sharmila et al., 2004)

The Edinburgh Duck (Desalination Version)

Another concept based on the use of wave power for desalination was introduced by Prof. Stephen Salter in 1985, following the work on the wave energy converter known as the Edinburgh Duck (Salter, 1985). This paper was recently extended (Salter, 2005), and a detailed description of the device is presented in this section. The motivation for the development of such a system is similar to the one of the Delbuoy: the direct use of wave energy to a run a desalination process in order to avoid the losses of all the intermediate electrical steps that are otherwise necessary. Desalination emerged within the Wave Power Group of the University of Edinburgh as a way of overturning the 1980's UK Department of Energy demand of a single $2\,GW$ wave energy plant, offering an alternative end product and directing the research to smaller units, an approach that is now proven to be more realistic.

Early research lead to a PhD thesis (Crerar, 1990) linked with mathematical and experimental modelling, which allowed measurements of pressures, temperatures and condensation/evaporation rates on a compressor system which simulated the inside of a desalination duck (Crerar and Pritchard, 1991).

The desalination duck uses the vapour-compression principle to extract the salt from sea water, a thermal method. This uses the partial evaporation of the feed falling downwards on a sheet of heat transfer surface due to the reduction of pressure induced by the action of a pump. Evaporation cools the surface while compression warms the vapour. The pressurised output is then fed to the other side of the cooled surface, which allows condensation to occur and recycle latent heat for further evaporation. Instead of the high-speed turbo-compressor used in most vapour-compression desalination systems, the pumping action is provided by the water motion in a partially filled (with water) duck body rather than with high speed compressors. The pressure across the vertical dividing wall of the duck will be proportional to the angular velocity and so gives the ideal linear damping. The inner water is not only an inertial referential but also a double-acting piston (see the 'Compression' and 'Suction' chambers in Fig. 6.64). This is the major conceptual difference between such a system and the common vapour-compression one. Similarly to the electricity production version, it is manly the pitching (also referred to as nodding) motion of the duck about an axis that will produce useful work. A new series of hydrodynamic modelling, both numerical and experimental, has been recently conducted at the University of Edinburgh, for both versions of the duck. One of the key results lead to the change of shape of the concept, loosing the front beak and moving the rotation axis away from the cylinder's own axis (Cruz and Salter, 2006).

Other relevant components of this particular device include the loop pumps and the heat exchanger. The first ones are responsible for both the pumping of all the working fluids and of the product ashore, being based on the principle that pressure is induced in a flow around a coiled pipe subjected to alternating angular accelerations in the same way as a column of fluid in a vertical tube induces pressure at the bottom. The use of several loops ensures the required capacity, as each 360 degree loop is expected to produce a pressure of $30\,kPa$.

The basic design motivation of this specific heat exchanger, developed by Maxwell Davidson also from the University of Edinburgh, is to ensure very large heat transfer areas at a cost similar to the one of the building material. A description of the heat exchanger is taken from Salter (2005): 'The Maxwell Davidson heat exchanger consists of sheets of the heat transfer material into which have been pressed corrugations that lie at 45 degrees to the sheet edge. The press tool is used twice to make two rectangles of corrugations separated by an area of plain sheet. Other holes can be punched out for pipe work and clamping rods. When a sheet is folded into a U shape the corrugations will contact one another and make the form of an X. A large number of U-folds are clamped together with short spacing tubes which can form feed sprayers and condensate collectors. The open ends are sealed by the inflation of an elastomeric bag to a pressure higher than the one chosen for the process. Sea water is fed to the gaps outside the U-folds. It then moves as a falling film down the sheets and falls through the bottom from where it is collected for recirculation. Vapour is drawn off and pumped through a demister stack to the inner surfaces of the U-folds where it condenses and collects at the bottom and is drained. The falling film is ideal for good heat transfer and there is lots of demisting taking place in the heat-exchanger itself. This design allows the construction of heat exchangers with very large areas, many thousands of square metres, at a cost not much greater than that of the raw material.

The concept has been built in small scale but not as a stand-alone desalination unit (only the hydrodynamic behaviour has been fully characterised). A comprehensive mathematical model has been developed, predicting an output linear with regard to the significant wave height. A $12\,m$ by $24\,m$ module would produce between 1,000 and 2,000 m^3/day in a moderate wave climates (10 to 20 kW/m).

Other proposed wave-powered desalination technologies

Many other wave-powered desalination technologies have been proposed, with details on mode of operation for many very sketchy. Of the technologies described in sufficient detail to be analysed, they all propose the use of reverse osmosis for desalination. In this respect they are all essentially copies or modifications of the Delbuoy concept; however two ideas proposed are worthy of further discussion.

Waterhammer has been proposed as a method of providing the pressure required for reverse osmosis, utilising the hydro-ram (Maratos, 2003). The hydro-ram converts the energy of a large volume of fluid at low pressure into a small volume of fluid at high pressure. It is thus proposed that low head wave energy converters such as the Tapchan and WaveDragon could be used to supply large volumes of low pressure water. The conversion efficiency of the hydro-ram is claimed to be 90 %, though little data is provided to substantiate this important data. The pressure pulses that are inherent in a device that utilises waterhammer may also be problematic for RO membranes, though it is likely that an appropriately sized pressure accumulator would be able to eliminate excessive pressure fluctuations.

The Wavemill wave energy converter uses the same reverse osmosis desalination technology as Delbouy, but is supplemented by an energy recovery system that recycles the energy in the brine stream, which was simply discarded in the Delbuoy system. The energy recovery technology proposed is the Clark pump (Thomson et al., 2002), which converts the pressure energy in the brine stream directly into the feed stream. The extensive use of energy recovery technology in reverse osmosis desalination is a recent development, which has only become standard since the mid-1990's. However, because of the large reduction in specific energy consumption it is likely that any wave-powered desalination technology that utilises reverse osmosis will also use some form of energy recovery technology.

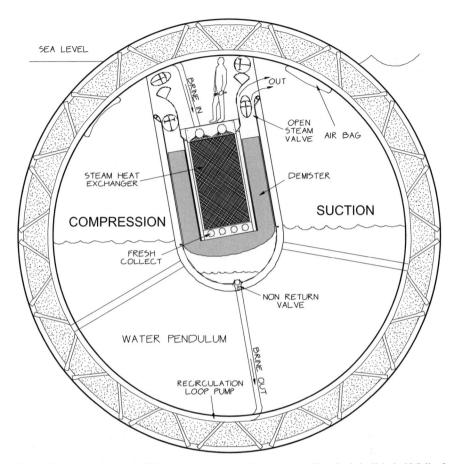

Fig. 6.64. A cross-section of the vapour-compression system. The duck hull is half full of water which acts like a double acting steam pump of enormous displacement but needing no accurate machined parts or sliding fits. All internal circulation and the pumping to shore performed using loop pumps (Cruz and Salter, 2006)

6.4.4 Challenges in Wave-Powered Desalination

The majority of the apparent challenges associated with wave-powered desalination are related to the variability of the power generated by a wave energy converter. Nevertheless, desalination plants have traditionally been designed to operate with a constant power supply; their ability to operate with a variable supply of power will differ, dependent on their basic operating principles and design.

One effect of a variable power supply is that at times the desalination plant will operating away from the optimal operating conditions that minimise the specific energy consumption of the desalination process. Ideally, the desalination plant would have a wide operating bandwidth, so that the specific energy consumption remains low for a wide range of input powers. Additionally, some form of energy storage, either inherent to the desalination process, or as a supplementary component, can help to maintain low specific energy consumption during short-term variations in the power supply. Configuring the desalination plant in modules that can be independently switched on can also help to match the power supply with the power demanded by the desalination plant.

In addition to maintaining low specific energy consumption, acceptable operating conditions for the desalination plant need to be sustained during the power variations to avoid damage to the desalination plant. For example, reverse osmosis membranes must operate between a maximum and minimum flow rate and below a maximum pressure. Manufacturers of reverse osmosis membranes also typically demand that their membranes experience a minimal fluctuation in pressure to avoid membrane fatigue. It is currently unclear whether this is a genuine problem or merely caution on the part of the manufacturers (Thomson and Infield, 2005). A particular operational scenario that must be considered is for periods of zero production, when the sea is calm, which though rare could last a number of days. Unless appropriate remedial measures can be taken, periods of zero production can result in irreversible fouling of reverse osmosis membranes and encrustation of heat exchangers.

If the only source of energy for the desalination plant is wave energy, then the variation in power means that at times the plant will be running at part-load, producing less water than its rated capacity. The economic effect of running at part-load is that the capital investment cost per unit of water produced is higher. This is no different from the economics when electricity is produced, except that the costs of desalination plants are typically much higher than electrical generator plants. A reverse osmosis plant, or vapour-compression plant typically costs about £1,000 per m^3/day of installed capacity. If the specific energy consumption of the desalination plant is $2-6\,kWh/m^3$, then this is equivalent to £4,000–12,000 per kilowatt of rated power. This is ten-times greater than the specific cost of an electrical generation plant, which is typically estimated to be £500–1,000 per kilowatt. If the costs of the desalination plant when coupled with wave energy are unchanged this implies that the wave energy converter used for desalination would have to be de-rated in comparison to its electricity-generating sister. Alternatively, the economics of wave-powered desalination can be made more attractive if the cost of a wave-powered desalination plant can be reduced with certain plant components becoming redundant in a wave-powered desalinator and/or exploiting inherent characteristics of the wave energy converter that can provide necessary functions of the desalination plant.

Additional challenges for offshore (not shoreline) wave-powered desalinators include: the pre-treatment of the feed water, the maintenance of the desalination plant in a location that is difficult to access and the provision of a pipeline to carry the fresh water to shore, which is likely to be much more expensive than the equivalent electrical cable.

6.4.5 Prospects for Wave-Powered Desalination

In considering the prospects for wave-powered desalination only concepts that utilise the wave energy directly will be considered, i.e. wave energy converters that generate electricity to be used for desalination are not considered. Whilst wave-powered desalination via the generation of electricity is a valid concept it differs little from the generation of electricity for feeding to an electrical network grid and can best be considered with reference to the other sections in this chapter which deal explicitly with the generation of electricity. An attractive prospect for wave-powered desalinators that utilise the wave energy directly is that they are likely to have a higher wave-to-water conversion efficiency than concepts that involve the generation of electricity due to a smaller number of energy transforms. For similar reasons they are also likely have less components and therefore have the potential to be cheaper and more reliable.

Wave-powered desalinators can be configured in two basic ways: autonomous and hybrid. A hybrid configuration means that an additional source of power is available to run the desalination plant, whilst autonomous means that wave-power is the sole source of energy used. The appropriateness of each configuration will depend on the circumstances where the desalinator is installed, together with the characteristics of the desalinator itself. By coupling a wave energy converter with an additional power source many of the problems associated with the variable supply of energy can be avoided (assuming that the alternative source of energy is controllable). On the other hand, hybrid designs are likely to add complexity and cost so may not be universally suitable.

Due to the mechanical nature of wave energy, the only desalination technologies that appear to be suitable for exploitation are mechanical vapour-compression and reverse osmosis. The prospects for utilising these two processes are discussed in the final two subsections.

Utilisation of the vapour-compression cycle

Vapour-compression is a desalination technique that can have some advantages over the other options. Salter (2005) presents a review on the principles of vapour-compression: seawater close to boiling point is dropped down one side of a hot heat-transfer surface and the pressure is reduced so that some is converted to vapour by drawing the necessary latent heat from the heat transfer surface. In the mechanical version of the technique (MVC; note that a pure thermal approach is also possible) a mechanical pump increases the pressure of the vapour by about 0.02 *MPa*, thereby increasing its temperature by a few degrees, and delivers it to the other side of the heat-transfer surface, which will of course have just been cooled by the loss of latent heat. The pure vapour condenses and releases its latent heat, which is needed for evaporation of more seawater. While the volumes of vapour are very large, the pressure drop is much smaller than for reverse osmosis. But more importantly the drop is from the product to the feed so that any leaks will lose small amounts of product rather than contaminating it. With an efficient demister to remove the small drops of water which are thrown from the boiling surface it is possible to produce an extremely pure output, better than 0.5 parts per million. As the output will have been taken to a high temperature for several minutes there is a further means for sterilisation. Pharmaceutically pure water has been produced from hospital sewage and the method has been used to dry the mash residues from whisky manufacture. The only intractable source of polluted output will be volatiles in sea water, which have boiling points close to the operating temperature.

The continuous recycling of the latent heat can make the process very efficient. The energy needed depends on the heat exchanger area and heat-transfer coefficients. The value is given by the product of the latent heat of steam (2.256 *MJ/kg* at 100°C) times the temperature difference across the heat exchanger divided by the absolute temperature of operation. A part of the temperature difference is the result of the boiling point of salty water being above the condensing temperature of pure water. For 3.5% *NaCl* sea water the elevation is 0.46°C but as soon as some has evaporated the strength of the remainder will rise, perhaps by a factor of two. The remainder of the temperature drop depends on the heat exchanger and, for economic reasons, is likely to be larger. With a perfect heat exchanger and 3.5% *NaCl* feed operating at 373.3 *K* the energy requirement would be only 2.78 *kJ/kg* (0.77 *kWh/m³*). For a more realistic brine strength the requirement would perhaps double to 5.56 *kJ/kg* (1.55 *kWh/m³*) for a very large heat transfer surface working at a very low rate. For practical throughputs the energy requirement would range from 9–35 *kJ/kg* (2.5–10 *kWh/m³*).

All hot desalination methods are bedevilled by the problem that some of the many materials dissolved in seawater are close to their limit of solubility and that this solubility falls with rising temperature. They will therefore come out of solution and form a hard scale on any suitable substrate. This will grow in thickness and will rapidly reduce the heat transfer coefficients, halving them in a few days of operation. A series of heat transfer materials with steadily improving transfer coefficients and resistance to fouling have been developed by Maxwell Davidson at the University of Edinburgh. The present choice is a metal mesh which has been sprayed with a continuous layer of polyvynilidene fluoride, usually known as PVDF, filled with flakes of carbon to improve heat transfer. The PVDF layer has excellent high temperature properties and resists fouling in the same way as a non-stick Teflon-coated frying pan. It is also used for the fibres of some reverse osmosis modules.

This text refers clearly to the application of a mechanical vapour-compression technique to a wave energy converter. Such coupling is only possible with devices that have sufficient inner volume. Furthermore, such thermal energy storage can help to overcome the problem of the random nature of the energy resource. In the example of the duck, all the pumping can be associating with its nodding motion in response to the incoming waves, making it particularly suited for vapour-compression. The duck remains the most studied concept when it comes to directly link wave energy conversion and this desalination method.

Utilisation of reverse osmosis membranes

The dynamics of many wave energy converters is such that power is extracted by applying a large force, opposing the movement of a slowly moving body; this is ideally suited to hydraulics. By using sea-water as the working hydraulic fluid the feed water for the desalination plant it can be pressurised directly using wave energy, minimising the required equipment and maximising the potential system efficiency.

Reverse osmosis (RO) membranes for the desalination of sea-water have progressed dramatically in the last decade, with them becoming significantly more durable, which means that they can more easily be integrated into a wave-powered desalination plant that operates with a variable feed pressure and flow rate through the membrane. Although manufacturers continue to specify relatively tight constraints on the desirable operating conditions for the RO membranes, this does not always appear to be necessary. In particular, fatigue of the RO membranes is often quoted anecdotally as a reason why RO cannot be used with a variable energy source; however recently results from a number of projects have cast doubt on this restriction (Paulsen and Hensel, 2005; Thomson and Infield, 2005). Undoubtedly, operating RO membranes with a "correct" and constant pressure and flow rate will maximise their longevity; but as the cost of RO membranes has reduced and their durability increased this has become less of a concern. RO membranes continue to improve and if a market existed for membranes that must operate in variable conditions, i.e. wave-powered desalination, then there is no clear reason to except that suitable RO membranes would not be developed.

With the use of energy recovery systems, which recycle the energy in the pressurised brine flow exiting the RO membranes, it has been possible to reduce the specific energy consumption to less than $7\,kJ/kg$ ($2\,kWh/m^3$). As with the RO membranes the majority of the energy recovery technologies have operating constraints that may not be suitable for operation with a variable supply of energy, without an auxiliary energy supply, or where the water is pressurised directly without the use of a rotary-dynamic pump. Consequently, the energy recovery technology used needs to be considered with respect to the characteristics of the wave energy converter and current technology modified where appropriate to provide the characteristics required. Although current energy recovery systems may not be suitable for use in wave-powered desalination, there appears to be no fundamental reason why they cannot be adapted to achieve the low specific energy consumption that is currently achieved by other RO plants.

The efficient recycling of the energy in the brine flow from the RO membranes means that it becomes economic to have a low recovery ratio, which reduces or eliminates the need for chemical pre-treatment of the feed water. This clearly reduces problems for RO plants integrated in a wave energy converter that is not located on the shoreline; increasing the prospects for wave-powered desalination.

If the wave energy can be coupled to an additional, controllable, energy source in a hybrid solution then the desirable operating conditions for the RO membranes and energy recovery technology can more easily be obtained. If the additional energy source is an electrical network then it may be possible to configure the hybrid solution so that depending on the sea-state and/or electrical network conditions, the power can either be drawn from the electrical network or fed into it. This flexibility would enable the best economic use of the wave energy to be achieved, although this system requires additional equipment, which may make it less attractive than an autonomous system.

Finally, although improvements in sea-water hydraulics, the RO membranes and modifications to the energy recovery systems are required to exploit fully the potential for wave-powered desalination by reverse osmosis, no fundamental problems appear to exist. Moreover, that many wave energy converters are ideally suited to the utilisation of a hydraulic power-take-off, the reduction in equipment and potential improvements in efficiency mean that the prospects for wave-powered desalination by reverse osmosis are very good.

6.5 Discussion

In this chapter the main alternatives with regard to the power take-off systems that can be implemented in a wave energy converter have been discussed. As Chapter 7 will illustrate, such alternatives can be at the core of many different technological solutions.

In section 6.1 the use of air turbines was addressed. This is directly associated with the oscillating water column principle, and corresponds to the most studied alternative until recently. Nevertheless the amount of study should not be confused with the suitability of the technology to the resource; in the air turbine case this is mostly due to the experience of academics in the aerodynamics field. The Wells and Impulse turbines were presented in detail as the best and most commonly used examples.

In 6.2 the use of directly driven linear generators was discussed. The idea of direct drive is appealing since it enables simple systems with few intermediate conversion steps and reduced mechanical complexity. The requirements for the generator have additional challenges, mainly related to the nature of the energy resource. Connection to the grid is also an issue that needs to be addressed at an early design stage. Direct drive in wave energy conversion is similar in principal to direct drive in wind energy, where the moving part of the generator is directly coupled to the energy absorbing part.

Section 6.3 was dedicated to hydraulic power take-off systems. Such systems have been envisaged for several wave energy converters, namely the Edinburgh duck and the Pelamis. Hydraulics ideally suits all concepts based on the principle of converting energy by applying a large force opposing the movement of a slowly moving body, and so seem immediately appealing for wave energy conversion. A review on the pros and cons of hydraulics and on the major components such as cylinders, valves and pumps, was conducted. Examples from the Wave Power Project at the University of Edinburgh and several spin-off companies were also given. One of the latter, Artemis Intelligent Power Ltd, is now pursuing a technology patented with the name Digital Displacement^{TM}. Such active-valve hydraulic machines could improve considerably the efficiency of a hydraulic power take-off system when operating at partial load.

The chapter was concluded by considering an alternative to electricity production: desalination. Examples of concepts that directly use wave energy to run a desalination process, either thermal of physical, were addressed in detail. Vapour-compression and reverse osmosis were identified as the ideal technologies that can be applied in wave energy converters. Although at an earlier stage than the electricity production versions, wave-powered desalination can prove to be a vital solution to tackle the growing issue of water scarcity worldwide, and positively contribute to the energy mix in remote areas.

References

References (6.1)

Abbott IH, Von Doenhoff AE (1959) Theory of Wing Sections, 2nd edn. Dover Publications
Alcorn RG, Beattie WC, Douglas R (1998) Transient performance modelling of a Wells turbine. Third European Wave Energy Conference, Patras, Greece, pp 80–87
Boake CB, Whittaker TJT, Folley M, Ellen H (2002) Overview and initial operational experience of the LIMPET wave energy plant. 12th International Offshore and Polar Engineering Conference, Kyushu, Japan, vol 1, p 586–594
Count B (1980) Power from Sea Waves. Academic Press, New York (ISBN 0-12-193550-7)
Curran R, Denniss T, Boake C (2000) Multidisciplinary Design for Performance: Ocean Wave Energy Conversion. Proc ISOPE'2000, Seattle, USA, pp 434–441 (ISSN 1098-6189)
Curran R, Gato LC (1997) The energy conversion performance of several types of Wells turbine designs. Proc Inst Mech Eng A J Pow 211(A2):55–62 (ISSN 0957-6509)
Curran R, Whittaker TJT, Raghunathan S, Beattie WC (1998) Performance Prediction of the Counterrotating Wells Turbine for Wave Energy Converters. ASCE J Energ Eng 124:35–53
Curran R (2002) Ocean Energy from Wave to Wire. In: Majumdar SK, Miller EW, Panah AI (eds) Renewable Energy: Trends and Prospects. The Pennsylvania Academy of Science, pp 86–121
Dhanasekaran TS, Govardhan M (2005) Computational analysis of performance and flow investigation on wells turbine for wave energy conversion. Renew Energ 30(14)2129–2147
Eves ARW (1986) The biplane Wells turbine. Master of Science Thesis, The Department of Aeronautical Engineering, The Queen's University of Belfast, UK

Falcão AF, Whittaker TJT, Lewis AW (1994) Joule 2, Preliminary Action: European Pilot Plant Study. European Commission Report, JOUR-CT912-0133, Science Research and Development-Joint Research Center

Finnigan T, Auld D (2003) Model Testing of a Variable-Pitch Aerodynamic Turbine. Proc 13th Int Offshore Mechanics and Arctic Engineering Conf, ISOPE, Vol 1, pp 357–360

Finnigan T, Alcorn R (2003) Numerical Simulation of a Variable Pitch Turbine with Speed Control. Proc 5th European Wave Energy Conf, Cork, pp 213–220

Folley M, Curran R, Boake C, Whittaker TJT (2002) Performance investigations of the LIMPET counter-rotating Wells turbine. Second Marine Renewable Energy Conference, Newcastle, UK

Folley M, Curran R, Whittaker TJT (2006) Comparison of LIMPET contra-rotating wells turbine with theoretical and model test predictions. Ocean Eng 33:1056–1069

Gato LMC, Falcão AF de O (1989) Aerodynamics of the Wells Turbine: Control by Swinging Rotor-Blades. Int J Mech Sci 31:425–434

Gato LMC, Henriques JCC (1994) Optimisation of Symmetrical Blades for Wells Turbine. EU Report for JOULE2-CT93-0333: Air Turbine Development and Assessment for Wave Power Plants

Horlock JH (1966) Axial Flow Turbines: Fluid Mechanics and Thermodynamics. Butterworths

Inoue M, Kaneko K, Setoguchi T (1987) The Fundamental Characteristics and Future of Wells Turbine for Wave Power Generator. Sci Mach 39(2):275–280

Jacobs E, Sherman A (1937) Aerofoil Section Characteristics as Affected by Variations of the Reynolds Number. National Advisory Committee for Aeronautics, Report No. 586

Justino PAP Falcão AF (1998) Rotational Speed Control of an OWC Wave Power Plant. Proc of Int Conf on Offshore Mechanics and Artic Engineering (OMAE), Lisbon, Portugal

X Kim TW, Kaneko K, Setoguchi T, Inoue M (1988) Aerodynamic performance of an impulse turbine with self-pitch-controlled guide vanes for wave power generator. Proceedings of 1st KSME-JSME Thermal and Fluid Eng Conf, Vol. 2, pp133–137

Maeda H, Santhakumar S, Setoguchi T, Takao M, Kinoue Y, Kaneko K (1999) Performance of an impulse turbine with guide vanes for wave power conversion. Renew Energ 17:533–547

Mamun M, Kinoue Y, Setoguchi T, Kim TH, Kaneko K, Inoue M (2004) Hysteretic flow characteristics of biplane Wells turbine. Ocean Eng 31(11–12):1423–1435

Mei CC (1976) Power extraction from water waves. J Ship Res 20(2):63–66

Raghunathan S, Tan CP, Ombaka OO (1985) The Performance of the Wells Self Rectifying Air Turbine. Aeronaut J, pp 369–379

Raghunathan S (1995) A Methodology for Wells Turbine Design for Wave Energy Conversion. J Pow Energ IMechE 209:221–232

Raghunathan S, Beattie WC (1996) Aerodynamic Performance of Counter-rotating Wells Turbine for Wave Energy Conversion. J Pow Energ 210:431–447

Salter SH (1988) World Progress in Wave Energy. Int J Ambient Energ 10:3–24

Sarmento AJNA, Gato LMC, Falco AF (1990) Turbine-Controlled Wave Energy Absorption by Oscillating Water Column Devices. Ocean Eng

Setoguchi T, Kaneko K, Taniyama H, Maeda H, Inoue M (1996) Impulse turbines with self-pitch-controlled guide vanes for wave power conversion: guide vanes connected by links. Int J Offshore Polar 6:76–80

Setoguchi T, Takao M, Kaneko K (1998) Hysteresis on Wells turbine characteristics in reciprocating flow. Int J Rotating Mach 4(1):17–24

Setoguchi T, Santhakumar S, Maeda H, Takao M, Kaneko K (2001) A review of impulse turbines for wave energy conversion. Renew Energ 23:261–292

Stewart T (1993) The influence of harbour geometry on the performance of OWC wave power converters. Ph.D. Thesis, The Department of Civil Engineering, The Queen's University of Belfast, UK

Thakker A, O'Dowd M, Slater S (1994) Computational Fluid Dynamics Study of Air Flow in a Wells Turbine and Oscillating Water Column Device. EU Report for JOULE2-CT93-0333: Air Turbine Development and Assessment for Wave Power Plants

Thakker A, Dhanasekaran TS (2003) Computed effects of tip clearance on performance of impulse turbine for wave energy conversion Renew Energ 29:529–547

Thakker A, Dhanasekaran TS (2005) Experimental and computational analysis on guide vane losses of impulse turbine for wave energy conversion. Renew Energ 30:1359–1372

Thakker A, Hourigan F (2005) Computational fluid dynamics analysis of a 0.6 m, 0.6 hub-to-tip ratio impulse turbine with fixed guide vanes. Renew Energ 30:1387–1399

Watterson JK, Raghunathan S (1997) Computed effects of tip clearance on Wells turbine performance. Proceedings of the 35th Aerospace Sciences Meeting and Exhibit, Reno, NV, Paper No: AIAA-1997-994

Wells AA (1976) Fluid Driven Rotary Transducer. British Patent Spec 1 595 700

Whittaker TJT, Thompson A, Curran R, Stewart T (1996) Operation of the Islay shoreline wave power plant as a marine test bed for turbine generators, project phase 5. Energy Technology Support Unit Report, ETSU Report No. V/02/0017/00/REP, Harwell, UK

Whittaker TJT, Beattie WC, Raghunathan S, Thompson A, Stewart T, Curran R (1997a) The Islay Wave Power Project: an Engineering Perspective. Inst Civil Eng Water Maritime Energ, pp 189–201

Whittaker TJT, Thompson A, Curran R, Stewart TP (1997b) European Wave Energy Pilot Plant on Islay (UK). European Commission, Directorate General XII, Science, Research and Development – Joint Research Centre, JOU-CT94-0267

Whittaker TJT, Beattie WC, Raghunathan S, Thompson A, Stewart T, Curran R (1997) The Islay wave power project: an engineering perspective. Inst Civil Eng Water Maritime Energ 124:189–201

References (6.2)

Baker NJ (2003) Linear generators for direct drive marine renewable energy converters. Ph.D. thesis, School of Engineering, University of Durham

Baker NJ, Mueller MA (2004) Permanent magnet air-cored tubular linear generator for marine energy converters. In: IEE Power Electronics and Electrical Machines & Drives Conference, Edinburgh

Boldea I, Nasar SA (1999) Linear electric actuators and generators. IEEE Trans Energ Convers 14(3):712–717

Chen Z, Spooner E, Norris WT, Williamson AC (1998) Capacitor-assisted excitation of permanent-magnet generators. IEE Proc Electric Pow Applic 145(6):497–508

Danielsson O (2006) Wave Energy Conversion – Linear Synchronous Permanent Magnet Generator. Ph.D. thesis, Acta Universitatis Upsaliensis Uppsala

Danielsson O, Eriksson M, Leijon M (2006) Study of a longitudinal flux permanent magnet linear generator for wave energy converters. Int J Energ Res 30(14):1130–1145

Danielsson O, Leijon M, Sjöstedt E (2006) Detailed Study of the Magnetic Circuit in a Longitudinal Flux Permanent-Magnet Synchronous Linear Generator. IEEE Trans Magnetics 41(9):2490–2495

Falnes J, Budal K (1987) Wave power conversion by point absorbers. Norwegian Maritime Res 6(4):2–11

Harris MR, Pajooman GH, Abu Sharkh SM (1997) The problem of power factor in vrpm (transverse-flux) machines. In: Eighth International Conference on Electrical Machines and Drives, no 440, pp. 386–390

Harris MR, Pajooman GH, Abu Sharkh SM (1997) Comparison of alternative topologies for vrpm (transverse-flux) electrical machines. In: Proceedings of the 1997 IEE Colloquium on New Topologies for Permanent Magnet Machines, Jun 18 1997, IEE Colloquium (Digest), pp 2–1

Leijon M, Bernhoff H, Agren O, Isberg J, Sundberg J, Berg M, Karlsson K, Wolfbrandt A (2005) Multiphysics simulation of wave energy to electric energy conversion by permanent magnet linear generator. IEEE Trans Energ Convers 20(1):219–224

Leijon M, Danielsson O, Eriksson M, Thorburn K, Bernhoff H, Isberg J, Sundberg J, Ivanova I, Ågren O, Karlsson KE, Wolfbrandt A (2006) An electrical approach to wave energy conversion. Renew Energ 31(9):1309–1319

McLean GW (1988) Review of recent progress in linear motors. IEE Proc B Electric Pow Applic 135:380–416

Mueller MA (2002) Electrical generators for direct drive wave energy converters. IEE Proc Generation Transmission Distribution 149(4):446–456

Neuenschwander VL (1985) Wave activated generator. US Patent (540602), 1985-09-03

Nilsson K, Danielsson O, Leijon M (2006) Electromagnetic forces in the air gap of a permanent magnet linear generator at no load. J Appl Phys 99(3):1–5

Polinder H, Damen MEC, Gardner F (2004) Linear pm generator system for wave energy conversion in the aws. IEEE Trans Energ Convers 19(3):583–589

Polinder H, Mecrow BC, Jack AG, Dickinson PG, Mueller MA (2005) Conventional and tfpm linear generators for direct-drive wave energy conversion. IEEE Trans Energ Convers 20(2):260–267

Prado MG, Gardner F, Damen M, Polinder H (2006) Modelling and test results of the Archimedes wave swing. Proc I Mech E Part A, J Pow Energ 220(8)855–868 (14)

Thorburn K (2006) Electric Energy Conversion Systems: Wave Energy and Hydropower. Ph.D. thesis, Acta Universitatis Upsaliensis Uppsala

Xiang J, Brooking PRM, Mueller MA (2002) Control Requirements of Direct Drive Wave Energy Coverters. Proceedings of IEEE TENCON'02

Waters R, Stålberg M, Danielsson O, Svensson O, Gustafsson S, Strömstedt E, Eriksson M, Sundberg J, Leijon M (2007) Experimental results from sea trials of an offshore wave energy system. Appl Phys Lett 90:034105

Weh H, Hoffmann H, Landrath J (1988) New permanent magnet excited synchronous machine with high efficiency at low speeds. In: International Conference on Electrical Machines, pp 35–40

References (6.3)

Chapple P (2002) Principles of hydraulic system design, Coxmoor Pubishing Company, Oxford (with the British Fluid Power Association)

Douglas JF, Gasiorek JM, Swaffield JA, Jack LB (2005) Fluid Mechanics, 5th edn. Pearson Prentice Hall, Harlow, UK, (ISBN 0-13-129293-5)

Edinburgh-SCOPA-Laing (1979) Report to the Department of Energy, UK

Ehsan MD, Rampen WHS, Taylor JRM (1995) Simulation and dynamic response of computer controlled digital hydraulic pump/motor system used in wave energy power conversion. 2nd European Wave Power Conference, Lisbon, November 1995, pp 305–311

Hägglunds (2006) Installation and maintenance manual. EN320-20h 2006, Hägglunds Drives AB, URL: http://www.hagglunds.com/Upload/20060809110542A_en320.pdf, accessed 1st July 2007

Henderson R (2006) Design, simulation, and testing of a novel hydraulic power take-off system for the Pelamis wave energy converter. Renew Energ 31:271–283

Hunt T, Vaughan N (eds) (1996) Hydraulic Handbook, 9th edn. Elsevier Science (ISBN 1856172503)

Hydraulics & Pneumatics (Monthly), trade journal of fluid power equipment and systems, published monthly by DFA Media Ltd, Tonbridge, free subscriptions are available to UK residents. URL: http://www.hydraulicspneumatics.com/

Ivantysyn J, Ivantysynova M (2000) Hydrostatic pumps and motors: principles, design, performance, modelling, analysis, control and testing. Akademia Books International, New Delhi (ISBN 1-85522-16-2)

Majumdar SR (2001) Oil hydraulic systems – principles and maintenance. Tata McGraw-Hill Publishing Company Limited, New Delhi (ISBN 0-07-463748-7)

Nebel P (1992) Maximizing the efficiency of wave energy plant using complex conjugate control. Proc Inst Mech Eng I J Systems Control Eng 206(14):225–236

Payne GS, Stein UBP, Ehsan M, Caldwell NJ, Rampen WHS (2005) Potential of digital displacement hydraulics for wave energy conversion. Proc 6th European Wave and Tidal Energy Conference, Glasgow, UK, 29th August – 2nd September, pp 365–371 (ISBN 0-947649-425)

Rampen WHS, Almond JP, Taylor JRM, Ehsan MD, Salter SH (1995) Progress on the development of the wedding-cake digital hydraulic pump/motor. 2nd European Wave Power Conference, Lisbon, November 1995, pp 289–296

Salter SH (1974) Wave power. Nature 249(5459):720–724

Salter SH (1980) Recent progress on ducks. IEE Proc A 127(5)

Salter SH (1982) The use of gyros as a reference frame in wave energy converters. Proc. second international symposium on wave energy utilization, Trondheim, 22–24 June

Salter SH (1993) Changes to the 1981 reference design of spine based ducks. Report to the UK Department of Trade and Industry, June 1992, reprinted as Renewable Energy Clean Power 2000, IEE Conference, 17–19 November, pp 121–130, IEE, London

Salter SH, Clerk RC, Rea M (1988) Evolution of the Clerk tri-link hydraulic machine. Proc 8th International Symposium on Fluid Power, Birmingham, UK, 19–21st April, Elsevier, Barking, pp 611–632 (ISBN 1-85166-201-4)

Salter SH, Rampen W (1993) The wedding cake multi-eccentric radial piston hydraulic machine with direct computer control of displacement. BHR Group 10th International Conference on Fluid Power, Brugge, Belgium, 5–7th April, Mechanical Engineering Publications, London, pp 47–64 (ISBN 0-85298-869-9)

Salter SH, Taylor JRM, Caldwell NJ (2002) Power conversion mechanisms for wave energy. Proc Inst Mech Eng M 216:1–27

Yemm R, Henderson R, Taylor C (2000) The PWP Pelamis WEC: Current status and onward programme. Proc 4th European Wave Energy Conference, Alborg Denmark

References (6.4)

Abdul-Fattah AF (1986) Selection of solar desalination system for supply of water in remote zones. Desalination 60(2):165–189

Al Suleimani Z, Rajendran Nair V (2000) Desalination by solar-powered reverse osmosis in a remote area of the Sultanate of Oman. Appl Energy 65:(1–4)367–380

Andrews W, Laker D (2001) A twelve-year history of large scale application of work-exchanger energy recovery technology. Desalination 138:201–206

Barlow M, Clark T (2002) Blue Gold. New Press, New York, USA

Belessiotis V, Delyannis E (2000) The history of renewable energies for water desalination. Desalination 128(2):147–159

Bouguecha S, Hamrouni B, Dhahbi M (2005) Small scale desalination pilots powered by renewable energy sources: case studies. Desalination 183:151–165

Clément A, McCullen P, Falcão A, Fiorentino A, Gardner F, Hammarlund K, Lemonis G, Lewis A, Nielsen K, Petroncini S, Pontes M, Schild P, Sjostrom B-O, Sorensen H, Thorpe T (2002) Wave energy in Europe: current status and perspectives. Renew Sust Energ Rev 6(5):405–431

Crerar AJ, Low RE, Pritchard CL (1987) Wave powered desalination. Desalination 67.127–137

Crerar AJ (1990) Wave powered desalination. PhD Thesis, The University of Edinburgh

Crerar AJ, Pritchard CL (1991) Wavepowered desalination: Experimental and mathematical modelling. Desalination 81(1–3):391–398

Cruz JMBP, Sarmento AJNA (2004) Wave Energy: Introduction to the Technological, Economical and Environmental Issues (in Portuguese). Portuguese Ministry for the Environment (ISBN: 972-8577-11-7)

Cruz JMBP, Salter SH (2006) Numerical and Experimental Modelling of a Modified Version of the Edinburgh Duck Wave Energy Device. Proc IMechE Part M. J Eng Maritime Environ 220(3):129–147

Davies PA (2006) Wave-powered desalination: resource assessment and review of technology. Desalination 186(1):97–109

Dempster WF (1999) Biosphere 2 engineering design. Ecol Eng 13(1–4):31–42

El-Dessouky H, Ettouney H (2000) MSF development may reduce desalination costs. Water Wastewater Int, pp 20–21

Einav R, Harussi K, Perry D (2003) The footprint of the desalination processes on the environment. Desalination 152(1–3):141–154

Engelman R, Cincotta RP, Dye B, Gardner-Outlaw T, Wisnewski J (2000) People in the Balance: Population and Natural Resources at the Turn of the Millennium. Population Action International, Washington D.C., USA

García-Rodríguez L (2002) Seawater desalination driven by renewable energies: a review. Desalination 143(2).103–113

García-Rodríguez L (2003) Renewable energy applications in desalination: state of the art. Solar Energy 75(5).381–393

Geisler P, Krumm W, Peters T (1999) Optimisation of the energy demand of reverse osmosis with a pressure-exchange system. Desalination 125:167–172

Harris C (1999) Energy recovery for membrane distillation. Desalination 125:173–180

Hicks DC, Pleass CM (1985) Physical Mathematical Modelling of a Point Absorber Wave-Energy Conversion System with Non-Linear Damping. In: Evans DV, Falcão de O AF (eds) Hydrodynamics of Ocean Wave-Energy Utilization. Springer-Verlag, Berlin

Hicks DC, Mitcheson GR, Pleass CM, Salevan JF (1989) Delbouy: Ocean wave-powered seawater reverse osmosis desalination systems. Desalination 73:81–94

Hicks DC (2004) Communication to the Horizon Solution Site (available online at http://www.solutions-site.org/artman/publish/article_60.shtml)

Höpner T, Windelberg J (1996) Elements of environmental impact studies on coastal desalination plants. Desalination 108:11–18

Husseiny AA, Hamester HL (1981) Engineering design of a 6000 m^3/day seawater hybrid RO-ED helio-desalting plant. Desalination 39:171–172

Isaacs JD, Seymour RJ (1973) The ocean as a power resource. Int J Environ Stud 4(3):201–205

Koklas P, Papathanassiou S (2006) Component sizing for an autonomous wind-driven de-salination plant. Renewable Energy 31(13):2122–2139

Pleass CM (1974) The use of Wave Powered Seawater Desalination Systems. Proc Int Symp Waves Energy. Canterbury

Maratos DF (2003) Technical feasibility of wavepower for seawater desalination using the hydro-ram (Hydram). Desalination 153.287–293

McCormick ME (1981) Ocean Wave Energy Conversion. John Wiley & Sons

Membrane Technology Newsdesk (2004) Desalination plant is wave-powered. Membrane Technol 2004(3):4

Miller JE (2003) Review of Water Resources and Desalination Technologies. Materials Chemistry Department, Sandia National Laboratories, Albuquerque, New Mexico

Paulsen K, Hensel F (2005) Introduction of a new Energy Recovery System—optimized for the combination with renewable energy. Desalination 184:211–215

Pontes MT (1998) Caracterização Energética das Ondas Marítimas e Estudo dos Problemas de Refracção no seu Aproveitamento. PhD Thesis (in Portuguese). Instituto Superior Técnico, Lisbon, Portugal

Pontes MT, Falcão AF (2001) Ocean Energies: Resources and Utilisation. Proc 18th World Energy Conf. Buenos Aires, Argentina (Paper 01-06-02)

Raluy RG, Serra L, Uche J, Valero A (2004) Life-cycle assessment of desalination tech-nologies integrated with energy production systems. Desalination 167:445–458

Salter SH (1985) Wave Powered Desalination. Proc 4th Conf Energy Rural Island Commu-nities. Pergamon Press, Inverness, UK

Salter SH (1989) World progress in wave energy – 1988. Int J Ambient Energy 10(1):3–24

Salter SH (2005) High Purity Desalination Using Wave-driven Vapour Compression. World Renew Energy Conf. Aberdeen, UK

Schiffler M (2004) Perspectives and challenges for desalination in the 21st century. Desali-nation 165:1–9

Sharmila N, Jalihal P, Swamy AK, Ravindran M (2004) Wave powered desalination sys-tem. Energy 9(11):1659–1672

Sommariva C, Hogg H, Callister K (2004) Environmental impact of seawater desalination: relations between improvement in efficiency and environmental impact. Desalination 167:439–444

Spiegler KS, El-Sayed YM (1994) A Desalination Primer. Balaban Desalination Publica-tions, Santa Maria Imbaro, Italy

Stover R (2004) Development of a fourth generation energy recovery device. A CTO's notebook. Desalination 165:313–321

Subiela V, Carta J, González J (2004) The SDAWES project: lessons learnt from an inno-vative project. Desalination 168:39–47

Thomson M, Miranda M, Infield D (2002) A small-scale seawater reverse-osmosis system with excellent energy efficiency over a wide operating range. Desalination 153:229–236

Thomson M, Infield D (2005) Laboratory demonstration of a photovoltaic-powered sea-water reverse-osmosis system without batteries. Desalination 183.105–111

7 Full-Scale WECs

In this chapter, some of the concepts that have reached the full-scale stage are described in detail. Other examples could be given, but the analysis was limited to four main technologies: OWC (Oscillating Water Column), AWS (Archimedes Wave Swing), Pelamis and Wave Dragon, which are presented in sections 7.1 to 7.4, respectively. Each illustrates one particular power conversion mechanism that was addressed in Chapter 6, namely air turbines, direct drive (linear generators) and hydraulics. A subsection regarding overtopping theory is also presented, as it was not addressed in the previous chapter and is linked with one of the technologies (Wave Dragon). Section 7.5 gives an account of the operational experience gathered by the several technology developers. Such detail provides valuable lessons to those interested in the field and also to a wide engineering audience. To conclude, section 7.6 gives a brief update on test centres, pilot zones, a review on the most relevant EU funded projects and a case study related to one of the technologies.

7.1 Oscillating Water Column

7.1.1 LIMPET

Tom Heath

Wavegen
Inverness
Scotland, UK

Since the inception of the UK Wave Energy Programme in 1974 the combination of an Oscillating Water Column (OWC) collector and a Wells turbine has been promoted as a reliable and cost effective combination for converting wave energy into useful power.

An OWC is formed by a chamber which is filled with air above the water line. Driven by wave action the water level inside the chamber rises and falls, alternately pressurising and rarefying the air within the chamber. As the water level inside the chamber rises pressurised air escapes from the chamber through a turbine-generator

Oscillating Water Column (OWC)   Wells Turbines

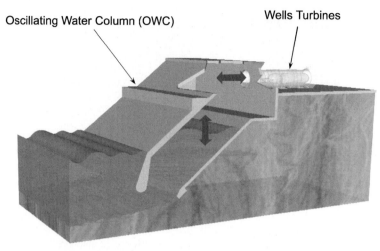

Fig. 7.1. Schematics of the LIMPET OWC

unit producing electrical power. As the water level in the chamber falls air is drawn back into the chamber through the turbine-generator assembly to continue power production. Wells turbines are self rectifying so that the direction of turbine rotation remains constant throughout the power cycle.

The technology has found commercial application in navigation buoys and numerous demonstration plants have been built across the globe in India, China, Norway, Japan, Portugal and in the UK, including Wavegen's LIMPET plant on Islay (N.E.: LIMPET is an acronym for Land Installed Marine Power Energy Transmitter). OWC breakwater units are under construction at Mutriku in Northern Spain and in Portugal at Porto. OWC's are not uniquely fitted with Wells turbines; a minority of developers favour impulse turbines whilst some systems with valve ducted flow feeding unidirectional turbines have also been proposed (see section 6.1). At present however the simplicity and flexibility of the Wells system make it the most attractive conversion option for OWC systems. The great virtues of the Wells turbine are its simplicity and effectiveness. The simplicity can we judged from Fig. 7.2 which shows, with the outer duct removed, the final stage of assembly of one of the pair of $250\,kW$ turbine rotor units originally fitted to the LIMPET plant. The seven blades of the $2.6\,m$ diameter unit are of a symmetrical airfoil section. The blades are bolted via a containment ring (which carries the centrifugal loads) to a plate which in turn fits directly on to the shaft of the generator (which is hidden under a cylindrical cover). The baseline configuration of the LIMPET plant used two of these assemblies, back to back, so that the combination formed a contra-rotating biplane turbine. Dependent upon the particular plant requirements other configurations may be used. For example an individual turbine-generator may be run as a monoplane turbine (with or without stator vanes) whilst a through shaft generator may be fitted with a turbine at either end to form a co-rotating bi-plane unit.

Fig. 7.2. Wells turbine used in the LIMPET plant

The principle of operation of the Wells turbine may be described with reference to Fig. 7.3. The plane of the rotating turbine is perpendicular to the air flow in the turbine duct. The forward motion of the blade combined with the perpendicular air flow means that relative to the blade there is an angle of attack of the air relative to the plane of the turbine blades. This angle of attack may vary

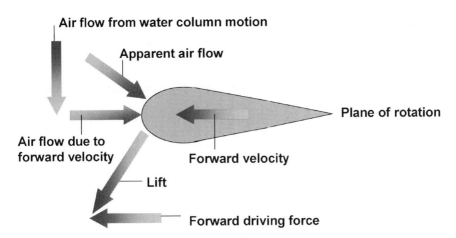

Fig. 7.3. Principle of operation of the Wells turbine

between zero (when the turbine is rotating but there is no airflow in the duct) to 90° (when there is airflow in the duct but the turbine is stationary). All airfoils generate lift at 90° to the angle of attack and, as is seen from Fig. 7.3, this lift has a component in the same direction as the rotation of the turbine. The great virtue of the Wells turbine is the fact that, irrespective of the direction in which the air is flowing, the turbine is driven in the same direction. In this regard the system is described as self-rectifying.

In practice there are drag forces on the turbine blades as well as lift and by comparison with some turbine forms the drag, by virtue of the large blade area, can be relatively high. This means that at small angles of attack (typically $<2°$) the component of lift in the direction of rotation is insufficient to overcome the drag force and the turbine will not drive. The same thing happens at high angles of attack (typically $>14°$) when the blade stalls and again the turbine will not drive. The blades are most effective at an angle of attack of around 7°.

Rather than describe the condition of the air flow on the turbine in terms of the angle of attack (θ) it is more usual to use a non-dimensional flow coefficient (Φ) which is defined as the tangent of the angle of attack at the blade tip. Thus

$\Phi = \tan(\theta) = \dfrac{V_x}{V_t} = \dfrac{V_x}{r\omega}$, where V_x is the axial flow velocity, V_t the turbine tip

speed, r the turbine radius and ω the rotational speed.

The detailed performance of the turbine changes with the form of the airfoil and the size of the turbine and peak efficiencies approaching 90 % are possible but the efficiency curve shown in Fig. 7.4 may be taken as representative for discussion purposes.

The peak efficiency shown is 75.5 %. As the water level in the OWC rises and falls in response to wave action the flow through the turbine will vary and hence the flow coefficient also changes. This means that the flow coefficient is constantly tracking up and down the efficiency curve with the effect that the whole cycle efficiency will be lower that the peak.

An estimate of the whole cycle efficiency can be made by noting that the instantaneous input pneumatic power is

$$P = cQ^2, \tag{7.1}$$

where Q is the instantaneous flow and $c = \dfrac{P}{Q}$ is the turbine damping (or pressure

drop per unit flow), where P is the instantaneous pressure. Similarly the instantaneous power converted by the turbine is $P_t = \eta(\Phi)cQ^2$ and integrating over a cycle the average efficiency is given by

$$\eta_{av} = \frac{\sum cQ^2}{\sum c\eta(\Phi)Q^2} = \frac{\sum Q^2}{\sum \eta(\Phi)Q^2}. \tag{7.2}$$

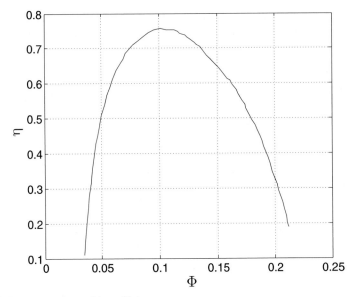

Fig. 7.4. Representative turbine efficiency curve

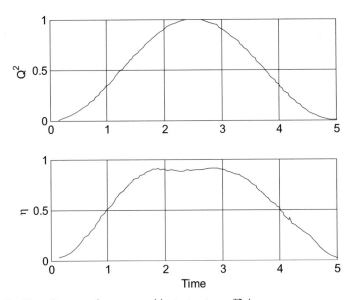

Fig. 7.5. Profiles of pneumatic power and instantaneous efficiency

If a sinusoidal flow profile is assumed and the turbine tip speed is optimised in relation to the assumed flow velocities then it is possible to develop the profiles of pneumatic power and instantaneous efficiency shown in Fig. 7.5. It is seen when there is significant pneumatic power the efficiency is close to the maximum and conversely when the efficiency is low there is little or no pneumatic power. For the example shown the average efficiency is 72% which is surprisingly close to the peak of 75.5%. This demonstrates that, in principle, the simple Wells turbine can deliver good conversion efficiencies even under random flow conditions.

The turbine-generator system does not work in isolation of the OWC capture chamber and the characteristics of the turbine and the influences of the power take-off damping and the motion of the water surface in the column are critical in determining the effectiveness of the primary power capture of the OWC.

As previously described in Chapter 6, the behaviour of the pneumatic power capture as a function of turbine damping for the LIMPET OWC was tested on a tank model for four different sea states. It became clear that there is an optimum damping which increases with the power in the sea. A turbine which over or under-damps the water column by a factor of two will cause the pneumatic power capture to fall by approximately 15%. The turbine design must therefore take full account of the influence of the interrelationship between primary power capture and turbine characteristics and it must also take account of the wide range of incident wave energies which will be encountered in a typical year. The power in the sea is not uniformly distributed. Figure 7.6 shows the distribution of pneumatic powers measured on the LIMPET plant between November 2001 and March 2003. It is seen that there are a large number of occurrences of low power and a small number of high power, or storm, events.

This is a typical power distribution with the skew increasing with the distance from the equator. In this data set approximately 70% of occurrences are below the annual average pneumatic power of $112 \, kW$. In designing the turbo-generation equipment a compromise must be found between the ability of the generating plant to extract the maximum power from storm events and the additional parasitic losses of larger plant generating at part load.

For the turbine to work most effectively it must simultaneously operate at the ideal flow coefficient whilst at the same time producing the right damping to maximise pneumatic power capture. To complicate the situation it must do this in all sea states. Fortunately the Wells turbine has the right characteristics to achieve this. Section 6.1 focus such characteristics in detail but a summary is presented here, along with an application exercise to the LIMPET case.

It can be shown that the damping of a Wells turbine (c) is proportional to turbine speed (ω) so that $c = k\omega$. Also: the axial flow velocity is $V_x = \dfrac{Q}{A}$ or $Q = V_x A$. Combining these relationships with Eq. (7.1) gives:

$$P_n = cQ^2 = k\omega (V_x A)^2 . \tag{7.3}$$

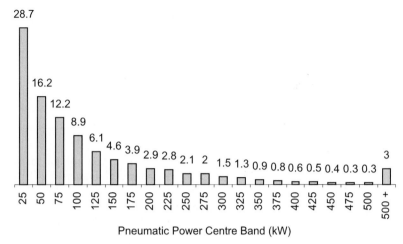

Fig. 7.6. Distribution of pneumatic power in percentage of applicability (Nov. 2001–Mar. 2003)

But if the objective is to maintain the flow coefficient in order to achieve peak turbine performance then $\dfrac{V_x}{V_t} = \Phi_{opt}$ or $V_x = \Phi_{opt}V_t = \Phi_{opt}r\omega$ (where r is the turbine tip radius). Substituting this in Eq. (7.3) gives:

$$P_n = k\omega\left(\Phi_{opt}r\omega A\right)^2 = k^*\omega^3,\tag{7.4}$$

where $k^* = k\Phi_{opt}^2r^2A^2$.

This means that for optimum turbine efficiency the turbine speed must vary in proportion to $\sqrt[3]{P_n}$. If the optimum damping measured for the LIMPET collector is compared to a plot of damping against $22P_n^{1/3}$ it can be concluded that there is a reasonable fit between the two, confirming that the Wells turbine has the required characteristics for good generation under a wide range of sea conditions. It is interesting that with inverter drive systems it is wholly practicable to vary the working speed of the turbine by a factor of 4 or more so that the optimum flow coefficient can be maintained for input powers varying over a $64:1$ range whilst keeping the collector damping close to ideal. This offers an excellent turn down ratio enabling generation from both the frequently occurring low sea states and the more intensive storm events.

Table 7.1 Optimum damping for different collectors (off resonance)

Type of Collector	Optimal Turbine Damping -15 kW/m input power ($N.s/m^5$)	Area (m^2)
4 m square breakwater section	600	16
7 m diameter Monopile	251	38.5
LIMPET	57	168

Theoretically it can be shown that in general terms the optimal damping of an OWC collector varies inversely with the water plane area. Table 7.1 shows values for optimum damping measured by Wavegen on three different collectors of widely different designs (but working off resonance). The data from Table 7.1 leads to a theoretical relationship of Optimal Turbine Damping $(N.s/m^5) = 9609 \times \text{Area}^{-1} \ (m^{-2})$.

Changing the input power will change the gradient of the line and OWC systems operating near resonance will not follow the same trend but an initial estimate of the optimum damping requirement can be based on the water plane area of the OWC.

Whilst the foregoing describes features of interest in OWC technology it makes no attempt to discuss the fine detail of the influence of detailed collector form or turbine design on the overall system performance. The technology is continuing to evolve and the overall performance improving with both experience and understanding.

The majority of practical experience with OWC systems has been obtained from shoreline systems where land access is possible but a number of floating OWC systems are also under development and will become more prevalent as the technology develops.

7.1.2 Pico – European Pilot Plant

António Sarmento, Frank Neumann, Ana Brito e Melo

Wave Energy Centre
Lisbon
Portugal

Until now only shoreline or nearshore bottom standing OWC plants have been developed, except the KAIMEI and Mighty Whale floating prototypes developed in Japan in the late eighties (JAMSTEC, 1998), a project without further continuation, and in Ireland with the OE Buoy project (a floating BBDB concept – Backwards Bend Duct Buoy): a 1 : 4 scale model underwent sea tests in the Galway Bay test site in January 2007. One of the pilot plants built so far has been the shoreline Pico plant in the Azores (Falcão, 2000). This is a $400\,kW$ rated plant, equipped with a $2.3\,m$ diameter Wells turbine coupled to an asynchronous generator. The turbine includes two fixed guide-vane stators, one at each side of the rotor. The generated electrical power is rectified and modulated with power electronics equipment (a hyper-synchronous cascade) before being fed into the electrical grid. The plant is also equipped with a pressure relief valve in parallel to the turbine as shown in Fig. 7.7. It also shows the sluice-gate isolation valve between the air chamber and the air duct and the fast reacting valve. The former is intended to be used whenever the plant is disconnected for a long time, whereas the latter is intended to be used to avoid turbine over-speed if in energetic seas an electrical grid fault occurs.

The monitoring equipment of the plant includes sensors to measure the:

- Rotational speed;
- Air pressure and water free-surface elevation in the air chamber;
- Static pressure both at the inner and outer covers of the air duct immediately upstream and downstream to the stators;
- Dynamic pressure at three radius averaged along three circumferential angles at the sections where the static pressure is measured;
- Vibrations and oil temperature at the turbo-generator bearings;
- Temperature, voltage and current at each of the three electrical circuit phases;
- Lubrication flow;
- Total, active and reactive power delivered to the grid;
- Cumulative active energy produced.

The plant was built with support from the European Commission through two Joule projects and a national funded project by EDP (Electricidade de Portugal) and EDA (Electricidade dos Açores), respectively the national and local utilities. The scientific responsibility of the European projects was of IST, Technical University of Lisbon, in particular of Prof. António Falcão. IST was also responsible for the conception and basic engineering of the plant and turbine aerodynamic design. Profabril was responsible for the civil engineering design and Marques Lda for the erection of the plant. The electrical, power electronic and monitoring equipment was supplied by EFACEC (Portugal) and ART (Scotland) supplied the mechanical equipment.

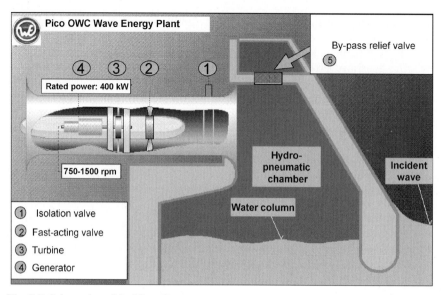

Fig. 7.7. Schematics of the Pico plant

Fig. 7.8. Back view of the Pico Plant

7.2 Archimedes Wave Swing (AWS)

Miguel Prado

Teamwork Technology BV
Zijdewind
The Netherlands

Among the main full-scale concepts, the Archimedes Wave Swing (AWS) is a u-
nique wave energy converter because it is completely submerged. This is extremely relevant for its design as it makes the system less vulnerable to storms.
Also, it is not visible, thus the public acceptance of a AWS wave farm is not as
problematic as, for example, a wind farm. It can be consider to be of the point-
absorber type (characteristic length, in this case the diameter, small when compared with the wavelength), and, as in the case of Pelamis (for example), it is envisaged to be deployed in arrays of devices unit rated in a few *MW*. This contribution aims to provide the historical background of the AWS and details regarding
its design, and it is based in two previous publications: Prado et al. (2006)[1] and
Cruz and Sarmento (2007)[2].

The AWS consists of an air filled chamber fixed to the seabed and open at the
top (the silo), closed by another cylinder (the floater). An air lock is created between the two cylinders and so water can not flood the silo. The floater can move
up (or down), due to the pressure increase (decrease) linked with the incoming
wave crest (trough) directly above the device. By adding a power take-off (PTO)
system this oscillation can be converted into electrical power. In the case of the
AWS, the PTO is a permanent-magnet linear generator (see 'Generator Design'
for more details). By tuning the system frequency to the mean wave frequency, the
stroke of the linear motion can be made larger than the wave height. In Fig. 7.9 the
device is moored to the seabed. For simplicity, the 2 *MW* pilot plant that was built
as a full-scale demonstration project was attached to a pontoon (Fig. 7.10). A brief
description of the evolution of the concept is therefore justified.

7.2.1 History

The inspiration for the AWS emerged in 1994 in a novel company called Teamwork Technology B.V, and one year later a 1:20 model was tested in regular
waves in a partnership that involved the Netherlands Energy Research Founda-
tion/Energy research Centre of the Netherlands (ECN) and WL Delft Hydraulics.

---

[1] Reprinted from the Journal of Power and Energy, Volume 220, 2006, Pages 855–868, with
permission from the Institution of Mechanical Engineers (IMechE).
[2] Reprinted from Ocean Engineering, Volume 34, Issues 5-6, 2007, Pages 763–775, with permission from Elsevier.

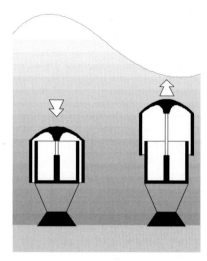

Fig. 7.9. Schematics of the AWS concept (Sarmento et al., 1998)

Fig. 7.10. The AWS at the Leixões harbour in 2004

Portugal was selected in 1995 as the location for the full scale-testing, with the north area being considered suitable not only from the energy resource point of view but also due to the proximity between the shore and the national electricity grid. The existence of several harbours and shipyards, which could be useful manly for maintenance purposes, was also seen as a vital aspect.

In the following years the target 2 *MW* prototype was assessed on the basis of overall system performance. This work was conducted by ECN, Instituto Superior Técnico (IST), WL Delft Hydraulics, Teamwork Technology and some industrial partners. A new set of 1 : 20 scale model tests were performed in the beginning of 1997, with irregular waves. By analysing the performance of the system and simulating it computationally, specifications for the regulation and transport of the electricity were made. By mid 1997 AWS B.V. was founded, joining NUON, ECN, Delft Hydraulics, Teamwork Technology and a private investor. Again a year later experimental tests, now on a 1 : 50 scale model, performed at the Hydraulics and Maritime Research Centre (HMRC) in Cork, Ireland, led to small changes on specific components like the power take-off (PTO) solution and confirmed the 2 *MW* rated capacity as the one to aim for the prototype to be installed offshore Portugal.

Fig. 7.11. Model tests (WL Delft Hydraulics)

Fig. 7.12. Model tests (HMRC Cork)

7.2.2 Design of the Full-scale Pilot Plant

By early 1999 the AWS design was ready, and components were ordered from several partners. The pilot plant was then assembled in Romania, with two main additions to the original concept: a pontoon used as support for the submergence operation and a vertical guidance structure to assist the floater's motion (see Fig. 7.10). Both these features are not present in the plans for future devices, and were intended to add additional control to operations with the prototype. In November 2001 the AWS was towed to the test site and the first submergence attempt was made offshore the Portuguese coast. Stability issues emerged and the operation was suspended. The AWS was towed back to the harbour of Leixões (near Porto). A second attempt one year later was also unsuccessful but the experience gathered proved valuable when the pilot plant was finally submerged in 2004 (for more details on the submergence operation see Prado and Gardner, 2005b). The pilot plant was then tested in a variety of sea states and operation conditions with the help of the several codes developed, including a time domain simulation of the whole system, by the team at the substation located at Aguçadoura beach, and by the team members at the support boat near the plant.

Table 7.2 Characteristic dimensions of the AWS pilot plant

Description	Length (m)	Diameter (m)	Height (m)	Width (m)	Weight (ton)
Floater	–	9.5	21	–	400
Pontoon	48	–	5.5	28	1200
Guidance Structure	–	–	33.5	–	120

Figure 7.10 shows the pilot plant while docked in May 2004 at Leixões. Both the guidance structure for the floater and the pontoon (with the four corner towers that would be filled with water during the submergence operation) are clear in the photo. Table 7.2 synthesises the main dimensions and weights of the AWS' components. For a detailed description see Prado et al. (2005a) and Rademakers et al. (1998).

The maximum peak power is 2 *MW*. The rated stroke is 7 *m* and the rated velocity is 2.2 *m/s*. A braking system had to be included in the device to damp the motion in case of generator's limitation or failure. This system consists on two cylinders sliding into each other, forcing the water trapped inside to flow through an orifice. By changing the area of the orifice the braking force can be adjusted to the desired level. During the tests, the area of the orifice was always kept at the minimum (water brakes were closed) and therefore the damping was quite high. The moving part of the device has a mass of around 400 *ton*, and the total mass of the device, including the pontoon, is approximately 7000 *ton* (from which around 5000 *ton* are just due to the sand ballast tanks). Most of the volume is reserved for the ballasting tanks necessary for the submerge/emerge operation. The air volume of the device at mid position is approximately 3000 m^3 and can be changed by pumping in/out water. The total volume of water that can be pumped is approx. 1500 m^3 and permits to tune the natural period of the device in the range of 7–13 *s*.

7.2.3 Linear Generator Design

To finalise the basic description of the AWS technology, some generic considerations about the design of the power take-off system are made. When designing the PTO for the AWS, the key driving forces were identified as the following:

- Maximum stroke: 7 *m*;
- Maximum speed: 2.2 *m/s*;
- Maximum force: 1 *MN*;

- Robustness;
- Maintenance as low as possible;
- Efficiency;
- Cost;

During the design process, the following choices were made in order to meet the requirements as good as possible.

1. Probably, a generator system consisting of a gearbox that converts the linear floater motion into rotating motion and a standard rotating generator would be a cheap and rather efficient solution. However, it appears to be extremely difficult to build one that is robust and maintenance-free. Therefore, a linear generator is used.
2. It is nearly impossible and extremely expensive to make the generator large enough to take all possible forces generated by waves. Therefore, the AWS also has water dampers that can absorb very high forces. This implies that the generator can be designed as a compromise between energy yield and cost.
3. The linear generator that converts the mechanical energy into electrical energy is a permanent-magnet (PM) generator because it has a rather high force density and efficiency at low speeds.
4. The magnets are on the translator that moves up and down, so that there is no electrical contact between the moving part (the translator) and the stator, which is important because such an electrical contact suffers from wear.
5. The generator is flat. Possibly, round generator constructions fit better in the construction of the AWS. However, for a single generator for a pilot plant, it was much cheaper to remain close to existing production technology and build a flat generator.
6. The number of slots per pole per phase is one. Increasing this number would lead to large pole pitches, resulting in thicker yokes and a higher risk of demagnetization. Decreasing the number of slots per pole per phase (using fractional pitch windings) would lead to additional eddy-current losses due to additional space harmonics.
7. The magnets are skewed to reduce cogging.
8. The translator with the magnets is only a few meters longer than the stator in order to reduce cost. This means that in the central position, the magnets of the translator are completely overlapping the stator so that maximum forces can be made, but in the extreme positions, the magnets only partly overlap the stator.
9. To balance the attractive forces between stator and translator, the generator is double sided, as depicted in Fig. 7.13.
10. For cooling the stator of the generator, a water cooling system was implemented.

11. The power electronic converter for the grid connection is placed on shore so that possible problems with the power electronics and the control could be easily solved. A 6 *km* long cable connects the generator terminals to the converter on shore.
12. A current source inverter on the shore is used for the utility grid connection. A voltage source inverter would have advantages of better control characteristics, better power factor, better generator efficiency and higher forces and energy yield (Leijon et al., 2005). However, it appeared to be easier and cheaper to use a readily available current source inverter.
13. Coatings are used to protect the generator against the aggressive environment.

Figure 7.13 depicts a cross section of the generator. When the translator with the magnets moves up and down, voltages are induced in the coils in the stator slots. Figure 7.14 shows a photograph of a part of the stator.

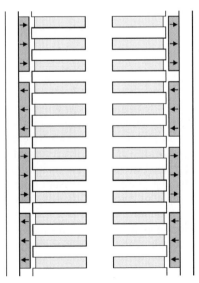

Fig. 7.13. Section of four pole pitches of the linear PM generator. The middle part is the stator with stator iron and coils in the stator slots in between. The left and right parts are the translators with the magnets with arrows indicating the magnetisation direction

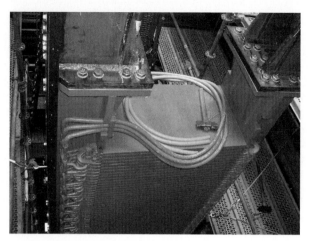

Fig. 7.14. Photograph of a stator in the AWS

7.3 Pelamis

Richard Yemm

Pelamis Wave Power Ltd
Edinburgh
Scotland, UK

The Pelamis wave energy converter (WEC) is a semi-submerged, articulated structure composed of cylindrical sections linked by hinged joints and is held on station by a compliant mooring system that allows the machine to weathervane to align itself head-on to incoming waves (it takes its 'reference' from spanning successive wave crests). As waves travel down the length of the machine they cause the structure to articulate around the joints. The induced motion of these joints is resisted by hydraulic rams that pump high-pressure oil through hydraulic motors via smoothing accumulators. The hydraulic motors drive electrical generators to produce electricity. Power from all the joints is fed down a single umbilical cable to a junction on the seabed. A number of devices can be connected together and linked to shore through a single seabed cable. Some key features should be highlighted:

1. A Pelamis wave farm can be installed in a range of offshore water depths and sea bed conditions to allow site developers flexibility in selecting installation sites.

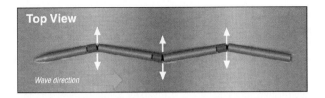

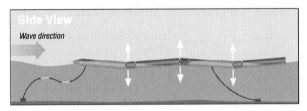

Fig. 7.15. Schematics of the Pelamis WEC (top and side views)

2. The Pelamis is constructed, assembled and commissioned off-site in safe conditions on land or in sheltered dock facilities, therefore requiring an absolute minimum of installation work required on-site. This avoids more expensive extended offshore construction activities, which are subject to longer weather delays.
3. The Pelamis has a rapid attachment / detachment electrical and moorings connection, which also allows the machine to be quickly retrieved and taken off-site to a safe quayside facility for any maintenance requirements, thus avoiding costly offshore operations with specialist equipment and vessels.

In this section details regarding the development of the Pelamis concept and a thorough description of the power take-off system are given. Some generic considerations regarding the hydrodynamics of the Pelamis are also presented.

7.3.1 Development Programme

Since its inception in 1998 it has been the sole objective of Pelamis Wave Power Ltd (PWP) to develop the Pelamis technology. At present, over 70 members of staff work daily from different locations to ensure that the Pelamis concept will become the market leader in the near future. It is fair to say that the Pelamis is one of the most studied WECs in history. Such claim is based on the several test rounds that were conducted at different wave tanks with several scale models, and the numerical modelling programme that was (is) followed in parallel. Table 7.3 presents the most significant test rounds, mentioning both the scale and the location of the tests. The experimental modelling programme is vital for validation of the numerical simulations, particularly when envisaging

Table 7.3. Detail of the experimental test round conducted to date

Model	Test Objective	Location	Date
80th Scale	Survivability	University of Edinburgh wide tank	May 1998
35th Scale	Numerical model validation	University of Edinburgh wide tank	July 1998
35th Scale	Alternative configurations	University of Edinburgh wide tank	July 1998
20th Scale	Survivability	London City University 55 m wave flume	Sept 1999
20th Scale	Numerical model validation	Glasgow University 77 m wave tank	Aug 2000
20th Scale	Power capture and mooring specification	Trondheim Ocean Wave Basin, Norway	Oct 2000
33rd Scale	Power capture	University of Edinburgh wide tank	Jan 2001
33rd Scale	Power capture	University of Edinburgh wide tank	Aug 2001
7th Scale	Digital control systems	Firth of Forth	Oct 2001
33rd Scale	Mooring response	Glasgow University 77 m wave tank	Mar 2002
50th Scale	Survivability	Glasgow University 77 m wave tank	Aug 2002
20th Scale	Control and survivability	Ecole Centrale de Nantes wide tank	Oct 2002
20th Scale	Control systems	Ecole Centrale de Nantes wide tank	Mar 2003
7th Scale	Mooring response and development	Ecole Centrale de Nantes wide tank	Apr 2003
20th Scale	Control and survivability	Ecole Centrale de Nantes wide tank	Mar 2005
21st Scale	Control and survivability	Ecole Centrale de Nantes wide tank	Feb 2007
21st Scale	Alternative configurations; Numerical model validation	Ecole Centrale de Nantes wide tank	Apr 2007

the next generation(s) of Pelamis machines. The role of numerical and experimental modelling in the development of the Pelamis WEC was addressed earlier in this book in section 5.4.

Among other significant milestones, the commissioning of a full-scale test rig for testing the power conversion module was pivotal for the development of the Pelamis. The test rig provided the final benchmark towards a full-scale prototype, by validating all steps of the power conversion chain. Two sets of actuation rams were mounted outside the module to simulate the wave action and drive the internal systems. Cycle testing of the power conversion module has provided confirmation of the suitability of each of the components and the reliability of the power conversion and electrical systems prior to the first offshore testing of the prototype.

Fig. 7.16. Full-scale test rig (view from control station)

The natural step forward was fulfilled with the construction of the first full-scale Pelamis wave energy converter, which can be seen as both a pre-production prototype as well as a technology demonstrator. It uses 100% 'available technology' as all system components are available 'off the shelf'. As with the full-scale test rig, the UK Department of Trade and Industry (DTI) have contributed to the funding of this demonstration machine. Verification of the design was performed by WS Atkins to ensure that it complies with offshore codes and standards (DNV). The machine was constructed entirely of steel using conservative safety factors. Steel has been chosen to simplify structural analysis, design and instrumentation and to ensure that modification and repair are straightforward. To reduce capital costs, the possibility of post-tensioned concrete tube elements is being investigated for subsequent devices.

The modular power-pack is housed in a second fully sealed compartment behind the ram bay so that in the event of seal failure only the hydraulic rams are immersed. Access to all system components is via a hatch in the top of the power conversion module. Maximum individual component weight is less than 3 tonnes to allow replacement using light lifting equipment.

The wave-induced motion of each joint is resisted by sets of hydraulic rams configured as pumps. These pump hydraulic fluid into smoothing accumulators which then drain at a constant rate through a hydraulic motor coupled to an electrical generator. The accumulators are sized to allow continuous, smooth output across wave groups. Output smoothness from the complete device will be comparable with that of a conventional thermal generator set. An oil-to-water heat exchanger is included to dump excess power in large seas and provide the necessary thermal load in the event of loss of the grid. Overall power conversion efficiency ranges from around 70% at low power levels to over 80% at full capacity.

Each of the three generator sets are linked by a common 690 *V*, 3 phase 'bus' running the length of the device. A single transformer is used to step-up the voltage to an appropriate level for transmission to shore, *HV* power is fed to the sea bed by a single flexible umbilical cable, then to shore via a conventional sub-sea cable.

Following construction the prototype machine underwent a comprehensive series sea trials in the North Sea (Fig. 7.17). After its success the device was towed to Orkney where it began a test programme at the European Marine Energy Test Centre (EMEC) in August 2004 (Fig. 7.18). When on site the machine's performance in supplying electricity into the grid is monitored and independently verified (see section 7.5.4). The Electric Power Research Institute (EPRI) presented a study in late 2004 which, in light of the success of this test round, concluded that the Pelamis technology was the only one available for immediate deployment.

Since the first sea trials and installation the prototype has been removed from site several times for both inspection and maintenance. The machine underwent a major refit at in 2006, when it was towed to a dry dock in Leith (Edinburgh) for a thorough check-up. This opportunity also allowed significant engineering improvements that tried to emulate the next stage: the P1A project.

Fig. 7.17. The Pelamis WEC full-scale prototype (150 *m* long, 3.5 *m* in diameter)

Fig. 7.18. The Pelamis WEC installed at the European Marine Energy Centre (Orkney)

Fig. 7.19. One of the P1A machines being assembled in Portugal

The P1A iteration of Pelamis machines corresponds to the first three machines that were ordered by a Portuguese consortium lead by Enersis in 2005. The three machines were towed to the Port of Peniche in 2006, for final assembly and commissioning (Fig. 7.19). The machines are to be installed in the north of Portugal, 5 km offshore Póvoa de Varzim, in what will become the first offshore wave energy farm. A letter of intent was already signed for a much larger project at the same location, involving a total of 30 Pelamis machines. The Enersis project is in the wake of the success of PWP's development programme, particularly the testing of the full-scale prototype. Such success was also recognised by the Scottish Executive, which recently (2007) announced that it would provide financial support to a Pelamis wave farm (four machines) at EMEC, in a project lead by Scottish Power.

7.3.2 Power Take-off

The power take-off system within the Pelamis is housed in Power Conversion Modules (PCM's) – as shown in Fig. 7.20 (previously presented in Chapter 6). There is a PCM positioned between each long tube section of the machine. There are four long tube sections in a Pelamis machine and three PCM's. The power take-off system uses hydraulic rams to resist the bending moments experienced about hinged joints at either ends of the PCM (1×Heave Joint and 1×Sway Joint). The bending moments are a result of the machine flexing under variable buoyancy forces from the local wave condition. The bending forces drive the hydraulic rams in extension and compression generating a high pressure flow of hydraulic fluid. This fluid is stored in pressure accumulators where it can be released under the control of a high pressure valve to flow across a variable displacement hydraulic motor which is coupled directly to a three-phase asynchronous generator with a rated output of 125 kW. Control of the high pressure fluid flow from the accumulator to the motor-generator set allows impulses from the absorbed wave energy to be smoothed within the accumulator so that the continuous out-flow of fluid at a constant pressure provides a smooth electrical output from the generator.

Once the fluid has passed through the hydraulic motor it is fed back into the low pressure hydraulic circuit where it is stored and recycled back into the high pressure circuit. For environmental reasons PWP only uses biodegradable hydraulic fluid so that in the unlikely event of a leak of fluid into the marine environment it will break down within days. Figure 7.21 shows a simple drawing of a single hydraulic circuit from the PCM (also previously presented in Chapter 6).

Within each PCM there are two separate hydraulic circuits similar to the one shown above. These circuits can be operated in combination or in the event of a component failure, or in an effort to increase system efficiencies, the circuits can be operated in isolation of each other. Each circuit contains one hydraulic ram resisting the heave bending moment and one hydraulic ram resisting the sway bending moment- this provides control on both axes in the event of a component failure within one of the circuits. Pelamis also has the ability to "dump" all absorbed power in response to grid constraints or cable connection failure; this is achieved through heat exchangers. As each of the hydraulic circuits has the capacity to generate 125 kW, each PCM therefore has the capacity to generate 250 kW. The three PCM's within one Pelamis machine give a rated capacity of 750 kW. The generated electricity is transported from the three PCM's down the length of the machine to the nose via medium voltage cabling where it then passes through a step-up transformer (11 kV in the prototype system) before being connected to the static sub-sea cable through a flexible umbilical electrical inter-connector.

The Pelamis benefits from a number of inherent characteristics and technological principles that gives it a unique and optimised balance of power absorption and survivability characteristics. To understand these it is essential to grasp the underlying principles behind the resource – wind driven water waves. Chapter 4 is fully dedicated to the wave energy resource, but particular aspects of its influence on the hydrodynamics of the Pelamis are detailed in section 7.3.3.

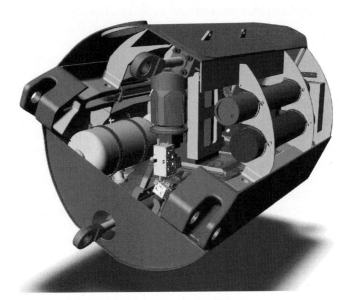

Fig. 7.20. Detail of the Pelamis Power Conversion Module

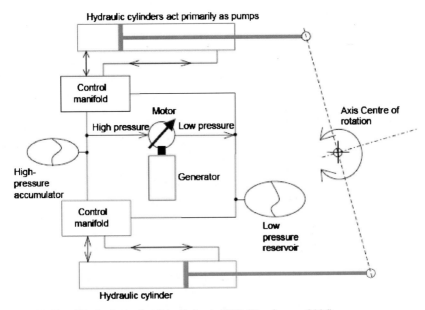

Fig. 7.21. Simplified schematic of the Pelamis PTO (Henderson, 2006)

7.3.3 Pelamis Hydrodynamics – Harnessing and Hiding

The primary driver

The ability of a wave energy converter to absorb power is directly related to its 'water-plane area' (for pressure activated systems, like Oscillating Water Columns or the Archimedes Wave Swing, this corresponds to the pressure-plane area). Generally speaking, the greater the water-plane area, the larger the forces that will be induced, and the larger the power capture potential. This concept is similar to that of the swept area of a wind turbine, or the swept volume of the pistons in a car engine. However, a large water-plane area would normally mean a very large volume. Volume equals weight, and weight is a good indicator of costs. Pelamis gets round this by creating a large water-plane area to volume ratio by distributing its volume thinly along its length. For reference - a spherical buoy of the same water plane area would be 20 *m* in diameter, and have thirteen times the volume

A source of reaction

One of Newton's laws of motion states that every force or action must have an 'equal and opposite reaction'. This fundamental principle must be obeyed by all physical systems including wave energy converters. In order to absorb power, the source of reaction must be able to provide solid reaction against the large wave induced loads. Providing an effective source of reaction is therefore a key design challenge for WECs. Offshore structures and wave energy converters have included a number of sources of reaction:

- Rigid coupling to the sea bed, either through self weight, direct fixture, or via a rigid mooring system is the most obvious and common choice of reaction reference. This requires either enormous mass or major insitu civil construction including extensive piling or rock anchors. As the system is essentially fixed in place, there is little or no scope for limiting hydrostatic or hydrodynamic loads in extreme conditions.
- Internal reaction mass. Here the relative acceleration of the WEC body and a coupled mass (either a solid lump or a constrained volume of water) is used. The problem with this system is that at the wave periods of interest the mass must be very large compared to the WEC body mass to provide sufficient reaction. Even with large masses there is significant 'lost motion' associated with 'dragging' the weight around with you, reducing the efficiency of the system. There is no scope for force limiting as wave height becomes larger.
- External reaction plate or body. Here a large mass is created using a large submerged plate, oriented at 90 degrees to the direction of motion, to push and pull against the surrounding water like a bi-directional parachute. Again, as inertial forces are weak relative to buoyancy forces at the wave periods of interest, the plate must be very large compared to the WEC body and waterplane area to provide even modestly effective reaction. Extreme loading on this type of system is typically even worse than for the case of an internal reaction mass.

All of the above systems require a separate reaction system to be supplied, often significantly larger or more massive than the WEC body itself. This separate system has to be strong enough to take extreme loads that may be an order of magnitude higher than typical operational forces. The direct and inescapable impact of this is significantly increased cost per unit power.

The Pelamis introduces the concept of 'self-reference' whereby the buoyancy forces applied by the waves are reacted against by buoyancy forces elsewhere on the machine. The length of the machine is chosen to be comparable to the maximum wavelength of interest, so that the system acts as a bridge between successive wave crests that pass by. The relative motion of the machine body to the water surfaces induces strong bending-moments along the machine while maintaining an overall vertical equilibrium with the wave. It is these bending-moments that drive the joints to absorb power. All forces are internal to the machine, no external source of reaction is required, and all components in the system both directly absorb energy from the wave and provide a source of reaction for the rest of the machine. Such direct link between power absorption and source of reaction make the Pelamis concept unique.

Tuning-up

Waves provide a cyclic or harmonic force input. It has long been understood that absorbing the energy from such an input is maximised by 'tuning' the natural frequency of the system to the frequency of the input force – very much like pushing a swing in time with its motion. This ideal match between the force applied and the response of the system at its natural frequency is known as 'resonance'.

Physics dictates that the frequency at which a mechanical system will naturally oscillate when pushed is a simple function of both its stiffness and its mass. The stiffer and lighter a system, the higher its natural frequency, and vice-versa. To be an efficient absorber, a WEC must therefore be given a stiffness and mass carefully chosen to provide a natural period matched to that of the incoming waves.

Most modern WEC concepts attempt to take advantage of this principle by carefully choosing their stiffness (buoyancy) to couple with their mass (weight) to produce a natural oscillation, or bobbing, period that matches the waves where most power is found. Some can, to some degree, 'tune' themselves to the incoming waves by varying their stiffness or mass. However, most WECs have this condition 'hard-wired' into their design - they are resonant, or near-resonant, whether the waves are small or large. In the latter case loads and motions become extreme at exactly the time when the WEC should to be reducing its response to survive.

The Pelamis achieves resonance in a creative way: the cross section of the machine is chosen to give a natural period far shorter than the waves of interest – typically, the Pelamis structure would 'bob' up and down once every 2 or 3 s, which is three or four times quicker for the most commonly encountered waves. This is because the system is hydrostatically too stiff for its own weight. It is as if we have artificially shortened the ropes on our swing to make it impossible to push quickly enough. To achieve resonance we must then find a way of either increasing the mass or reducing the stiffness. The former is not possible without incurring a large cost penalty. However, the geometry of the Pelamis joints is designed to dramatically reduce the stiffness of the system when desired through the following patented mechanism.

Each Pelamis power conversion module allows motion about two independent axes; these axes are arranged at 90° to each other. However the axes are not horizontal and vertical – they are, instead, biased 25–30° away from this to give an orthogonal pair of inclined axes of motion. The axis that allows motion in a more vertical direction is known as the heave axis, the axis that allows motion in a more sideways direction is known as the sway axis. Together these axes allow motion in any direction, like a universal joint on a car transmission shaft.

The challenging part comes with controlling the restraint applied to each axis depending on the waves encountered. If equal or similar restraint is applied to both joints the machine will respond as a simple articulated raft. All motions will be vertical as if the system only had simple horizontal hinges along its length. If, however, a much larger restraint is applied about the heave axis than the sway axis a very different response is stimulated. The joints will preferentially respond about the sway mode, giving an amplified sideways 'snaking' response along an inclined path. This response is analogous to the traditional child's toy snake that when held by the tail weaves its body and head vigorously back and forth under the action of very small motions of the tail. However, the twist in the tail for Pelamis is that the inclined axis of motion has a dramatically lower hydrostatic stiffness than the vertical mode, corresponding to what is need to achieve resonant response at the wave periods of interest. The key is that this cross-coupled, resonant response is only a result of a deliberate control application – the default mode is equal joint restraint and a safe, non-resonant response.

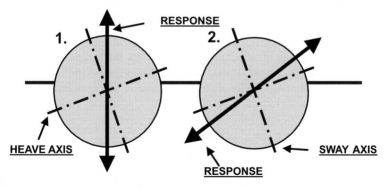

Fig. 7.22. Schematics of the Pelamis response under non-resonant (1) and resonant (2) conditions

Focusing your efforts

Engineers and mathematicians are always looking for ways to simplify the analysis of complex systems. In the case of wave energy, a very powerful analysis technique is to look at the waves a WEC can generate in still water if it was driven through a given motion cycle. That is to think of the machine as a wave-maker rather than a wave-absorber. It is an incontestable fact that a good wave-absorber must be capable of being a good wave-maker (Falnes, 2002). Some of these radiated waves can be used to cancel out the incident wave stream through destructive interference thereby absorbing power. This abstract technique allows the fundamental power capture capability of different WEC types to be studied more easily.

Another mathematical definition created to assess a wave energy converters performance is its 'capture-width' (see Chapter 3). This is the width of wave front absorbed, defined as the power output from the machine divided by the incident wave power level per metre of wave front width.

WEC's currently under development are, almost exclusively, of the heaving buoy 'point-absorber' type. Water is displaced by the vertical motion of a buoy on the surface, or by expansion/contraction of a submerged or semi-submerged volume. This action generates a symmetrical, circular set of waves propagating equally in all directions – similar to dropping a stone into a pond. Because of this, only a proportion of these radiated waves can cancel out the incident waves we are trying to absorb. Even so the theoretical limit of wave absorption capture-width turns out to be the wavelength divide by 2π, which for typical waves works out to be 15–25 m, irrespective of the frontal width of the machine. The maximum power of the machine is limited by the volume of the unit as this determines the maximum size of wave that can be radiated, and therefore absorbed.

Pelamis' radiation-pattern is fundamentally different. Imagine the Pelamis as a long line of dynamically linked, small heaving buoys, spaced along a line in the direction of propagation of the wave; hence the description of *line absorber*, mentioned previously in Chapter 3. When bobbed up and down, each of these produces a circular pattern of waves as before. However, if the relative phase of their motion is chosen carefully, these individual waves will reinforce in one chosen direction, while cancelling each other out in all other directions. This produces a focused wave pattern like a torch beam rather than the diffuse light from a lantern. The phenomenon is exactly the same as used by modern 'phased-array' radars to scan the radar beam without moving the antenna. The phasing of these motions can be actively controlled to maximise power capture in small seas, or minimise response under survival conditions. An example of the interaction of the directional radiation pattern with a regular wave field is shown in Fig. 7.23.

This inherent, and unique, hydrodynamic property of Pelamis means that the theoretical capture width of the system rises up to half a wavelength, three times higher than a simple heaving point absorber system. It follows directly that, for a given volume, the Pelamis is capable of absorbing up to three times as much power as the same volume configured as a heaving buoy.

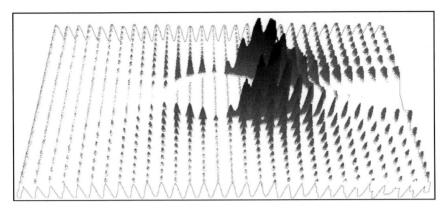

Fig. 7.23. Interaction of the directional radiation pattern with a regular wave field

Put more practically – on a good wave energy site, with an average annual incident power of 50 kW/m a WEC will be designed to reach rated power at an incident power level of approximately 50–60 kW/m to make most effective use of its power take-off system and electrical tie-back to shore. This inherently limits the economic rated power of any heave only buoy, irrespective of size, form or mode of operation, to approximately 1 MW. The corresponding ultimate limit for Pelamis is three times this, 3 MW.

This kind of conceptual advantage is analogous of the kind of advantage enjoyed by modern wind turbines, which use lift to drive the rotor, over old forms that used drag. Another analogy is the example of the torch versus the unfocussed bulb. The torch makes very efficient use of the bulb and batteries energy and power, compared to the omni directional lantern.

7.3.4 Shedding the Load

Power capture effectiveness is only one desirable characteristic of a WEC. Another is limiting loads and motions once rated power has been reached as the wave height increases. Figure 7.24 is a typical 'probability-of-exceedance' plot for wave height at a typical site.

For most of the year ($>90\%$) wave heights are less than ~4 m. It is in these conditions that the system must generate as much power, as efficiently, as possible. For the remainder of the year during storms (less than 10% of the time), wave heights increase to much higher levels, greater than 15 m in the worst storms. It is critically important that the fundamental hydrodynamics of the system limit the power that has to be absorbed by the onboard power take-off systems once the rated power sea state has been reached. Without this characteristic the power-take-off system would have to be rated for a much higher level with the dual penalty of higher capital cost, and significantly reduced efficiency at the normal operating power levels.

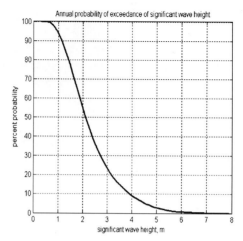

Fig. 7.24. Typical annual probability of exceedance plot (significant wave height)

Pelamis achieves this characteristic on account of its finite local volume down the length of the machine. As the wave height and steepness increases, the tubes locally fully submerge under, and emerge from wave crests, limiting the buoyancy and weight forces that generate bending moments in the machine. This process, known as hydrostatic-clipping, means that as wave height rises further the bending moments and joint angles (and therefore absorbed power) limit close to the chosen rated power level. This effect is analogous to the self-limiting characteristics of stall-regulated wind turbines whose blades stall above the rated power wind speed, limiting power capture without the need for complex control.

The effect described above also gives rise to desirable survivability characteristics that are at the heart of the Pelamis concept. Limiting the primary hydrostatic driving forces is an essential first step. However, most extreme loads in the marine environment arise from hydrodynamic effects – drag loads and inertia (or slamming) loads.

Drag loads arise from the turbulence created by the high speed flow of a fluid past a solid obstruction, in our case the body of the WEC machine. Loads rise with the square of velocity with the result that they are almost insignificant in small seas but rise rapidly to dominate loading in very large waves. For example, drag loads in 15 *m* high winter storm waves will be over 200 times those on a gentle 1 *m* summer swell. It is essential that the form of the machine is as 'slippery' and low drag as possible to minimise these loads, just as a racing car must be as streamlined as possible to increase speed.

Slamming loads can be even more destructive. These arise when a fast moving volume of fluid hits a solid body whose shape prevents its easy 'escape'. The most painful analogy is 'belly-flopping' off the top diving board at the swimming pool. The pressure spike that leads to red skin and a red face can crush or tear marine structures. The worst slamming loads occur on structures that have large flat panels that stop the fluid flow rapidly. Non-slamming forms are small, rounded or pointed, thereby allowing easy diversion of the flow rather than deceleration.

Research shows that load spikes induced in the vessel structure by the bow slamming are an order of magnitude higher than the basic hydrostatically induced bending loads (Cook, 1998). If these loads cannot be minimised or eradicated, the structure will have to be designed to withstand them, with an associated increase in material and cost.

A survivable WEC concept must be near invisible to hydrodynamic loading to prevent loads rising uncontrollably in storm seas. Pelamis tries to achieve this by presenting the 'slipperiest' form and minimal cross sectional area to the oncoming waves, and by buoyancy so that the machine can dive cleanly through the crest – similar to a surfer swimming out against breaking seas.

7.3.5 Implementation – Technology Transfer

Hydraulics

Wave loading is characterised by high forces and low velocities, continuously changing with time. An appropriate technology is required to effectively harness this and extract the energy. High pressure hydraulics are universally specified where similar characteristics are encountered or required in engineering. Everyday examples of this include:

1. Heavy lifting or pulling equipment
2. Metal presses and rolling machines
3. Extrusion machines
4. Excavators
5. Large traction vehicles

In addition, the lumpy, unsteady nature of energy input to a WEC means that its power take-off system has to cater for high instantaneous input power levels, and provide a significant amount of short-term energy storage to deliver a steady output power. Smooth output will be essential allow a large number of machines to work together and supply to the land electricity grid. Hydraulic accumulators are a cost effective technology currently available that offer the required balance of energy storage and high instantaneous power capability.

Optimising capture of wave energy also requires reliable, accurate control of forces and motions. Hydraulic systems can be highly effective for this due to the absence of significant control delay, the rigid coupling between the load and structure, the absence of inertia, and the very high instantaneous powers available. Hydraulics are used extensively for accurately controlled, high load, safety critical applications throughout engineering such as:

1. Aircraft control surface actuators
2. Power steering on road vehicles
3. Ship steering and stabilizer systems
4. Robot arms and industrial manipulators

This combination of proven high force, high power, and accurate control makes hydraulics one of the best-fit technologies for wave energy (if not the best).

Available technology – no prototypes-within-prototypes

The central rule of avoiding prototypes-within-prototypes has been learnt the hard way over and over again in the world of engineering. All WEC concepts are complex assemblies of structure and systems and many early designs fell foul of breaching this maxim. PWP set out to build the first prototype machine and all early production machines from 100% available technology. All of the machine elements are proven components with a service track record. Wherever possible, subassemblies of components, such as the motor-generator sets, are tested under operational conditions prior to deployment in Pelamis.

Capable and flexible control system

The importance of control to successful wave energy extraction has already been described. However, it is worth re-iterating that the ultimate ability of a WEC to absorb energy and generate smooth power across a wide range of conditions, whilst surviving extreme conditions, depends on the effectiveness of the control system. The flexibility and capability of the Pelamis power take-off and control system means that as better control methods are developed, they may be implemented on existing hardware (potentially hundreds of machines) with a remote software upgrade at little cost. Operation and monitoring must be automated to allow many machines to run without human intervention. This functionality has already been developed as part of the prototype programme.

Redundancy and robustness

All components crucial to the safe operation of Pelamis have a level of redundancy within each power module to avoid safe operation being compromised due to a single component failure. For example, multiple transducers, of negligible cost relative to the entire machine, are installed and can be switched between automatically if one fails. The entire hydraulic circuit of each power module may be split in half to avoid a fault on one side, such as a leak, from affecting the other (analogous to dual circuit brakes in a car, or redundant controls in an aircraft). All communications are dual routed and any failures are handled by distributed intelligence and ultimately a mechanical failsafe system of joint restraint that can operate without electrical power for indefinite periods. The benefit of this approach lies not just with the safety of the machine, but also with the reduced requirements for intervention and maintenance. The Pelamis power take-off and control system has been designed from the outset not to be vulnerable to single-point failures. It can continue to operate after a number of individual components have failed.

Deployment, operations and maintenance

Major cost is associated with the installation and retrieval of offshore systems. It is essential that the mooring system of a WEC is designed to minimise time spent on-site, minimise equipment costs (size and classification of vessel required), and maximise the range of conditions under which operations can take place. The Pelamis mooring system requires no sub-sea work or divers, all connections are made at the surface, and it can be installed using a single vessel.

In addition, PWP is committed to the policy of off-site maintenance. As long as the system can be removed and reinstalled quickly and safely in a wide range of conditions it is much more cost effective to take the machine to the equipment, rather than the equipment to the machine. A simple example of this is a 200 tonne crane on wheels costs less than £1000 per day, an equivalent waterborne unit costs upwards of twenty time this. The Pelamis mooring and connection system minimises the amount of work on-site, and maximises the range of operable conditions.

Volume production

At every stage of the design process, the engineers at PWP consider the implications of their design on mass production. Every aspect of the machine lends itself to volume manufacturing, notably the modular nature of the tubes and power modules, and the identical subassemblies within those modules. Effort is ongoing to reduce material costs and use innovative methods of production to make a more effective machine for less money.

7.4 Wave Dragon

James Tedd[1], Erik Friis-Madsen[1], Jens Peter Kofoed[1], Wilfried Knapp[2]

[1]*Wave Dragon ApS, Copehagen Denmark*
[2]*Technical University of Munich, Munich, Germany*

Wave Dragon is one of the foremost technologies within the field of wave power. Unlike most other devices is does not oscillate with the waves; it gathers the wave energy passively by utilising the overtopping principle. The front face of the device is a curved ramp, oncoming waves surge up it, as if it were a beach. Behind the crest of this ramp lies a reservoir which gathers the water "overtopping" the ramp which now has higher potential energy than the surrounding water. The effect of Wave Dragon is amplified by long reflector wings. Mounted to the reservoir, they channel the waves towards the ramp. The energy is extracted as the water drains back to the sea through low head hydro turbines within the reservoir.

Fig. 7.25. The Wave Dragon prototype in Northern Denmark. The reservoir and ramp are situated in the middle, with the reflectors concentrating the waves towards the ramp

The Wave Dragon is designed as a floating offshore device to be placed in water depths above 20 *m*. These areas are where the greatest wave energy is, and also where it is easiest to gain the permission to deploy. Over three years of sea testing have been conducted on a prototype in Northern Denmark.

This section will introduce the technology and the history of the Wave Dragon. A short introduction to overtopping theory is presented. The rationale behind the design is considered in detail for the wave reflecting wings, the low head hydro turbines and the control strategies of the Wave Dragon.

7.4.1 A Floating Overtopping Device

The Wave Dragon is built on the concept: use proven technologies when going offshore. It consists of three main elements:

- Two patented wave reflectors focusing the waves towards the ramp, linked to the main structure. These wave reflectors have the verified effect of increasing the wave height substantially and thereby increasing energy capture.
- The main structure consisting of a patented double curved ramp and a water storage reservoir built in concrete. This is very comparable to concrete boats, many of which were built during the first world war and are still floating to this day.
- A set of low head propeller turbines for converting the hydraulic head in the reservoir into electricity. These are similar to turbines which have been used in low head river hydro plants for generations.

The challenge is to put these sub-systems together to work in this novel method and survive in the extreme environment present in offshore conditions.

Wave Dragon is by far the largest envisaged wave energy converter today. Each unit will have a rated power of 4–11 *MW* or more depending on how energetic the wave climate is at the deployment site.

This will be a device with a displacement of approximately 30,000 tonnes and dimensions as shown in Fig. 7.26. This size brings many advantages. The device will respond minimally to waves, reducing fatigue problems. Also as it is

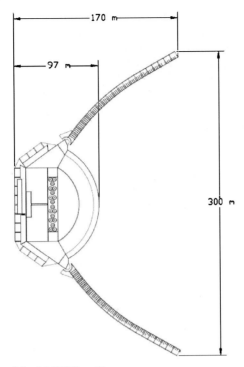

Fig. 7.26. Top view of the 7 *MW* Wave Dragon

large and stable it will be possible to work on board the device, which will dramatically reduce maintenance costs and downtime. As an overtopping device there are also many advantages to robustness of the design, in particular there are no end-stop problems as in larger seas the waves will wash over the platform harmlessly.

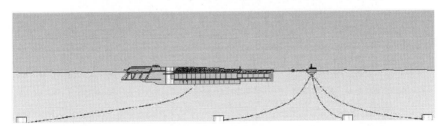

Fig. 7.27. This figure shows the CALM mooring system envisaged to be used on the demonstrator unit

The front mooring of the device will be a Catenary Anchored Leg Mooring (CALM) buoy, with a single rear mooring to restrain the device to a given rotation about the front mooring. The choice of anchor will depend on the sea bed. For the demonstrator unit concrete buckets filled with ballast rocks will be used as gravity anchors.

7.4.2 History of the Wave Dragon

The inventor, Erik Friis-Madsen, initiated the development of the Wave Dragon in 1986. In the early years he developed the principle of the Wave Dragon, and in 1994 an application for patent was submitted. Both Danish and European patents for the Wave Dragon have been obtained since then.

During the period 1995–1999 a number of studies of structural layout, overtopping of a fixed model, reflector efficiency, financial aspects, geometry, optimal choice of turbine configuration, and movements of the Wave Dragon were carried out. The findings are described fully in the report EMU et al. (2000) and Kofoed et al. (2000). Using the findings of these studies, the Wave Dragon design was slightly modified and the first test programme formulated.

The first test programme, financed by the Danish Wave Energy Program, consisted of thorough laboratory investigations of movements of the floating structure, mooring forces, forces in the reflectors, overtopping/amount of captured energy and survival in extreme wave conditions. These tests were carried out at the Hydraulics and Coastal Engineering Laboratory, Aalborg University, 1998–1999, using a floating 1 : 50 scale model of the Wave Dragon built by the Danish Maritime Institute, see Fig. 7.28.

As a continuation of the work performed under the Danish Wave Energy Program and on the basis of the research feasibility study, from 2000 to 2002 the EU granted funding for the project "Low-pressure Turbine and Control Equipment for Wave Energy Converters" under the Non-Nuclear Energy RTD Program. The total budget for this project was 1 *M€* of which the EU funds contributed 50 %.

Fig. 7.28. Testing of 1 : 50 scale Wave Dragon in Aalborg University wave basin

The objective of this project was to optimise the hydraulic performance, structural design, design of configuration, control and regulation of the turbines and design of electrical components and connection to grid. The laboratory model was modified, and tested at Aalborg University and at the facilities in University College Cork-HMRC (Ireland). The modifications showed significant improvements to energy capture and hydraulic behaviour. In parallel the configuration and regulation of the turbines was designed by the following companies: Ossberger Turbinenfabrik (Germany) / Kössler GmbH (Austria), Hälleryd Turbiner AB (Sweden) and Veteran Kraft (Sweden) together with turbine tests and computer simulations conducted at the Technical University Munich (Germany). The structure was designed by the inventor Erik Friis-Madsen and adjusted to shipbuilding standards by Armstrong Technology (UK). Electrical components and connection to the grid was designed by Balslev (Denmark), Belt Electric (Denmark) and Elsamprojekt / Eltra (Denmark).

After these successful tests, the next stage for Wave Dragon was to build, deploy and operate a field prototype. Funding from the Danish Energy Authority and European Commission led to deployment of this prototype in April 2003 in The Nissum Bredning (Broads), an inland sea, connected to the Danish North Sea (see section 7.6 for complete details). Testing has been ongoing since then on the many aspects of the device, from control systems and hydraulic performance to behaviour in extreme sea states (during a 100 year storm) to effects on bird life to subsea acoustic behaviour and more. As the world's first grid-connected floating WEC this Wave Dragon has been providing power to the grid of Northern Denmark consistently. Section 7.5 gives the full details of this deployment, including results and operational experiences learnt. Progress was reported in the conference papers of Sørensen et al. (2003, 2005).

The current focus for the Wave Dragon technology is to build and deploy a multi *MW* unit. A European Commission project was begun in May 2006 to implement the design for this commercial size. Off the coast of Pembrokeshire in South West Wales much work is ongoing to gain the relevant permissions to deploy a unit there. This is mainly focused on the Environmental Impact Assessment (EIA), studying the flora, fauna, geology, shipping and other aspects.

After the first successful deployment of a Wave Dragon unit there are several projects in the pipeline to construct farms of Wave Dragons in a variety of countries.

7.4.3 Overtopping Theory

The theory for modelling overtopping devices varies greatly from the traditional linear systems approach used by most other WECs. A linear systems approach may be used with overtopping devices. This considers the water oscillating up and down the ramp as the excited body, and the crest of the ramp as a highly non-linear power take-off system. However due to the non-linearities it is too computationally demanding to model usefully. Therefore a more physical approach is taken. The time series of the overtopping flow is modelled, thus relying heavily upon empirical data.

Figure 7.29 shows the schematic of flows for the Wave Dragon. Depending on the current wave state (H_S, T_p) and the crest freeboard R_c (height of the ramp crest above mean water level, MWL) of the device, water will overtop into the reservoir ($Q_{overtopping}$). The power gathered by the reservoir is a product of this overtopping flow, the crest freeboard and gravity. If the reservoir is over filled when a large volume is deposited in the basin there will be loss from it (Q_{spill}). To minimise this, the reservoir level h must be kept below its maximum level (h_R). The useful hydraulic power converted by the turbines is the product of turbine flow ($Q_{turbine}$), the head across them, water density and gravity.

Within the field of coastal engineering there is a considerable body of work looking at the overtopping rates on rubble-mound breakwaters, sea walls and dykes. The studies of Van der Meer and Janssen (1994) provided the basis of the theory on the average expected overtopping rate. Gerloni et al. (1995) investigated the time distribution of the flow. However this work was focused on structures designed to minimise the rate of overtopping, counter to the aims of the Wave Dragon. Kofoed (2002) performed laboratory tests on many permutations of ramp angles, profiles, crest freeboard levels in a variety of sea states, all with heavy overtopping rates. These studies showed the Wave Dragon's patented double curved ramp to be highly efficient at converting incident wave power.

When comparing results between different scales of model testing it is very useful to use Non-Dimensional figures to describe the variables. Results from the model scale can then simply be used for any size of device. In coastal engineering the average flow $\overline{Q}$ is converted into non dimensional form by dividing by the breadth of the device b, gravity g and the significant wave height H_S:

$$Q_{ND} = \frac{\overline{Q}}{b\sqrt{gH_S^3}}. \tag{7.5}$$

In the case of the floating Wave Dragon it has been seen that there is a dependency on the wave period. The dominant physical explanation for this is the effect of energy passing beneath the draft of the structure. Figure 7.30 shows a typical distribution of wave energy in the water column, with the left side showing the portion influenced by the ramp of Wave Dragon and therefore available to be exploited.

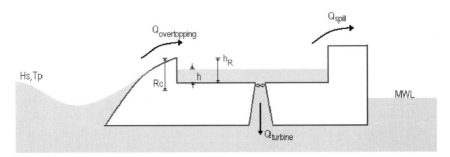

Fig. 7.29. Schematic of flows on the Wave Dragon

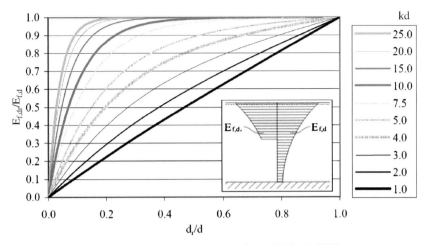

Fig. 7.30. Vertical distribution of energy in water column (Kofoed, 2002)

Shorter period waves have their energy concentrated in the upper part of the water column so Wave Dragon will absorb proportionately more energy from these. For Wave Dragon the following non-dimensional form has been used to include this period effect:

$$Q_N = \frac{1}{\lambda_{d_r}} \frac{\overline{Q}}{b\sqrt{gH_S^3}}. \tag{7.6}$$

This uses the coefficient λ_{d_r}, the ratio of energy between free surface and device draft ($E_{f,dr}$) to incident wave energy ($E_{f,d}$). This is based on linear wave theory and defined as following equation:

$$\lambda_{d_r} = 1 - \frac{\sinh(2k_p d(1-\frac{d_r}{d})) + 2k_p d(1-\frac{d_r}{d})}{\sinh(2k_p d) + 2k_p d}, \tag{7.7}$$

where k_p is the wave number at peak period, d is the water depth and d_r is the draft of the device.

To analyse the overtopping flow performance the Non-Dimensional overtopping rate is compared to the relative crest freeboard R, as shown in Eq. (7.8). This allows scale test results to be scaled to a full sized device:

$$R = \frac{R_c}{H_S}. \tag{7.8}$$

Time variation of the overtopping flow is also very important for modelling the power produced by the Wave Dragon. To make the model overtopping events are assumed to be random and independent, with a Weibull distribution. This has been confirmed by comparisons with data from a prototype Wave Dragon (section 7.5).

With this good understanding of the overtopping flows a simulation programme was designed and has been extensively used to optimise and model the Wave Dragon behaviour. This programme provides as an input a randomly generated time history of waves overtopping the ramp according to a mean rate and a specified distribution. This allows modification of many attributes (such as: reservoir depth and area, crest freeboard height, turbine number and type and turbine operational strategy) in order to pick the configuration which will produce the most electricity for each sea condition present at a location.

7.4.4 Wave Reflector Wings

One of the most distinctive aspects of the Wave Dragon are the long slender wings mounted to the front corners of the reservoir platform. These are designed to reflect the oncoming waves towards the ramp. A wider section of wave is available to be exploited with only a moderate increase in capital cost. The overtopping volume in a wave is very dependent on the wave height, therefore by providing only a moderate increase in height, much more energy can overtop the ramp.

In order to choose the correct lengths, angles, and position of these wings Kramer and Frigaard (2002) did extensive computer modelling of many combinations of these. The computer modelling used a 3D boundary element method. The meshing requirement is reduced to the structure's boundary surface, so as a result it provides a fast, efficient and accurate frequency domain solution for linear wave structure interaction problems. The method modelled the wings in isolation. The energy flux through the central gap (where the ramp and reservoir would be) gave

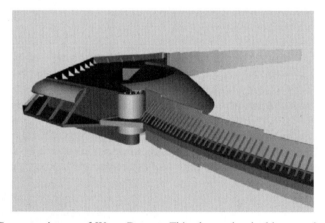

Fig. 7.31. Computer image of Wave Dragon. This shows the double curved ramp facing waves approaching from the right. The reservoir is behind and would have the turbines placed roughly in the centre. The slender wave reflectors are shown, with a flat side facing the waves, and stiffeners on their rear side

a performance coefficient for each setup. A smaller selection of sea-states and configurations was tested in a similar manner in a wave basin. Here again the wings were fixed, the platform removed and wave probes measured the waves passing through the central gap. A very good correlation was found between the two methods.

Reflectors which are floating and have finite draft will reflect higher frequency waves better than lower frequency. Typically these waves are also the smaller waves. Therefore increasing their input is very advantageous, improving the bandwidth of the device. The results show an increase in energy approaching the ramp of 85 % in these smaller operational wave states. When averaged over the sea states expected within the Danish part of the North Sea during one year, this translates to an increase in power of around 40 %.

Secondary bonuses of the presence of the wave reflector wings include: better weather-vaning performance to face the waves, lower peak mooring forces, and improved horizontal stability of the main platform.

As the aft and rear mooring attachment points are separated further, the yaw of the platform is more stable. Therefore the device will not turn away from the predominant wave direction, and will also realign itself faster as when the wave direction changes.

The peak mooring force is decreased as there is an internalisation of the wave loading forces within the structure. As the length of the device is comparable to the length of the waves, at some instants the wave force will push the platform, while pulling the wings or vice versa. Therefore the net peak force on the mooring links will be lower. This does however demand careful consideration for the design of this joint to resist these forces. Initial work and prototype work has concentrated on strain gauges mounted in the shoulder. Corona and Kofoed (2006) were able to show this internalisation of forces in practice by analysing the frequency spectrum of the strains close to the joint. Two force peaks were evident, one corresponding to wave frequencies and a secondary lower frequency peak at the slow surge frequency of the device.

Lastly the reflectors wings act as stabilisers to the device. As they float under their own buoyancy they counteract any list of the platform. This is important as the more horizontal the platform is kept the less water is spilt and so the more efficient the device operation.

Fig. 7.32. Computer mesh of the reflector wings and laboratory set up

7.4.5 Low Head Turbines and Power Train

Water turbines which are suitable for this purpose have been used in low head river water power plants for many decades and have been developed to a high level of efficiency and reliability. In France the 240 MW La Rance tidal power station has used such turbines in a salt water environment since 1967. Thus, in contrast to most of the other WEC principles, a proven and mature technology can be used for the production of electrical energy.

Turbine operating conditions in a WEC are quite different from the ones in a normal hydro power plant. In the Wave Dragon, the turbine head range is typically between 1.0 and 4.0 m, which is on the lower bounds of existing water turbine experience. While there are only slow and relatively small variations of flow and head in a river hydro power plant, the strong stochastic variations of the wave overtopping call for a radically different mode of operation in the Wave Dragon. The head, being a function of the significant wave height, is varying in a range as large as 1 : 4, and it has been shown by Knapp (2005) that the discharge has to be regulated within time intervals as short as ten seconds in order to achieve a good efficiency of the energy exploitation.

From a river hydro power installation which is properly maintained, a service life of 40–80 years can be expected. On an unmanned offshore device, the environmental conditions are much rougher, and routine maintenance work is much more difficult to perform. Special criteria for the choice and construction of water turbines for the Wave Dragon have to be followed; it is advisable to aim for constructional simplicity rather than maximum peak efficiency.

Figure 7.33 shows the application ranges of the known turbine types in a graph of head H vs. rotational speed n_q. The specific speed n_q is a turbine parameter characterising the relative speed of a turbine, thus giving an indication of the turbines power density. Evidently, all turbine types except the Pelton and the cross flow type are to be found in a relatively narrow band running diagonally across the graph. Transgressing the left or lower border means that the turbine will run too slowly, thus being unnecessarily large and expensive. The right or upper border is defined by technological limits, namely material strength and the danger of cavitation erosion. The Pelton and the cross-flow turbine do not quite follow these rules, as they have a runner which is running in air and is only partially loaded with a free jet of water. Thus, they have a lower specific speed and lower power density. Despite its simplicity and robustness, the cross flow turbine is not very suitable for wave power applications:

- Its operating principle entails a 'lost head' in the order of one runner diameter. This leads to a very low efficiency at very low heads.
- Due to the typically very narrow blade passages this turbine cannot cope very well with debris like seaweed and fishing net parts.
- Due to its low specific speed, the turbine itself is rather bulky, and it needs a gearbox to drive the generator.

This type of turbine has thus not been further evaluated.

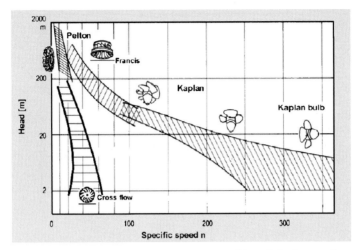

Fig. 7.33. Head range of the common turbine types (Voith and Ossberger characteristics; see Knapp et al., 2000)

The Kaplan type is the only turbine suitable for the head range in question. The shape of a turbine's guide vanes and runner blades is designed to give an optimal energy conversion in its design point, which is defined by optimum values of head and flow (H_{opt} and Q_{opt}) at a given speed. For every other operating point, there will be a discrepancy between the flow angles and the blade angles, decreasing the efficiency of the turbine. Whenever a turbine is required to operate in a relatively wide head and flow range it is important that the efficiency curve is flat and widely spread. This criterion is best fulfilled by the double regulated Kaplan type.

In this type of turbine, both the guide vanes and the runner blades are adjustable, thus making the turbine very adaptable to varying operating conditions. This is only achieved by a relatively complex construction which implies an oil-filled runner hub with a number of critical bearings and oil seals and a great number of joints and bearings in the guide vane operating mechanism. There is an immediate reflection in higher manufacturing costs, but also in a higher demand for maintenance, especially when the turbine is operated in an aggressive environment i.e. saltwater with possible silt contents. For these reasons single regulated variants of the Kaplan turbine have been conceived, namely the Propeller type with fixed runner blades, the Semi-Kaplan type with fixed guide vanes and the unregulated on/off turbine with fixed runner blades and fixed guide vanes. These turbines are simpler in construction, but they have a narrower efficiency curve, see Fig. 7.34.

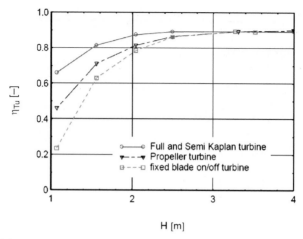

Fig. 7.34. Efficiency vs. Head for variants of the Kaplan turbine

With a projected plant of a larger size, it should be considered to use a number of smaller turbines instead of a single larger turbine. This has the following advantages:

- By stopping a number of turbines at lower flow rates, the flow rate can be regulated over a wider range without sacrificing efficiency, see Fig. 7.35.
- Single units can be taken out of service for maintenance without stopping production.
- Capacity demanded for hoisting and transport equipment to perform repair and maintenance work is greatly reduced.
- The smaller turbines have shorter draft tubes, and are thus easier to accommodate in the whole device.
- The smaller turbines have a higher speed, which reduces the cost of the generator.

Depending on the location of a production Wave Dragon it is envisaged that there will be between 16 and 24 turbines mounted.

In normal hydro power stations, the turbines are operated at constant speed, as they are coupled directly to generators feeding into a fixed frequency grid. However, if the generator is connected to a frequency converter, the turbine can be operated in a relatively wide speed range. This is very advantageous in situations where a large variation in turbine head occurs. By adapting the speed to the actual turbine head, the efficiency of the turbine can be kept almost constant, see Fig. 7.36.

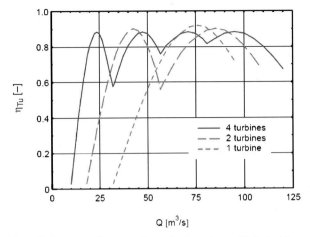

Fig. 7.35. Turbine efficiency vs. flow rate for a single and a multiple turbine configuration

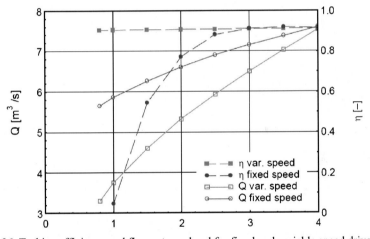

Fig. 7.36. Turbine efficiency and flow rate vs. head for fixed and variable speed drive

In the development of the Wave Dragon, different turbine regulation strategies have been evaluated by means of simulation software. Maximum overall plant efficiency was obtained when the turbine flow was reduced along with the emptying of the reservoir. The variable speed turbines adapt well to this, naturally reducing flow at the lower head. Therefore by using enough variable speed on/off turbines a good efficiency can be delivered, with smooth power delivery and a high load factor.

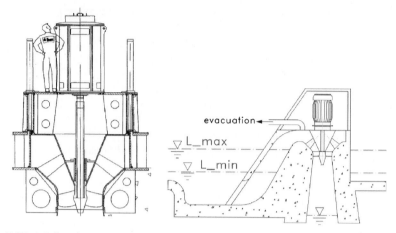

Fig. 7.37. A full scale Wave Dragon cylinder gate turbine and a Siphon on/off turbine

Two alternative methods for interrupting the flow have been analysed, the first one using a large mechanically operated cylinder gate, the second one using a siphon principle, see Fig. 7.37. The siphon type has no gate, it is stopped by simply admitting air into the top of the inlet duct; the turbine is started again by partly evacuating the air until the flow starts again and takes the rest of the air along with it. The cylinder gate type has the advantage of shorter start-up time and slightly better efficiency. If suitably designed it has been found that it has low maintenance requirements.

7.4.6 Control

Improved control algorithms are very valuable, as for no extra capital cost or maintenance cost they can improve the performance of a device, effectively free extra energy capture. Therefore this is an area which currently is a major focus of research work. On the basic level, as with several other WECs, Wave Dragon has two control loops. A slow acting control loop is used to tune the device to the current sea state. A much faster acting control strategy is used to extract the maximum energy from wave to wave.

The long period control's main aim is to regulate the floating height of the Wave Dragon to the optimal level for the sea state. This aims to maximise the power flowing over the ramp. A lower floating level will have more flow but at a lower head, and a higher floating level will have lower flow but a higher head – the optimum must be found. The time scale of the seas state's increase is of the order of a few hours, therefore the platform can also change its buoyancy at a similar rate. The input to this control strategy is the current, or future sea state which can be measured directly in the region of the Wave Dragon, or predicted based on weather forecasts.

The method for controlling the floating level of the platform is by blowing air into, or venting air from, open compartments beneath the reservoir. Due to the free surface of the reservoir this can be compared to balancing a tray full of water. The layout of these compartments and the detailed strategy for filling them is crucial to maintain stability. For example if there is a large central compartment filled with air, and low buoyancy at the edges the device will be quite unstable. In general the more stable the platform is, the closer to full the reservoir can be, and so the more power will be generated.

The fast acting control is to maintain a suitable water level within the reservoir. If the water level is too high, then large waves will not be able to be accommodated in the reservoir so there will be considerable spill from it. However if the water level is lower the head across the turbines is also low so less power can be produced from the same water overtopping the ramp. Again an optimal level must be found.

The reservoir level is controlled by the turbines. They are controlled on and off in a cascade fashion using the cylinder gates as explained earlier. At a minimum reservoir set point, the first turbines cut-in. As waves fill the reservoir, the remaining turbines progressively start. At a maximum level all turbines are operational. The input for this can either come from pressure transducers within the reservoir itself giving the level, or from direct measurements of the power generated by the generators, from which the head can be inferred. An area of development here is in the use of predictive algorithms, to control the turbines dependant on the expected overtopping in the next few waves. By lowering the reservoir level when some large waves are expected, spill would be minimised. Also by maintaining a higher reservoir level when smaller waves are expected, less water would be discharged at a lower head. Initial studies have shown that this small work alone could increase performance by 5 to 10 % (see Tedd et al, 2005).

7.4.7 Summary

Wave Dragon is a large floating overtopping type WEC. Its natural broad-banded behaviour and the use of established components make it one of the leaders in the wave energy field. It is a challenge to operate in the uncompromising environment where wave energy is greatest. Progress has been made in implementing existing technologies in this new manner. More development continues to improve the Wave Dragon device.

7.5 Operational Experience

Section 7.5 is effectively a compilation of contributions from technology developers (the same that were responsible for sections 7.1 to 7.4) which aim to gather the operational experience that has been acquired so far. The lessons that are presented are extremely relevant for the wave energy industry, and also interesting to a wider engineering community, particularly at a time when the deployment of the first large scale wave energy farms is being envisaged. For the first time the experiences from several of the key technology developers are presented in the same publication, a fact that greatly enhances the appeal of this section.

7.5.1 Oscillating Water Column - LIMPET

Tom Heath

Wavegen
Inverness
Scotland, UK

The most extensively studied and reported OWC plant in operation is Wavegen's LIMPET plant on the Scottish island of Islay. The grid connected plant has been in operation since November 2000 and serves both as a generator and research unit. The original configuration of the turbo-generation equipment gave an installed capacity of $500\,kW$ but this was reduced to $250\,kW$ after it became clear that the pneumatic power capture of the plant had been overestimated at the design stage. The LIMPET project was coordinated by the Queens University of Belfast supported by the EU under the JOULE III programme. Wavegen are now owners and operators of the plant.

Fig. 7.38. LIMPET OWC operating on Islay since 2000

Collector Form and Construction

The LIMPET OWC is inclined at an angle of 40° to the horizontal (Fig. 7.39). The column inclination affords and easy entry for water in surge and also facilitates tuning to longer periods.

The structure was designed to withstand either a frontal wave pressure of 6 *bar* across the full 21 *m* width of the device or a peak internal pressure of 1 *bar*. This necessitated internal supports for the roof dividing the water column into three equal chambers each 6 *m* wide.

To avoid the uncertainties associated with rock quality at the site a rear wall was added to the structure so that the internal pressure forces were totally contained within the collector structure. In this way the requirement for anchoring the structure to the site rock was minimised. With the exception of the lower section of the roof structure the collector was built primarily entirely of in-situ cast reinforced concrete. The lower section of the roof was formed from beams pre-cast at site which provided a firm foundation for an in-situ capping without the need for complex scaffolding. The major problem faced when building the collector was the need to protect the construction site from wave activity during the construction phase. This was achieved by excavation a hole behind the cliff edge and building in the lee of the protective bund (Fig. 7.40).

In times of storm waves overtopped the bund and in bad weather conditions it was necessary to cease working at the base of the structure. The degree of protection was however sufficient to limit lost time to 25 % during the summer months when construction activity had been planned. In general terms the construction team was able to use 10 day weather forecasts to predict weather down time and could plan accordingly.

On completion of the excavation the construction commenced with the casting of the rear wall. This was followed by the erection of the side walls and finally the roof of structure to leave the structure complete but isolated from the sea by the wave wall (Fig. 7.41). The final stage of the construction was then to remove the

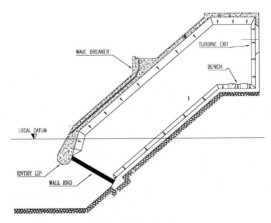

Fig. 7.39. Schematic of the LIMPET plant (side view)

Fig. 7.40. Details of the construction of the LIMPET plant

wave wall. This was achieved by using explosives. A series of four rows of holes were drilled along the length of the wall and the rows blasted sequentially, with millisecond separation, with the row to seaward being the first to be fired. The result of the sequential firing was to throw the blasted rock towards the sea and away from the structure.

Prior to the blast the excavation had been pumped full of water with the objective of creating a differential head to apply an outward force to the final pillar of rock to be fired. The rock was overcharged in relation to a normal quarrying operation in order to ensure that the shattered rock was in small enough pieces to be

Fig. 7.41. Completion of the structural work

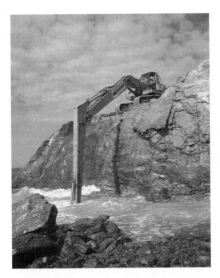

Fig. 7.42. Rock removal operations

removed by a long reach excavator (Fig. 7.42). Prior to blasting the wave wall valves had been fitted to the two outlets on the landward side of the collector to facilitate the fitment of power take-off systems. The turbo-generation equipment, which had previously been assembled and tested by Wavegen was transferred to the LIMPET site and connected to the collector. Figure 7.43 shows the turbo generation system connected to the outlet at the centre of the collector back wall. A second outlet, which is available for testing alternative power take-off systems, is also visible. After installation the turbo generation equipment was enclosed in a simple building to give protection against the worst vestiges of the weather.

As well as generating to the grid via the main plant a second outlet is available on the collector chamber which facilitates the continuing development of

Fig. 7.43. Turbo-generation system

the turbine-generator system. These developments have centred on a 750 *mm* diameter unit rated at 18.5 *kW* (Fig. 7.44). The focus of the development is on ensuring the long term reliability of the system whilst maintaining or improving performance. It is only when the high availability of the generation system necessary to justify economic power production have been demonstrated that, other than for research purposes, turbines will be installed on the floating systems now under consideration.

Figures 7.45 and 7.46 show typical site measured curves for non dimensional turbine torque and non-dimensional pressure drop measured on the LIMPET plant. These curves combine to give an overall conversion efficiency from pneumatic to electrical power of approximately 50 %.

The non-dimensional parameters are defined as follows:

- Non-dimensional Torque $T^* = \dfrac{T}{\rho\omega^2 r^5}$ where T is torque, ρ air density, ω angular speed and r turbine radius.

- Non-dimensional pressure $P^* = \dfrac{P}{\rho\omega^2 r^2}$ where P is the pressure drop across the turbine.

The immediate prospects for OWC technology appear excellent with OWC breakwaters under construction in Portugal and Spain for commissioning in 2008 and a number of other schemes worldwide in the pipeline. Once the baseline technology has been firmly established there will be opportunities for significant step increases in performance by, for example, introducing advanced turbine technologies such as the variable pitch Wells units under development at Wavegen. There is every indication therefore that OWC technology, which is perhaps the longest established concept for extracting power from the waves, will continue at the forefront of wave energy development and make a major contribution to the harnessing of ocean energies.

Fig. 7.44. 18.5 *kW* turbine-generator system

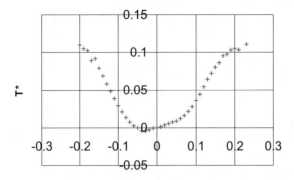

Fig. 7.45. Non-dimensional torque vs. flow coefficient

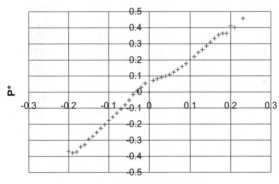

Fig. 7.46. Non-dimensional pressure drop vs. flow coefficient

7.5.2 Oscillating Water Column – Pico Plant

António Sarmento, Frank Neumann, Ana Brito e Melo

Wave Energy Centre
Lisbon
Portugal

The Pico plant was built from 1995 to 1999 under sometimes difficult conditions. These resulted from a number of reasons: Pico is a small island (15000 inhabitants) with limited infra-structures and qualified man-power about two thousands kilometres away from Lisbon; no direct flights were available at the time and the access to the island could be difficult in the winter due to rough weather and in the summer because of shortage of available seats in the flights due to tourism; a storm destroyed the breakwater protecting the in situ construction of the plant in the early phase of the plant erection and a second storm destroyed a small part of the machine room and was responsible for flooding the room with the electrical equipment a few weeks before the completion of the plant. The main lesson learned with the erection of Pico plant is that in situ construction must be avoided whenever possible and that the plant should be as much accessible as possible.

The initial tests in the Pico plant started in 1999. Since the beginning a number of problems with the mechanical equipment were found: vibrations in the guide-vanes of the stator due to very fragile guide-vanes; vibrations on the turbo-generator support structure (due to a resonance frequency excited by the turbine rotation); fatigue in the housing of the sluice-gate isolation valve produced by pressure fluctuations propagated from the air chamber (about 3.000.000 cycles per year); significant loss of oil in the turbine lubricating circuit. Due to the limited technical support available at Pico and the difficulty in accessing the plant in waves higher than $3\,m$, the planning of the actions to repair these faults and problems was difficult. On top of these the cost of all the repairs/actions also created problems. These problems affected and delayed the testing programme of the plant, which in reality had a negligible amount of operational hours in these early days. Problems of different nature also occurred related to leakage of rain water and sea water under big waves to the machine room through the cover of the opening at the machine room ceiling used to install or remove the turbo-generator unit. Along the years this was responsible for the damage of some of the auxiliary electrical, electronics and monitoring equipment and for the need of significant repairs of some auxiliary mechanical components, namely the pneumatic actuators of the sluice-gate and blade valves.

As a result of the described conditions at the plant, the Wave Energy Centre submitted and coordinated a project financed by the Portuguese DEMTEC/PRIME program to refurbish the plant and proceed with the plant testing (Neumann et al., 2006). The co-funding of the project was provided by EDP, EDA, EFACEC, IST, Irmãos Cavaco and INETI. This project is now briefly described and the main results from several months of plant testing are described.

The first phase of the project consisted in the refurbishment of the plant: this included the removal and renovation of the auxiliary electrical, power electronics and control equipment by EFACEC and its installation in two containers outside the plant, about $100\,m$ away. The electrical transformers were also placed in one of these containers, and so new electrical connections were established between the plant and the containers and between these and the electrical grid. This phase also included the cleaning and repair of the pneumatic actuators of the two valves mentioned earlier, the repair of the turbine lubrication system, the reinforcement of the turbo-generator support structure, the connection between the housing of the sluice-gate valve and the outside atmosphere by an air duct in order to maintain approximately atmospheric pressure in the housing and thus reduce the fatigue problems, the proper isolation of the machine room opening at the ceiling and the installation of a slow relief-valve (supplied by Kymaner, Portugal) to allow the air chamber to escape directly to the atmosphere in very rough seas.

The second phase of the project consisted in the testing of the plant with the equipment as indicated above. The third phase of the project was intended to allow the testing of new blades with optimised profiles for this application and a fast-reacting relief-valve to be operated in a wave-to-wave basis. However due to problems encountered during phase two, and the additional costs involved, phase three shifted from the original program to the testing of the plant with the original turbine blades and no fast-reacting relief valve, but now without the guide vanes stator on the atmospheric side as will be explained later. The problems mentioned above were basically two: i) one of the blades of the stator on the atmospheric side of the turbo-generator broke due to fatigue and destroyed this stator, affecting the turbine ring and the other stator – these components were repaired and new and stronger stators (with blades), supplied by Kymaner, were built and installed in October 2006; ii) one of the ball-bearings of the turbine broke due to deficient assembly after inspection (new bearings were installed after this occurrence).

The experience gained with roughly 56 tests of 20 minutes duration along five months in two years can be summarised as follows (see also Figures 7.47 to 7.49):

- Fatigue on the mechanical components is a critical issue that needs to be properly addressed at the design stage.
- The vibrations on the support structure could not be completely removed and so the turbine could not operate above $1100\,rpm$ – note that maximum turbine rotational speed is $1500\,rpm$ and that the power increases with the cube of the rotational speed.
- As a result of the previous limitation the power levels at the turbine shaft are significantly less than what could be attained: $1,172\,kWh$ of pneumatic energy (average power of $62.8\,kW$) were available to the turbine, of which $617\,kWh$ were delivered to the grid (average of $33.1\,kW$), this representing an average efficiency from pneumatic to electric energy delivered to the grid of 52.6%.

- The analysis of the three test periods (October-November 2005; October 2006 without guide vane in the atmospheric side of the turbine; November 2006 with the new set of guide vanes) shows that:

1. The wave conditions were less energetic in 2006, leading to significantly smaller pneumatic energy available to the turbine;
2. The turbine average efficiency is about the same in the two first sets of tests, respectively 58.6% and 57.5% and a bit lower in the third set (49%) – these values are computed from the data in Figures 7.47 to 7.49.

- The inertia of the turbine allows smoothing the electrical power output as can be seen in Fig. 7.50 for a particular test. Other tests show similar results.
- Due to the limited water depth in front of the plant (about 8 m) in many cases the wave propagates in very shallow waters as a second order wave: shorter and more intense crests than troughs. This originates much higher upward velocities (and outward flows) than downwards velocities (and inwards flows) as seen in Fig. 7.51.
- Large outward flows mentioned in the previous point produces turbine blade stall with decrease of energy production and an increase of fatigue (typically in the stator on the atmospheric side) and noise generation.
- The very careful design of the air ducts originated very uniform flow velocities in the turbine entrance, in particular for outward flows (see Fig. 7.52). They are also responsible for the very good turbine average efficiency mentioned above and for the excellent agreement with the numerical estimations (see Figures 7.53 and 7.54).
- In energetic seas, when the relief-valve in the top of the air chamber (see Fig. 7.7) is open, an air jet is produced from the atmosphere into the chamber during the downward motion of the internal water free-surface. The air jet impinges directly in the free-surface pulverising the water. The resulting water droplets are subsequently transported through the turbine by the outward air flow thus reducing the turbine rotational speed and shaft power and eroding the turbine blades.
- The tested control laws produce a very stable and efficient operation of the power take-off equipment.

The efficiency of the first two sets of data (Figures 7.47 and 7.48) is very close, which is surprising as we would expect to have much higher efficiencies when the plant operates with the stator on the atmospheric side (Fig. 7.47). The efficiency of the third set of data (Fig. 7.49) is unexpectedly smaller than the ones for the two previous sets of data and seems to indicate that the new stator is less efficient than the original one. This is very surprising since they were built with the same aerodynamic design. Also surprisingly is to notice that the efficiency is smaller than the one measured without stator (data of October 2006). Research is ongoing to provide further clarification.

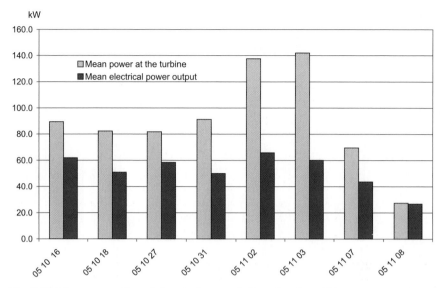

Fig. 7.47. Mean pneumatic and electrical power delivered to the grid in sea tests conducted in October and November 2005 (original stators)

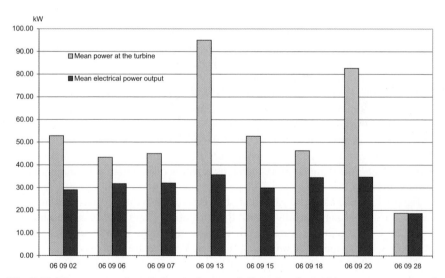

Fig. 7.48. Mean pneumatic and electrical power delivered to the grid in sea tests conducted in September 2006 (with no guide vane in the atmospheric side)

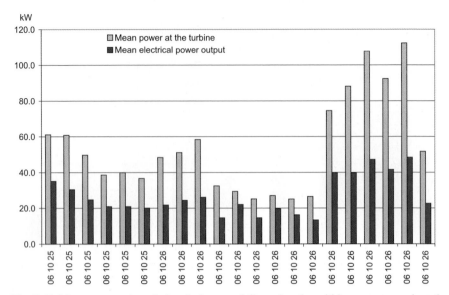

Fig. 7.49. Mean pneumatic and electrical power delivered to the grid in sea tests conducted in November 2006 (new set of guide vanes)

It is also strange that in two of the tests (051108 and 060928) the electrical and pneumatic energy are almost the same, resulting in turbine and generator efficiencies close to 100 %. This results from having the Pitot tubes that measure the dynamic pressure upstream the turbine blocked with dust, thus measuring dynamic pressures (and air flows) smaller than the real ones. As the pneumatic power is the product of the air flow and the static pressure drop in the turbine, smaller estimates of the real pneumatic power result in higher turbo-generator efficiencies.

Figure 7.50 shows that with the relief-valve closed the pneumatic power available at the air chamber is almost equal to the one available to the turbine, meaning that aerodynamic losses are very small. It is also seen that the electrical power delivered to the grid is much more stable as a result of the control strategy and the turbine inertia.

Figure 7.51 shows for a particular, but representative test, that positive (outwards) flows are significantly larger than negative flows due to wave shoaling produced by the reduced water depth close to the plant site (about 8 m). It also shows that, as expected, the pressure drop and flow rate are in phase.

Figure 7.52 shows the dynamic pressure measured at both sides of the turbine for three radii: inner, mean and outer radius. The uniformity of the measurements is outstanding and shows a very uniform velocity profile, in particular if it is noted that the differences in the dynamic pressure are the double of those in the velocity. Uniform velocity profiles are critical for good turbine efficiency as otherwise stall can be anticipated.

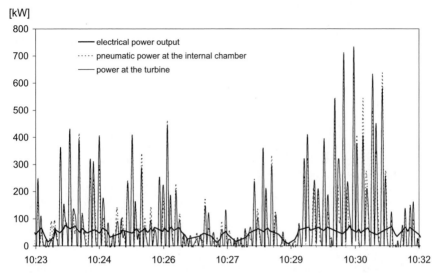

Fig. 7.50. Sample of time-series of pneumatic power in the air chamber, pneumatic power available to the turbine and electrical power delivered to the grid

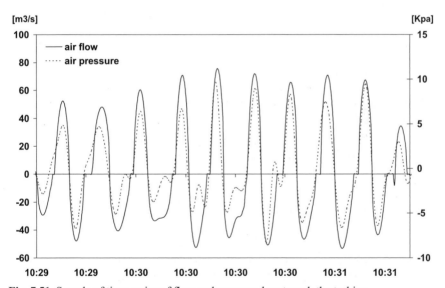

Fig. 7.51. Sample of time-series of flow and pressure drop trough the turbine

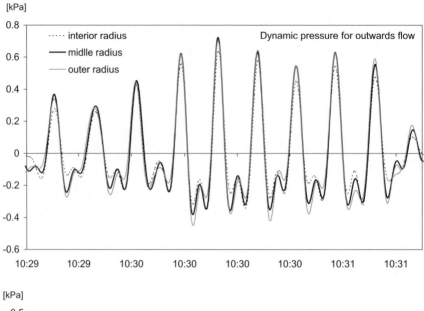

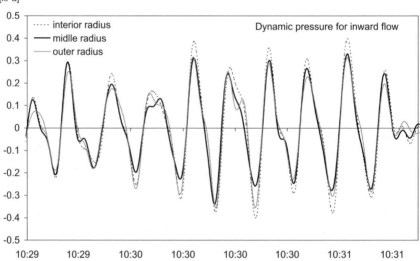

Fig. 7.52. Sample of time-series of dynamic pressure measured upstream (top) and down-stream (bottom) the turbine. Only positive dynamic pressures are relevant

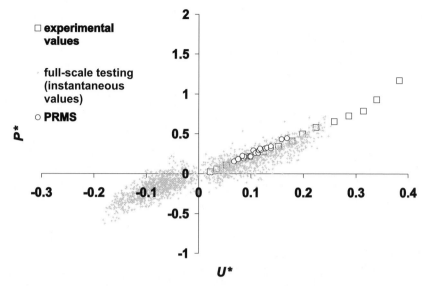

Fig. 7.53. Comparison of dimensionless flow rate versus pressure drop on the turbine

Figure 7.53 shows the comparison between the experimental curve measured in a steady flow wind tunnel (Gato and Falcão, 1990), the instantaneous and the *rms* values measured at the full-scale plant. The spread of the instantaneous values is expected due to measurement uncertainties. The root mean square values coincide extremely well with the experimental ones, showing that, as expected, the turbine flow is quasi-steady.

Finally Fig. 7.54 shows comparisons between numerical results and measured data relating the average power delivered to the electrical grid and root mean square values of the air pressure in the pneumatic chamber. A very good agreement is seen to occur indicating the good quality of the numerical model used to design the aerodynamics of the turbine. Again it is noted that the agreement is only possible because of the very uniform flow velocity approaching the turbine.

The results and conclusions presented here could only be possible with the contribution from a very large number of people and institutions, both in relation to the construction of the plant and its reimbursement and test. Prof. António Falcão was already mentioned as the scientific responsible for the two European JOULE projects that allowed the construction of Pico plant. Prof. Luís Gato, also from IST, was the responsible for the aerodynamic design of the turbine, guide vanes and air ducts. The technical contributions from Consulmar, Kymaner, EFACEC, Irmãos Cavaco, INETI, IST and EDA were also fundamental and must be acknowledged.

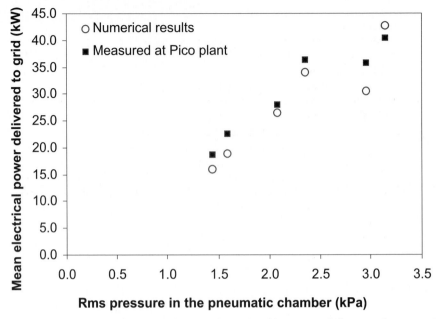

Fig. 7.54. Comparison of numerical estimations of turbine power delivery and measured values at Pico plant

7.5.3 Archimedes Wave Swing (AWS)

Miguel Prado

Teamwork Technology BV
Zijdewind
The Netherlands

The 2 *MW* pilot plant of the Dutch Archimedes Wave Swing (AWS I) was suc-cessfully submerged 5 *km* offshore Póvoa de Varzim (Northern Portugal) in May 2004, and decommissioned after testing approximately six months later. Despite resulting in a short operational phase, the AWS is thus among the first devices with operational record in significant dimensions, having tested components and proven the technology at full-scale.

Results of the AWS dynamic analysis process were presented and correlated with the theoretically expected results in Prado et al. (2005a). A short summary of the main milestones that were reached is presented in this section, but firstly some considerations regarding the submergence operation are given.

Submerging the AWS pilot plant

The controlled submergence of an offshore device requires a detailed stability study in order to predict and avoid problems that may occur during operation (like general instability, resonance effects, etc).

To guarantee a robust design of the submergence operation of AWS, a theoretical analysis of AWS dynamics was developed for all the steps of the operation. A quasi-linear model in the frequency domain was developed for the six degrees of freedom of the device (surge, sway, heave, roll, pitch and yaw) based on linear wave theory. Non-linearties due to drag damping forces were included in the model by equivalent linearisation. Critical aspects like natural periods and dynamic response to wave excitation can be evaluated for each degree of freedom.

One of the main problems of submerging a device in a controlled way is related to the fact that the dynamic behaviour of the device is not constant during all the operation and changes with the submergence depth. This results mainly from changes in the inertia and buoyancy of the device, inflicted by the ballasting procedure and the changing underwater geometry. Critical points of the operation could be identified with the described calculation methodology, which were confirmed later during operation.

The theoretical analysis used for evaluate AWS dynamics proved to be a vital tool in the design process of the submergence operation, leading to its success after two previous failed attempts. As described in 7.2, the first AWS full-scale prototype was composed by a generator (central cylinder) installed on top of a pontoon (see Fig. 7.55). The pontoon itself is not necessary for wave energy

Fig. 7.55. View of AWS at the harbour of Viana do Castelo, Portugal

conversion and it was only built with the purpose of facilitating the submergence operation during the test phase. In future versions the pontoon will be suppressed.

The pontoon is composed by several tanks that are ballasted during the submergence operation (Fig. 7.56):

- Four ballast tanks (BT), with a total capacity of $2200\,m^3$.
- Four Air & Water tanks (AW), with a total capacity of $1000\,m^3$.
- Four towers (T), with a total capacity of $150\,m^3$.

The total mass of the device at an unballasted state is $7200\,ton$, from which $5500\,ton$ are due to the sand tanks (SD). The purpose of the sand tanks is to decrease the gravity centre of the device in order to increase its stability.

Taking into account that the total ballast capacity of the pontoon is $4700\,m^3$, at 100% ballasted state the device mass will be $11900\,ton$.

The submergence operation can be described in the following eight steps:

1. Starting point, no ballast, depth $5.25\,m$;
2. Deck underwater, ballast $340\,m^3$, depth $5.5\,m$, fill AW tanks;
3. AW tanks full, ballast $750\,m^3$, depth $7.5\,m$, fill BT tanks;
4. Towers underwater, ballast $1870\,m^3$, depth $17.5\,m$, fill BT tanks;
5. Ballast tanks full, ballast $2980\,m^3$, depth $21\,m$, fill towers;
6. Lift floater by pumping air inside, ballast $3260\,m^3$, depth $25\,m$, fill towers;
7. Floater underwater, install buoys, ballast $4070\,m^3$, depth $31.5\,m$, fill towers;
8. Device at seabed, ballast $4150\,m^3$, depth $43\,m$.

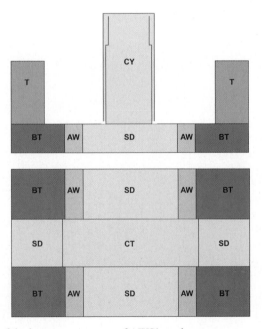

Fig. 7.56. Sketch of the inner arrangement of AWS's tanks

During the submergence operation, the device dynamics changed with submergence depth. It is therefore very important to analyse for each step the dynamic behaviour of the device.

The submergence operation took place between the 16th and the 18th of May 2004, 6 km offshore Póvoa de Varzim in Northern Portugal. The maximum recorded sea state during that period of time was $H_s = 1.6\,m$ with T_z between 7 and 9 s. Figure 7.57 shows the different steps of the operation. More details can be found in Prado et al. (2005b).

Fig. 7.57. Several stages of the submergence operation: deck submersion (top left), towers half submerged (top right), towers submerged (mid left), floater lifted to upper position (mid right), floater submersion (bottom left), pilot plant installed (bottom right)

Testing of the AWS pilot plant: results

Results from the testing round are valuable in order to validate the numerical simulation and to gather experience in a true operational environment. A $15\,min$ set of data is particularly interesting for detailed analysis. The measurements presented were performed on the 3rd of October of 2004, around $14:30$. At that time, the following conditions were observed:

1. Sea

- Significant wave height: $H_s \approx 2.3m$;
- Wave periods: $T_P \approx 11.7s$, $T_e \approx 10.5s$;
- Tide level: $\eta_T \approx 1.43m$.

2. AWS

- Water volume inside AWS: $V_W \approx 1180\,m^3$;
- Average air pressure inside AWS: $\overline{p}_a \approx 32.3\,mwc$;
- Water brakes closed (hence additional damping affecting the motion).

3. Landstation

- Converter input resistance: $R_{AC} = 2.75\Omega$.

    The following signals were measured (see Fig. 7.58):

- Water pressure on top of the AWS, to identify the waves;
- Air pressure inside AWS, to identify the air spring and the motion of the device;
- Electrical power at the converter, to quantify the energy output.

    By visual inspection of the peak values in Fig. 7.58 it is possible to see that a correlation between the three signals exists. However it is necessary to quantify this correlation in order to use the numerical models which describe the dynamics of the machine (more details in Prado et al., 2005a).

    One way of evaluating the consistency and correlation of the measured signals is by calculating a common signal. The electrical power produced by the generator P_{GEN} was calculated from the measured signals. The generated power can be related to the converter power P_{CONV} by

$$P_{GEN} = \frac{R_{AC} + R_{GC}}{R_{AC}} P_{CONV} , \tag{7.9}$$

where $R_{GC} \approx 1\Omega$ is the resistance of the generator plus the power cable and $R_{AC} = 2.75\Omega$ is the input resistance of the converter.

    The velocity of the device for small displacement is related to time derivative of the air pressure

$$\dot{x} = -\frac{S_f}{k_a} \dot{p}_a , \tag{7.10}$$

with S_f being the surface area of the floater and k_a the air spring coefficient.

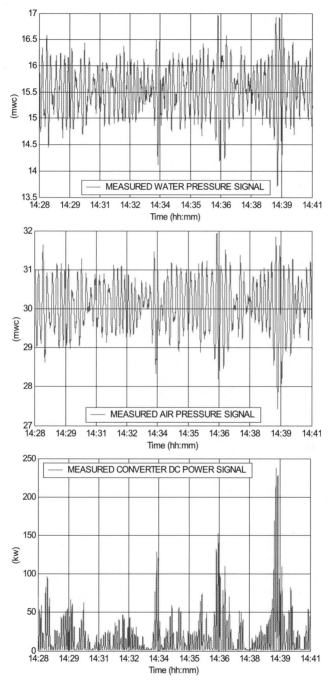

Fig. 7.58. Measured signals (3rd of October 2004)

The velocity is also related to the water pressure through the time domain model:

$$(m_f + m_{add\infty})\ddot{x} + \beta_L\dot{x} + \beta_{NL}\dot{x}|\dot{x}| + kx = F_{WAVE}, \tag{7.11}$$

where m_f is the mass of the floater (including all moving parts of the device), $m_{add\infty}$ is the added mass at infinite frequency, β_L is a linear damping coefficient due to the generator and the radiation force, β_{NL} is a nonlinear damping coefficient due to the drag forces and the water brakes, k is the total spring coefficient (air + nitrogen + hydrostatic) and F_{WAVE} is the wave exciting force.

The generator power may be computed using either air or water pressure velocity estimate by

$$P_{GEN} = \beta_{GEN}\dot{x}^2, \tag{7.12}$$

where β_{GEN} is the mechanical damping due to the generator:

$$\beta_{GEN}(R_{AC}) = \frac{\partial F_{GEN}(\dot{x}, x, R_{AC})}{\partial \dot{x}}\bigg|_{\dot{x}=x=0} \approx \frac{1.83 \times 10^6 (N.s.m^{-1}.\Omega)}{R_{GC} + R_{AC}}. \tag{7.13}$$

The estimated generator powers from the measured signals are shown in Fig. 7.59. The signals are quite similar, especially the generator powers estimated from the converter and the air pressure. The generator power estimated from the water pressure signal differs considerably with regard to the peak values. It is however interesting to see that the average powers derived from all three signals are very similar. To better evaluate the correlation between the estimated signals, a XY plot of the different estimates can be made (Fig. 7.60).

From Fig. 7.60 it can be seen that the estimated generator power from the converter and air pressure correlate quite well. However the estimated generator power from the water pressure deviates on average from the estimated generator power from the converter for higher values. For lower values the fit on average is reasonably good, although the dispersion is considerably higher than in the case of the estimated power from the air pressure. This difference in the correlation was expected since the relation between water pressure and electrical power is more complex the one between the air pressure and the electrical power.

To better evaluate the effect that the water brakes and the settings of the device have on its performance, the motion and the generator power were computed from the model for different scenarios. Figure 7.61 illustrates the scenario observed during the testing day and Fig. 7.62 shows what would happen for the same day if the water brakes were open and the device was optimally tuned. A very significant difference in performance by releasing the brakes and tuning the device with the optimal settings can be observed. The motion and the average generator power both increase by approximately a factor of eight.

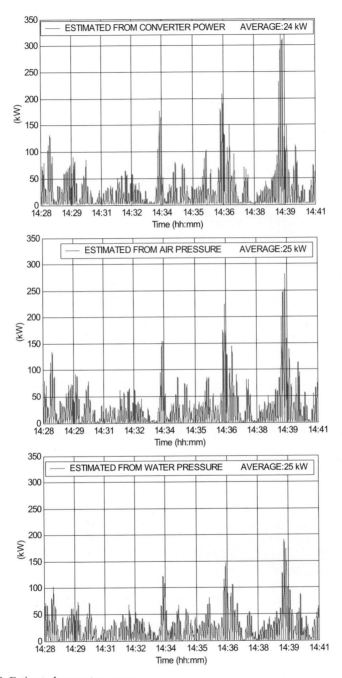

Fig. 7.59. Estimated generator powers

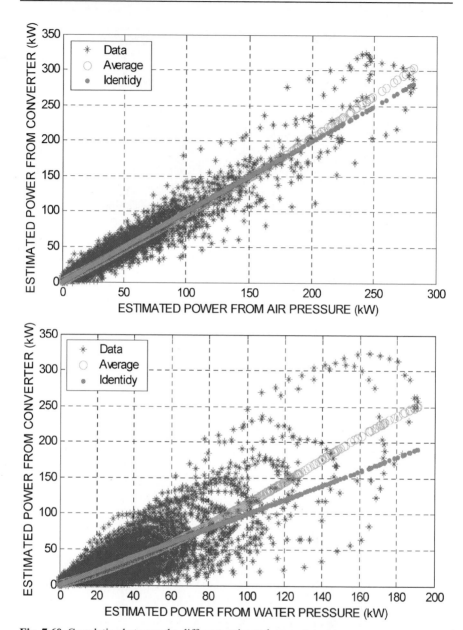

Fig. 7.60. Correlation between the different estimated generator powers

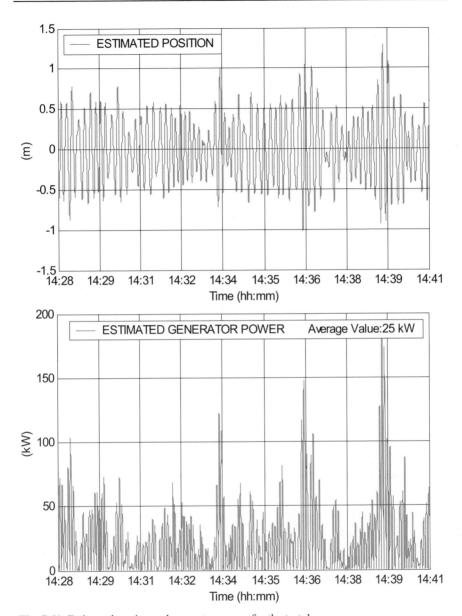

Fig. 7.61. Estimated motion and generator power for the test day

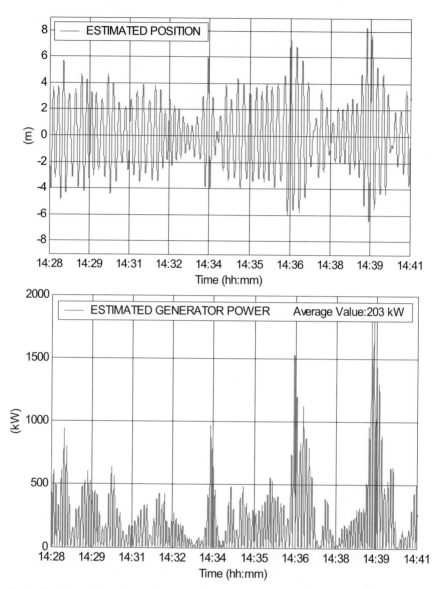

Fig. 7.62. Estimated motion and generator power for the test day, with water brakes open and optimal settings

The effect of the settings of the device can be better appreciated by envisaging the maximum position and average generator power versus the natural period and the setpoint resistance of the converter. There is a drastic effect caused by the action of the water brakes, resulting in excessive damping and thus reduction of the power output. Note that the water brakes were used at this stage given the nature of the tests. Correct tuning of the settings reveals a big impact on the performance. Average power levels of $200\,kW$ can be obtained for the same sea state. It should be noticed however that for this kind of power, the expected maximum stroke should be around $16\,m$, which is higher than the stroke of $9\,m$ available. For a maximum $9\,m$ stroke, average power levels of $150\,kW$ are estimated. In practice using a dynamic adjustment of the converter resistance, it is s possible to reach average power levels higher than $150\,kW$.

In summary, the experience with the pilot plant results shows that a simple 2nd order mass-spring system fits reasonably well to the measured data, which leads to the conclusion that the main structure of the AWS dynamics is relatively simple. Average power outputs can be estimated with high accuracy but peak values are underestimated by the model. A higher-order non-linear model should be used to get better fits to the measured data. Since water brakes were always closed, other forms of damping like drag and radiation could not be identified. From the model it could be seen the significant impact that the water brakes have on the overall performance of the device. By opening the water brakes and correctly adjusting the settings of the device, the average power level could reach levels in the range $150–200\,kW$. All the lessons learned in Portugal, both in the submersion and when in operation, are expected to be vital when developing the next generations of AWS machines.

7.5.4 Pelamis

João Cruz*, Ross Henderson, Richard Yemm

Pelamis Wave Power Ltd
Edinburgh
Scotland, UK
* Now at Garrad Hassan and Partners Ltd, Bristol, England, UK

As previously mentioned in 7.3, Pelamis Wave Power's (PWP) development programme reached a significant milestone in August 2004, when the first full-scale prototype (FSP) was installed at the European Marine Energy Centre (EMEC), in Orkney, and the first electricity generated from an offshore wave energy converted was delivered to the UK grid. The ongoing FSP programme provides operational knowledge which is pivotal for the first offshore wave farms, such as the Agucadoura wave farm in Portugal (three Pelamis machines at a first stage). This short contribution aims to briefly describe the production and assembly knowledge, the

O&M procedures and operational experience that PWP has gathered over the past years. Representative data from a recent round of tests at EMEC is also presented.

Production and Assembly

A production programme was initiated in 2005, following the first commercial order for three Pelamis machines to be installed in Portugal. The new P1A Pelamis machines have a few design improvements with regard to the full-scale prototype, including a rapid mooring connection system.

In June 2005 the first construction contracts were issued with the majority of the fabrication contracts being awarded to Scottish manufacturers. The main tube fabrication was given to Camcal, for completion at the Arnish manufacturing site on the Isle of Lewis. The Stonehaven based company, Ross Deeptech, secured the order for the fabrication of the nine power conversion modules. After completion of the first module, it was transported to the PWP site in Methil, Fife. Based in the ex-Kvaerner yard, now the Fife Energy Park, the PWP production team have established an assembly facility. Drawing upon the local skills base, twelve members of the production team were given the task of populating the power conversion modules with their internal components (Fig. 7.63). Further orders will allow PWP to develop their assembly facilities and work has already been initiated with the objective of reducing costs for the next generation of Pelamis machines.

Fig. 7.63. Seven of the nine power conversion modules of the P1A machines lined up at the Fife Energy Park during initial assembly

The power modules, tubes and all remaining components for the first three P1A machines were transported to Portugal for final assembly by PWP staff (Figures 7.64 and 7.65). The assembly of the machines at the quayside of the Peniche shipyard uses a new 'habitat' system, a localised floating 'dry dock', tested in accordance with DNV standards and certifications. It allows maximum flexibility for operations whilst ensuring work is carried out in a safe and dry environment.

Fig. 7.64. Arrival of the P1A machines at the Peniche (Portugal) for final assembly

Fig. 7.65. Detail of the power conversion modules at the Peniche shipyard (Portugal)

Once assembled, the three machines will undergo commissioning and sea trials prior to installation of the wave farm, which will take place 5 *km* off the Portuguese coast, near Póvoa de Varzim. The project will have a rated capacity of 2.25 *MW*.

O&M

Suitable maintenance strategies are vital to ensure the long term reliability of any technology, in particular for wave energy converters. The design for reliability and the initial maintenance schedules must be cautious because systems failures can be aggravated and there may be no suitable weather window immediately available to retrieve the machines in the event of any given failure.

The Pelamis is designed to be resistant to single point failures through a combination of inherent robustness and redundancy where required. It is cost effective to avoid loss in generation or integrity through the inclusion of redundant systems. It is also cost effective to avoid any on-site maintenance and make use of the ability to connect and disconnect quickly and cheaply across a wide range of conditions. The Pelamis mooring system has been repeatedly demonstrated on the refitted prototype machine allowing connection in under two hours and removal in around 15 minutes in seas of $H_s > 2\,m$. Further enhancements will allow even faster operations over a greater range of conditions.

PWP acknowledged the importance of inspections and maintenance from the start of its operational programme. The offsite maintenance policy means that if and when necessary the Pelamis can be removed from site and docked at the quayside; for the FSP this was in Lyness (Orkney). All electrical systems are kept alive while the machine stays in a default control mode, thus allowing continuous monitoring of all systems. The easy access to the prototype ensures that if any modification to a given component needs to be made then it can be done promptly and in conformity with strict health & safety policies. In order to maximise the knowledge of issues like fatigue and to proceed with technological updates, inspections and the opportunity for refits also need to be programmed. Figure 7.66 shows the refit of the FSP which took place in mid 2006 at one of the dry docks in the Port of Leith (Edinburgh). Along with the introduction of engineering innovations such as the quick moorings connection system (note that two years had passed since the launch of the full-scale prototype) inspections were carried out on major structural elements such as the main bearings and attachment points, and other systems such as hydraulic rams, accumulators, cable transits, and bellows seals.

After this operation, the full-scale prototype was ready for a new set of sea trials (which took place in early 2007) and for a further installation round at EMEC. Results from one of the installation periods are presented in the following subsection.

Full-Scale Prototype – Results

The full-scale prototype Pelamis has been in Orkney since 2004. Following a set of sea trials in the North Sea (see Fig. 7.17) the first installation took place in

Fig. 7.66. The full-scale Pelamis prototype while being refitted in Leith (Edinburgh), 2006

August 2004 (Fig. 7.18). As previously described the machine then underwent a major refit operation in Leith (Edinburgh) in 2006. New systems where installed to reflect the design enhancements of the P1A machines due to be installed in Portugal and to test other novel components.

The success of this refit operation lead to a new round of sea trials which were conducted by early 2007 in the North Sea. Subsequently the Pelamis FSP was towed up to Orkney for a further installation round at EMEC, which spanned between April and July.

In analogy to other subsections of this chapter, a significant 30 minute window was selected for detailed examination (the period between 14:00 and 14:30 on 08/07/2007). During that half-hour the sea state was characterised by $H_{m0}=2.8\,m$ and $T_{-10}=6.7\,s$.m. The average absorbed power over the half hour was $147\,kW$, while the average generated power was $95\,kW$, resulting in a 30 minute average wave-to-wire conversion efficiency of 65%, as expected for the FSP. The peak (instantaneous) values for absorbed and generated power over that period were $842.2\,kW$ and $201.2\,kW$, respectively. The *rms* equivalents were $182\,kW$ and $101.9\,kW$. The performance assessment software developed by PWP allows a thorough check of selected variables. For example, plots of averaged time series can be produced with an arbitrary buffer size (Fig. 7.68 shows the results for a 10 minute window, now for an hour period stating at 14:00). In this hour the corresponding levels of wave power dropped from $28\,kW/m$ to $19\,kW/m$ at 14:30, keeping this lower bound roughly for 20–25 minutes until 15:00. The efficiency however remained well above 60% throughout the entire one hour period. It is interesting to zoom in the absorbed and generated power time series: Fig. 7.69 shows a three minute window of the instantaneous values of both. The effect of the sea state is clear in the absorbed power plot, while the hydraulic power take-off system allows the generated power to be much smoother with respect to time. In fact, this plot would present a flat line if the control settings were not so conservative (as previously mentioned), even in a sea state such as this one, where the effects of different wave groups were particularly clear. All power measurements are independently verified by EMEC. Testing of the FSP has followed a cautious approach with steady implementation of the most robust control strategies with conservative settings relative to design constraints. Power absorption can be significantly higher with bolder control settings and future designs include constraints adjusted to take full advantage of potential absorption.

All machine data can be included in automatically generated summary reports, which also include the relevant wave statistics and the characterisation of the frequency and of the directional spectrum. Figure 7.70 shows the corresponding frequency spectrum for this half hour, where the continuous curve represents the Pierson-Moskowitz (PM) spectrum with the same H_{m0} and T_{-10}. Early Pelamis simulation work assumed a model PM spectrum but more recent models allow the use of buoy data directly. Thus the state of the art simulations include non-linear hydrodynamics, accurate modelling of the control and hydraulic PTO, moorings system, and wave excitation generated from buoy measurements in real seas. When such methodology is used, the outputs of the numerical simulation are typically within 10% of both the absorbed and generated power. This is a reassuring example of numerical models extending from the recreating tank tests and hypothetical situations into the real world with good agreement with a full-scale machine operating in real seas.

Another significant milestone achieved during the 2007 FSP operations is demonstration of rapid installation / removal times using small work boats in conditions up to $2\,m\ H_s$. This proves that future Pelamis wave farms can be serviced cost effectively. Greater advances in Pelamis operations technology are currently underway to further streamline the installation and removal of machines and extend the range of conditions available for intervention.

Fig. 7.67. The Pelamis full-scale prototype installed at EMEC (June 2007)

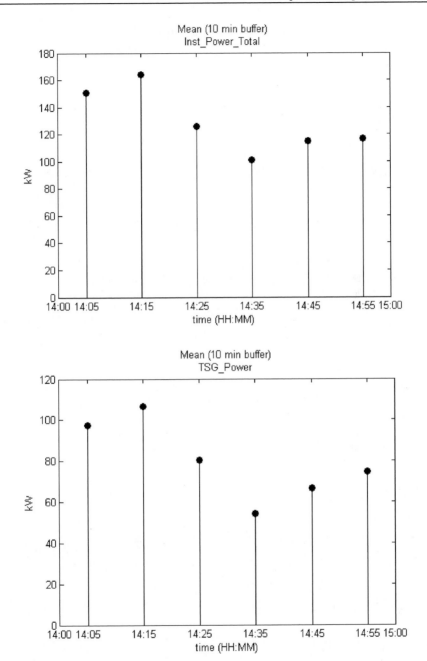

Fig. 7.68. 10 minute averaged absorbed (top) and generated power (bottom) – 08/07/2007

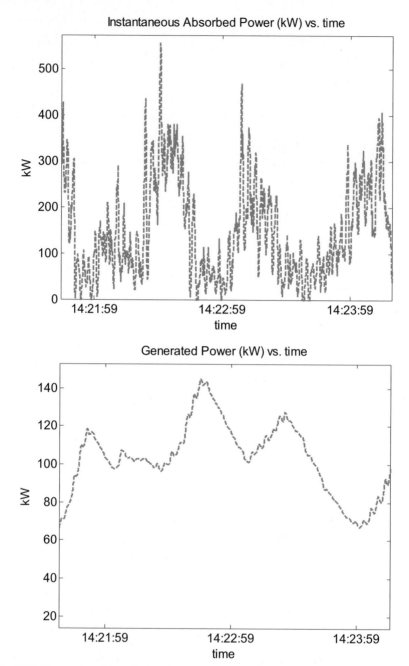

Fig. 7.69. Time series of absorbed and generated powers – 08/07/2007 14 : 21–14 : 24

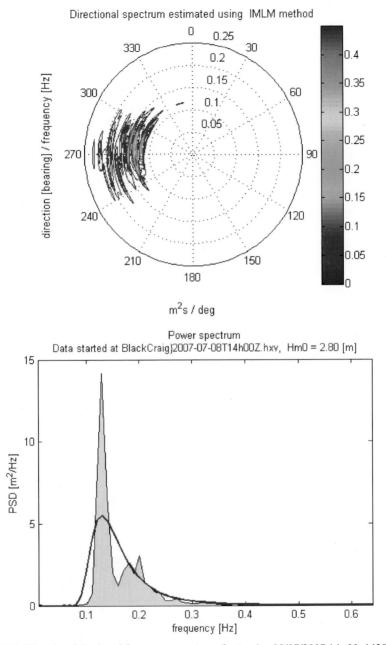

Fig. 7.70. Directional (top) and frequency spectrum (bottom) – 08/07/2007 14 : 00–1430

Control

Control of Pelamis is achieved by means of a bespoke real time system with a custom graphical user interface (GUI). Every output in the machine can be controlled in real time with respect to every input. This allows immensely powerful control algorithms to be implemented and provides the flexibility to improve the power absorption capability of existing machines through the upgrade of control algorithms. The potential for performance improvements through control upgrades is immense and the associated costs marginal.

While advances in the user interface are being made as the technology progresses towards less user intervention and greater autonomy. The FSP interface allows low level control of every aspect of the machine and continuous monitoring of all systems. Figure 7.71 shows the main screen of the FSP GUI. The left hand side represents the machines dynamic response and the right hand side represents the power generation systems. Each side is further split into three parts, each corresponding to one of the power conversion modules. Such flexibility and independence allows the PWP control team to safely implement new control algorithms which can maximise the generated power without compromising the survivability of the machine.

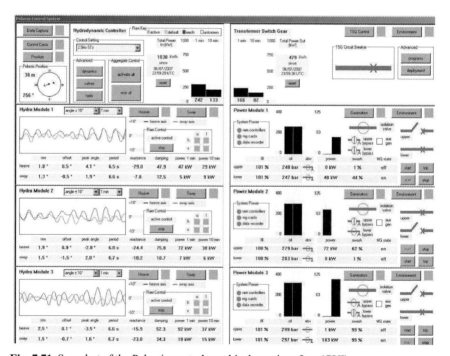

Fig. 7.71. Snapshot of the Pelamis control graphical user interface (GUI)

All changes to the control system in service are carried out as part of a robust development and testing programme. A key element in this is the testing of all software changes on a purpose built test-bench which includes all the control hardware of the real machine along with emulation hardware to provide real electronic signals to the various devices and model the physical systems of the real machine. Thus the software can be bench tested for errors in the context of the real hardware with real representative signals. This bench test approach also allows any issues that appear on the real machine to be recreated and examined safely.

Summary

This subsection has provided a short account of the operational experience that PWP has gathered from its full-scale demonstration programme in Orkney. The direct results of the lessons are now being applied in new designs and in other projects, like the Aguçadoura wave farm in Portugal. The flexibility and power of the control system is likely to provide dramatic improvements in performance in the newer generations of machines. PWP's numerical simulation package has been demonstrated its ability to model detailed full scale machines in real seas. The monitoring and post-processing packages are also a continuous reassurance of the quality of the methodology and these are now undergoing integration into the control and analysis suite for the new farm interface.

7.5.5 Wave Dragon

James Tedd, Jens Peter Kofoed, Erik Friis-Madsen, Lars Christensen

Wave Dragon ApS
Copenhagen
Denmark

Introduction

Wave Dragon have deployed and run a successful prototype wave energy power plant in Northern Denmark for over 3½ years. The device is located in an inland sea and so is scaled appropriately, while maintaining all automatic control and power take-off system and grid connection which will be seen in larger open sea power plants. This section will introduce the device and the Europe-wide consortium who built it. The operational experience gained from running the device will be shared by considering a selection of the subsystems on board. The power production results will be presented showing how the prototype has shown Wave Dragon to achieve an efficiency of 18 %, close to the long term goal.

A scale prototype in a 1 : 4.5 scale sea

Nissum Bredning, a sheltered sea connected to the Danish North Sea, was chosen as the site to deploy the first Wave Dragon prototype (see Figures 7.72 and 7.73). By deploying in a less energetic site many of the components and subsystems could be tested and developed at a reasonable cost. Also the expense of a site visit was greatly reduced. The available wave energy in the site makes it suitable for a 1 : 4.5 scale machine (relative to the North Sea). Therefore the original concrete construction drawings were scaled down and the construction material replaced with steel, allowing small changes to the structure to be made on board.

The main aspects to be tested at this site were those which can not be successfully modelled in the laboratory, like the low head hydro turbines controlled by variable speed Permanent Magnet Generators, and the hydraulic response of the platform with its open buoyancy compartments. In addition, control strategies for optimal power production and buoyancy regulation have been focused. Operating in the harsh offshore environment has taught many additional lessons.

The device itself has been very highly instrumented with over 100 sensors. These include: pressure sensors to measure the incoming waves, floating height and water in the reservoir; strain gauges and force transducers to measure the

Fig. 7.72. The platform at the dockside in 2003 before being launched. On the right the open air compartments used for buoyancy regulation can be seen

Fig. 7.73. Wave Dragon in Nissum Bredning. Wave Dragon was deployed at two sites in the sea, initially at the Northern site, and then later at the more energetic southern site

forces in the structure; a wind station; accelerometers and inclinometers to view the positions of the device; electrical sensors within the power take-off system and many more. Amongst the most used of these have been the five web-cameras mounted on board, allowing 24 hour visual checks of the situation on the platform.

Wave Dragon consortium

An early decision by the Wave Dragon team was to have the majority of work performed by experts in the field. This philosophy has led to a very dynamic company structure with a small core element, and a broad selection of trusted partner companies with years of experience in their own field. This structure is favoured by many public funding bodies, and allowed the Wave Dragon project to qualify for support from the European Commissions R&D targeted funds. The partner companies provided match funding for their own work and were rewarded with an option of a share in the Wave Dragon core company. With the modern age of telecommunications there was no problem with the wide spread of work across the European continent. The partner companies involved and their respective roles are:

Wave Dragon ApS is the development company for the wave energy converter technology.

Löwenmark Consulting Engineers FRI – A small Danish company which is home to the inventor of Wave Dragon Erik Friis-Madsen. This company has worked closely with all the engineering companies in the project to ensure the best possible design for the device. Most work is now performed by the Wave Dragon company.

SPOK ApS – This small Danish company provided the project management for the consortium and developed the Wave Dragon corporate side and initiated further projects. In addition to this work they had the responsibilities for the environmental surveys and Life Cycle Analysis work conducted for the prototype. Most work is now performed by the Wave Dragon company.

Balslev A/S – As a larger Danish electrical and control engineering company they have worked on the active control of the Wave Dragon. This has

involved implementing the control strategies into the Programmable Logic Controller (PLC) and development of the System Control and Data Acquisition (SCADA) user interface for the operation of the Wave Dragon.

ESBI Engineering Ltd – This large Irish consultancy (once part of the Electricity supply board of Ireland) have had responsibility for the design and monitoring of the grid connection of the Wave Dragon.

NIRAS AS – As a larger structural engineering consultancy based in Denmark this company has had a broad brief of responsibilities. The core of their work was to define the design criteria for the device and perform some detailed design. In addition they worked with Finite Element Modelling of the internal forces within the structure and prediction of waves from the wind states.

Promecon A/S – A subsidiary of MT Højgaard this company is one of the largest steel constructors in Scandinavia. They constructed the main structure in Steel and were responsible for the successful deployment of the device.

Kössler Ges.m.b.H. – This Austrian company has decades of experience in construction of hydro turbines. As a partner in the consortium they brought this experience into play when constructing the turbines used on board the device.

Aalborg University – As the local university in northern Denmark, and one of two major technical universities in the country, Aalborg University have been core to the Wave Dragon project since the earliest days. The first scale testing was conducted in their laboratories. In this project Aalborg University specialised in monitoring the prototype. This has led to many published research papers on all aspects of the prototype and countless students have been inspired from working on related projects.

Technical University Munich – The turbine group at this university joined the project with the aim of designing and testing the scaled turbines, and analysing their control strategies in cooperation with Veterankraft AB. As the project developed, they have also been crucial in many more practical matters, mainly with maintenance of the turbines.

Armstrong Technology Associates – A structural engineering company from the UK. Initially in the consortium to provide naval engineering expertise to convert original concrete 1 : 1 construction drawings to 1 : 4.5 scale steel equivalents. Unfortunately after initial design work the company changed direction and withdrew, their responsibilities were taken on primarily by NIRAS AS.

Performance in storms

Wave Dragon has had survivability built into all systems from the start. Overtopping devices are naturally adapted to perform well in storm situations, where the wave will pass over and under the device with no potential end-stop problems. However as the prototype has been deployed for more than three years in an offshore situation it comes as no surprise that the platform has had to survive some severe weather. With the web cameras on board some dramatic moments have been caught, as in Figures 7.74 and 7.75.

The two incidents shown on camera did not cause any damage to the Wave Dragon. Unfortunately this has not always been the case. On one occasion damage was suffered to the main platform and a large fixed buoyancy tank was punctured. This was a reminder of the need for redundancy in the design, here the other fixed buoyancy tanks were sufficient to allow the Wave Dragon to continue operating. During the next calm period the breached tank was emptied, the puncture welded tight, and the tank separated into two smaller units to add extra redundancy.

Fig. 7.74. View from web camera mounted to control container of large wave surging down reflector wing of Wave Dragon

Fig. 7.75. View from web camera mounted to the device's shoulder of waves crashing into turbine area

In 2005 the largest storm for 100 years hit northern Denmark. Average wind speeds were in excess of 110 *km/h* with gusts above 150 *km/h*. In addition to widespread structural damage to buildings in the region the storm broke a force transducer mounted in the mooring system of the device. On the prototype the force transducer was selected to give accurate readings of mooring forces in even small waves. It therefore did not have the safety factors of the other sections of the mooring system. Small defects in the steel could be seen at the nucleus of the fracture shown in Fig. 7.76.

After loosing the mooring connection the Wave Dragon soon drifted onto a sandy shore, where it suffered little damage. The device was soon refloated and taken to harbour for some scheduled maintenance. Thus Wave Dragon became the first WEC to be successfully recovered and redeployed at sea.

Fig. 7.76. Fractured force transducer after Force 11 storm and view of transducer in position taken by a diver

Control of the device

All of the control devices for the Wave Dragon prototype are housed in a container mounted to the back of the Wave Dragon platform. Clearly this is oversized on the 1 : 4.5 scale device as any operational staff will not scale.

The core of the control system is a Programmable Logic Controller (PLC). The PLC controls the blowers and valves to regulate the buoyancy of the device, aiming for a horizontal and stable platform. A series of regulation strategies have been implemented within the PLC with implementing aspects of PID control giving more stable behaviour. The regulation is able to change the floating height of the platform by around 20 *cm* every hour, allowing the crest freeboard to rise with increasing storms, and thus improve the power capture. The PLC is also responsible for controlling the action of the turbines, in order to extract the maximum energy from the water which has overtopped. The PLC controls a hydraulic pump and solenoid valves to allow the water hydraulic system to operate the turbine cylinder gates or evacuate the siphon. Again several strategies have been tested on board, with much improvement.

One of the PC's in the control room is running a System Control and Data Acquisition (SCADA) system allowing the primary human interface with the device. In Fig. 7.77 some screenshots from the system are shown. In addition to giving the current status of the device the system allows set points to be altered and records historic values for the power production system. In parallel with this a more typical data acquisition system is run recording the measurements not used in the active control of the device, but of interest for scientific and development uses, such as strain gauge readings and wind records.

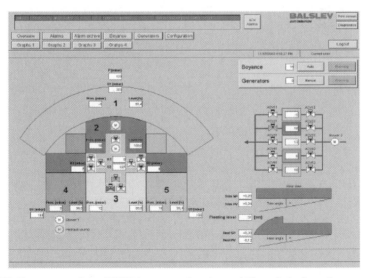

Fig. 7.77. Front page of the SCADA system showing stability of platform, buoyancy tank status, and turbine status

The SCADA PC is connected to the internet, and therefore allows control of the device remotely. This is aided by the use of five web cameras as a visual check on any abnormal behaviour. However on occasion this link has been lost; either due to grid power failures in the district, power failures on board, or simply issues with the internet provider. In these cases a backup system which uses the GSM network becomes operational. In the case of serious abnormalities the PLC will send SMS messages to two nominated people. These people are able to control the PLC in a basic way by sending coded messages to the device. If both of these lines of communication are all down then the PLC will continue to operate in a safe manner. If power is lost to the PLC, the structure of the Wave Dragon will stay afloat without any active control system.

Moorings

The design for a full sized Wave Dragon has the device moored to anchors by several catenary chains, a CALM buoy system. In Nissum Bredning the site is too shallow (at around $6\,m$) to allow this system. In the prototype the compliance of the mooring is provided by using elastic nylon and polyester ropes, a schematic of this is shown in Fig. 7.78. The anchor is provided by a steel bucket filled with local sediment. A pile is attached to this anchor which is of great use, for mounting wind measuring devices, a webcam, and a pressure transducer to measure the incoming waves. A second smaller back anchor is present in the prototype to prevent the device colliding with the mooring pile. This restricts the movement of the device to $\pm 60°$, which is acceptable close to a coast.

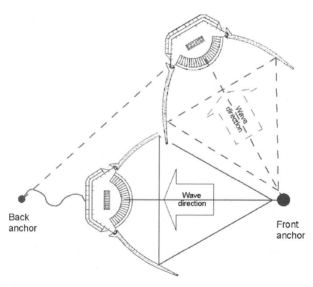

Fig. 7.78. Mooring Schematic

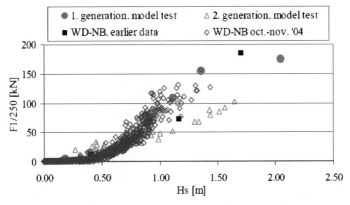

Fig. 7.79. Comparison of mooring forces in terms of F 1/250 (average of the 1/250 largest peaks) measured during 1 : 50 scale tests and prototype measurements (1 : 4.5). H_s is given in prototype scale

A lot of data has been recorded by the force transducers in the mooring system. As reported in Kofoed et al. (2006) and shown in Fig. 7.79, the mooring forces show good correlation to the scale testing in the laboratory. There was a noticeable increase in the stiffness of the mooring lines due to marine growth during the period. This will not be an effect on catenary chains in a full scale deployment.

Experience with hydro turbines

Power simulation investigations show that a power take-off system consisting of 16 to 20 variable speed on/off low-head turbines of the Kaplan type with fixed runners and guide vanes is optimal for a production Wave Dragon. For the prototype platform a choice was made to replicate these with a selection of turbines, 10 in all. Each generating turbine is directly connected (no gear box) to a permanent magnet generator (PMG). Each PMG is connected to a frequency converter, which is used for control of the speed of the turbine and for supplying the power from each turbine onto a common DC rail. The power is then via another frequency converter put onto the grid with grid frequency and voltage. The turbines used are shown in Fig. 7.80 and detailed below.

- A Kaplan turbine with siphon inlet. This is the original turbine which had been tested at the Technical University of Munich. The runner diameter of the siphon turbine is $0.34\,m$, rated flow is $0.22\,m^3/s$ (at $0.5\,m$ head) and rated power is $2.6\,kW$ (at $1\,m$ head) corresponding to $350\,kW$ in a full scale North Sea Wave Dragon.
- Six Kaplan turbines with cylinder gates. These turbines have the same runner diameter and performance data as the siphon turbine. The turbines were fabricated in Austria by Kössler GmbH, their runners were made by TUM.

Fig. 7.80. Turbines in the reservoir of the prototype: The structure in the background is the siphon turbine, the grey towers are five of the six cylinder gate turbines, and the small devices in unpainted steel are two of the dummy turbines

- Three dummy turbines. These turbines are not able to produce power; they are merely calibrated valves which let the overtopped water run back into the sea. The diameter of the valves is $0.43\,m$, and discharge is about twice one of the real turbines. They permit accurate simulation of the discharge from a further six real cylinder gate turbines at a fraction of the cost.

  In order to avoid debris in the turbines the turbine area was enclosed by a trash rack, see Fig. 7.81. There is also a powerful water pump on board to enable testing and calibration of the turbines in calm situations.

Fig. 7.81. Trash rack with $5 \times 5\,cm$ grid openings enclosing the turbine area

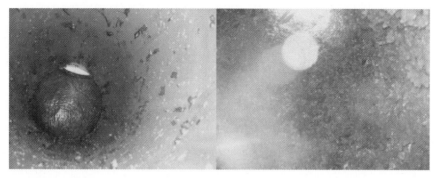

Fig. 7.82. The insides of two turbine draft tubes, on the left one painted with silicone paint, and on the right painting with conventional epoxy paint

Experience has shown that the aggressiveness of the salt water environment had been underestimated. A design proven in many river hydro power turbines has been used for the main turbine bearing, but the bearings failed after a few months of operation due to problems with the shaft seals.

The bearings of four out of the seven turbines have been modified and rebuilt during the 2004 summer, and the turbines have been operated without further problems from October to December 2004. The bearings of the remaining turbines have been modified and rebuilt during the 2005 summer. It was possible to conduct the majority of this work onboard the floating platform.

Dismantling the turbines during the above mentioned repairs has also shown that marine growth is a factor that may not been taken lightly. The draft tubes of the turbines had been made from different materials: uncoated stainless steel, black steel protected with conventional epoxy paint and black steel protected with a recently developed silicone-based anti-fouling paint. The draft tubes painted with the conventional paint system were found heavily overgrown (mainly by the acorn barnacle, *balanus crenatus*, and sea squirt, *ciona intestinalis*), which were almost impossible to remove. The stainless steel tubes as well as the ones coated with the silicone paint had only a few small barnacles on them which could be swept off very easily, as can be seen in Fig. 7.82.

Power Production

There are many ways to show the power production of a WEC, depending on where you measure the power. In several published papers (e.g. Tedd et al., 2006) some of these methods are explored. Figure 7.83 shows a typical time series for a record.

An enormous quantity of data has been collected during the testing period, which has not yet been fully analysed. However, the work done up to now has confirmed that the performance predicted on the basis of wave tank testing and turbine model tests will be achieved in a full scale prototype.

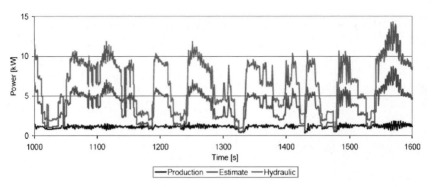

Fig. 7.83. Time series of a typical record, showing: Production, power delivered to the grid; Estimate power, which supposes the turbine set points were correct and the dummy turbines produced as real turbines; and Hydraulic power the power of the water passing through the turbines

Looking at the period May 2003 to December 2004 and scaling the energy production (1 : 4.5) to a typical 16 *kW/m* wave climate as found in the North Sea the prototype Wave Dragon would have produced from 50 to 500 *MWh/month*. Taking into account down periods and testing periods the real production would have been approximately 3.2 *GWh/year*. The latest tests have shown that an optimal setting of the set points of the PM generators has increased the power production with a factor of two. Therefore it is assumed that the real production easily could have been 6.5 *GWh/year* or equal to an 18 % average wave-to-wire efficiency.

This result should be compared to the 16 % prototype goal and the 21 % long term goal for the Wave Dragon technology. Measurements of the hydraulic power indicate that it will be possible to reach this value of energy production. Some of the discrepancies are believed to be due to the scaling which will cause extra energy losses in bearings etc.

Summary

Since March 2003 a prototype of Wave Dragon has been tested in an inland sea in Denmark. This has been a great success with all subsystems tested and improved through working in the offshore environment. The project has proved the Wave Dragon device and has enabled the next stage, a production sized version.

7.6 Test Centres, Pilot Zones and European Cooperation

António Sarmento

Wave Energy Centre
Lisbon
Portugal

The design, construction, deployment and test of pilot plants and prototypes are of fundamental importance in the development of the technology as they provide an opportunity to assess design methods and criteria, to work out engineering solutions, to identify the real costs of building, deploy and operate the pilot plant or prototype and, most importantly, to learn from experience, given that no matter how much effort is put in theoretical, numerical and experimental models and in engineering considerations, there are aspects that are only properly addressed in this phase.

However, prototype construction and testing are just the first step in the industrial and commercial development of a given technology, which corresponds to the demonstration of the concept, i.e., of the technological possibility of producing energy with such technology. The other important phases are the demonstration of deliverance, proving that the risks involved in the costing, quality of fabrication and time required to supply the device. The following step is the demonstration of performance: showing that the risk related to energy production, maintenance and operation costs and the resulting cost of energy are controlled. Only then a project-finance can be set on the basis of which wave farms projects are financed due to their intrinsic economic value.

The first phase of development, the demonstration of concept, either via a 1 : 2 or 1 : 4 scale pilot plant or a full-scale prototype, requires a test centre, like the test site in Galway Bay in Ireland[3], EMEC in Scotland[4], or, up to a certain extent, the former AWS site in Northern Portugal, to be used now for the deployment of three Pelamis units. The demonstration phase is usually very limited in time, will involve very low power levels and small energy production and, as it concerns an experimental phase, will require easy access to the device and a significant monitoring program. The impact of a single demonstration device during a limited period of time will be very small and should allow simple and fast licensing procedures, in particular if the test site is chosen in a low environmental sensitivity area. The wave resource and other environmental data (wind and current climatology, bathymetry, seabed properties, etc.) should be known and, in the case of the wave resource, it should be appropriate for the scale of the pilot plant. The test site should have one or more offshore connection points, monitoring facilities related

---

[3] http://www.marine.ie/home/aboutus/organisationstaff/researchfacilities/Ocean+Energy+Test+
   Site.htm
[4] http://www.emec.org.uk/

to the device itself and the environment. Proximity to commercial harbours should be strongly encouraged for both O&M and safety.

The second phase, demonstration of deliverance, requires a small wave farm of three to five devices. The Wavehub[5] project in Southwest England is intended to facilitate the deployment of these first wave farms. With a total connection power of 20 *MW* it will allow up to four small wave farms to be deployed, again with a simplified licensing procedure. However the Wavehub initiative does not seem to provide the possibility of these wave farms to expand and it is not obvious if the demonstration of performance will be possible to achieve with such small wave farms.

The demonstration of performance is just the initial phase of pre-commercial development. If successful, the three to five devices wave farms will expand to medium size farms of around 20 units in a first phase and to commercial wave farms sizes of 50 or more devices in subsequent phases. If several wave energy technologies are able to attain this stage, we will get very easily to the level of several hundreds of *MW*. The development may then be frozen by the lack of available connection points or by the time required to obtain the required licenses. It thus seems important that pilot zones with simplified regulations and licensing procedures, with significant grid connection availability and subsidised tariffs for the produced energy are developed. Like in the case of the test sites, these pilot zones should be set in areas of low environmental sensitivity and conflict of use, easy access from nearby ports, well served by nearby shipyards and proximity to the electrical grid, small distance between the 50 *m* contour and the shoreline, etc.

The pilot zone recently announced in Portugal has been thought to provide the conditions indicated above for continuous experimentation and expansion of the technologies, from the demonstration to commercial phase. It is an area of around 400 *km²* in waters from 30 *m* to 90 *m* depth, with a connection point up to 80 *MW* to the distribution grid (medium voltage) and, at a later stage, 350 *MW* to the transport grid (high voltage). The area has low environmental sensitivity, low conflicts of uses, appropriate wave resources, no significant currents, good meteorological conditions, sandy bottom including along the route to the shoreline and is well served by the nearby electrical grid and several ports and shipyards. The pilot zone will be managed by a company that will be responsible for the licensing of the wave farms, the environmental characterisation of the area and the availability of environment data, and the development of the electrical infrastructure. It is desirable that other pilot zones are created in different countries to allow the development and testing of the technologies in different environments and legislations and to reduce the risk of shortage of connection points. The company will also be responsible for the monitoring of the wave farms. This will allow the identification of the resulting positive and negative environmental and socio-economic impacts of wave energy in order to foster the firsts and mitigate the lasts and to correct the regulations and legislation where necessary.

---

[5] http://www.wavehub.co.uk/

Wave energy will be still in a long term process before it reaches the degree of technological maturity and economic competitiveness that will in turn allow it to compete with other forms of well established technologies. The pilot zones may provide an easier route for this development, but do not avoid the need for international collaboration in many areas and through partnerships of different kinds.

The areas of collaboration include the identification and removal of technological and non-technological barriers, the development of best-practices and standards, the availability of adequate research and test facilities and the training of young technicians and researchers. The support of the European Commission (EC) to enhance this collaboration has been very important, either by funding R&D and demonstration projects, or by funding appropriate networks, such the Coordinated Action of Ocean Energy Systems (CA-OE)[6] or the Marie Curie Research Training Network Wavetrain (Sarmento et al., 2006). The Coordinated Action is a forum to coordinate the research and development of about 50 partners with its work scheduled along five areas: i) Modelling of Ocean Energy Systems; ii) Component technology and power take-off; iii) System design, construction & reliability; iv) Performance monitoring of ocean energy systems; v) Environmental, economics, development policy and promotion of opportunities. The Wavetrain RTN is a research and training network involving 11 European partners (9 R&D partners and 2 technology developers) intended to train 14 young and experienced researchers in the area of wave energy, covering about the same topics as the CA-OE.

The collaboration involves or should involve different types of partnership: companies providing different expertise, R&D institutes and universities and governments and public administration departments and should be developed both at national, European and international levels. Indeed, if Europe is the leading region in developing ocean energy, there is no doubt that the involvement of countries with ocean resources from other parts of the globe is fundamental to create the proper dynamics by providing extra funding, expertise and grid connection points. This last point may be a major factor affecting the development of ocean energy.

7.6.1 Case Study: Wave Dragon

James Tedd, Hans Christian Sørensen, Ian Russel

Wave Dragon ApS
Copenhagen
Denmark

Introduction

As one of the pioneers of the wave energy field, Wave Dragon has been instrumental in developing Denmark's test centre for wave energy in Nissum Bredning.

---

[6] www.ca-oe.net/home.htm

This has spurred development of Wave Energy in the area with more devices be-ing tested in the vicinity. These later deployments have been much easier as Wave Dragon had already mapped the resource, performed environmental surveys, and liaised with local contractors to ensure reliable maintenance.

The next move for Wave Dragon is to build a device in Pembrokeshire, West Wales in the UK system. As a new technology deployed within a UK national park, Wave Dragon must provide a detailed Environmental Impact Assessment, before gaining consent for the deployment. This new pilot site is an early test case for the authorities of England and Wales, where it is only the second such applica-tion to be submitted.

Support from the European Commission together with the Danish and Welsh governments has been vital to the development of Wave Dragon. Support was forthcoming in the early tank testing stages, through the prototype development and now for development of a full sized pre-commercial demonstrator in Wales. Therefore Wave Dragon has operated a very open policy, publishing as much and as often as possible, supporting all Europe-wide initiatives, as well as hosting ex-change students. This has proved to be of great value to the company.

Nissum Bredning Test Centre

Nissum Bredning (broads) is a sheltered sea in the Danish mainland separated from the Danish North Sea by two thin tongues of sand, see Fig. 7.84. The sea covers an area of approximately $200\,km^2$ with the longest continuous fetch in the area of $29\,km$. Therefore the waves in the area are solely driven by the local wind, with no long period swell waves. The water depth is mostly around $6\,m$, although there are shallower regions in the west.

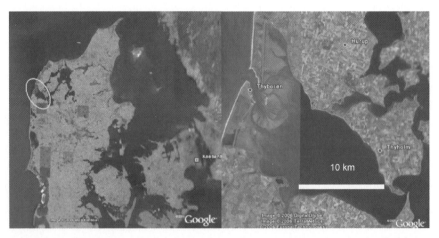

Fig. 7.84. Satellite images of Nissum Bredning, from Google Earth. The detailed image shows shallower sandbanks to the West

In 1998 the Danish Energy Authority and the Danish Wave Power Association collaborated with the local Nordic Folkecentre for Renewable Energy to construct a pier for initial testing of Wave Energy devices. This has been used by several inventors to test small wave energy devices, by putting them in the water at the end of the pier.

In preparation for the Wave Dragon test deployment in Nissum Bredning several studies were undertaken. The most important was to analyse the wave resource within the broads. As the waves are fetch limited the Shore Protection Manual method (SPM) could be used to determine the waves. Wind speeds to be used were well defined from four weather stations in the vicinity and the bathymetry is also well known. The study of Svendsen and Frigaard (2001) produced maps showing the mean wave power per unit width.

The Wave Dragon deployed here was built in scale 1 : 4.5 of a Wave Dragon for the North Sea. Wave power per meter of crest scales with a power of 2.5. The Northern site therefore has an equivalent power of $16\,kW/m$. The Wave Dragon was first tested here, as it is very accessible for early teething maintenance issues and a grid connection was available close to the test pier. The Southern site was chosen to test the device in the largest waves available within the broads; here there is an equivalent power of $24\,kW/m$. However it is less accessible, as a boat must travel from the harbours of Oddesund or Remmerstrand, both of which are around $5\,km$ from the site.

Before deploying, Wave Dragon sent a consent application – including a description of the device – to the Danish Energy Authority. They performed a consultation process with all possible authorities (environmental, H&S, regional, local and military), and with relevant associations (fishermen, environmental, etc.) and announced the project in local newspapers and called for public objections. The consent was granted with these objections converted into requirements. The authorities and others found that there would be no or close to no negative effects from the project given that it was temporary. All found that it was a good idea to test new technologies. The full process is described well by Hansen et al. (2003).

Fig. 7.85. The Nordic Centre for Renewable Energy's test pier, seen at sunset, with Wave Dragon floating to the right

Fig. 7.86. Bird life in Nissum Bredning. An Arctic Tern perches on the tip of the reflector wing, and a Herring Gull has made its nest in some spare cable on the roof of the control room

The broads area is designated an EU Bird Protection Area (due to three diving ducks species: Common Goldeneye, Red-Breasted Merganser and Goosander); the northern site is also protected by the Ramsar Convention (on wetlands and water-fowl habitat of international importance). In response Wave Dragon agreed to not perform intrusive on-site work during the spring breeding season. Only a small area of seabed is affected by the device, so it has only a small impact on mussel fishing. A non-toxic anti-fouling paint was applied to the turbine draft tubes to prevent poisoning of shellfish. The Wave Dragon has been grid connected at both the northern and southern locations.

In 2006 another Danish device, the Wave Star, a multi-point absorber was installed, mounted to the end of the test pier at the northern site. There are future plans for the southern site to be used for a prototype of a heaving buoy device. This would be using the offshore grid connection provided by Wave Dragon. Both of these devices have benefited from the Wave Dragon experience in the broads area, showing that wave energy can be environmentally benign.

Pembrokeshire Pilot Zone

Wave Dragon is progressing through the next stage of development: to build and deploy a device at a power production scale in order to show the technology to be commercially attractive. The project aims to deploy a 4–7 *MW* Wave Dragon, for 3–5 years in the Irish Sea, close to Milford Haven, Pembrokeshire, Wales. After the period of initial testing the device will be towed to a more energetic site, around 19 *km* offshore, to join a larger array. As this will be the first Wave energy device in Wales, and one of the earliest in the UK, a very stringent planning process must be observed.

Wave Dragon decided to make the pilot zone in Wales for practical and economic reasons. For a project to be economically feasible there must be possibilities to move the first pre-commercial demonstrator into a farm of devices. This farm requires good wave conditions, a large port for construction and maintenance and good access to strong grid connection. Considering Fig. 7.87, the areas of good wave resource in the UK are the islands of Western Scotland and the South West of Wales and England. Unfortunately the Western Islands of Scotland have very weak grid connections. Pembrokeshire was chosen as the Welsh Development Agency (now the Welsh Assembly Government) has been very supportive of the project since 2001, and also all the facilities such as large deep harbours and skilled workforce exist within Milford Haven.

In Pembrokeshire there were several factors in choosing the precise site for the deployment. These varied from a desire to be close to shore and near a major port to reduce cabling and maintenance costs, to avoid major shipping lanes, and fishing areas, to avoid military training sites and munitions dumps, to minimise environmental impacts, and of course to choose a location with good wave climate. After initial considerations in the region, a $6\,km$ square box was chosen with the intention to deploy the Wave Dragon within it. A scoping study (May and Bean,

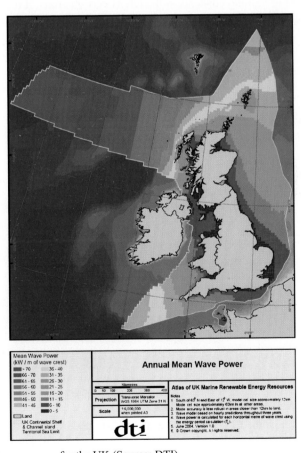

Fig. 7.87. Wave resource for the UK (Source: DTI)

2006) and consultation with all local, regional and national bodies and organisations enabled a full design for the Environmental Impact Assessment. This is required as the coast is a National Park, and the seabed itself is a Marine Reserve
and a Site of Special Scientific Interest (SSSI).

Currently a wide variety of specialist consulting companies are working on
the project, each looking at the impact within their own area of expertise.
These vary from the wildlife including seabed, inter-tidal, marine mammals,
birdlife, and more. The hydraulics of the device are studied for analyses of the
effect on coastal processes such as erosion, sediment travel and surf waves. The
human aspects are also covered including the effects on shipping, fisheries and
navigation risk and any archaeology which may be disturbed. Some results
from these studies are shown in Fig. 7.88 and 7.89. From the Geophysical survey, an area of coarse sediment could be seen to the north east of the survey
zone. The pilot plant will be deployed here, with the best mooring option, the
minimising of cabling costs, and limiting possible damage to some more sensitive rocky habitats.

This system of gaining consents in the UK is very rigorous; it must also be followed by other offshore projects, such as wind farm developments. This is advantageous as it clearly proscribes what needs to be done. However it is also a time

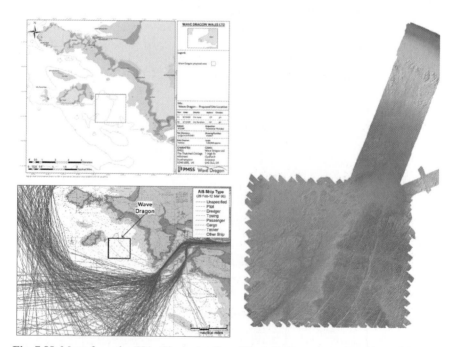

Fig. 7.88. Maps from the EIA, Clockwise from Top left, these show: Location of the surveyed $6\,km^2$; Bathymetry of the site and cable approaches taken during the Geophysical
survey; Ship paths recorded by AIS during a 14 day period

Fig. 7.89. Visual impact: A 7 *MW* Wave Dragon seen from the Pembrokeshire Coast Path above Marloes Sands. Photomontage by Enviros

consuming and expensive process, two luxuries thinly spread in the field of wave energy. A full discussion of the UK system, and comparisons to other country procedures, can be found in Neumann et al. (2006).

In parallel with the consents procedure the engineering design work is being conducted. Wave Dragon has started a new European Research Project to produce a baseline design for a *MW* sized Wave Dragon, and to instrument and monitor its deployment. This partnership is very similar to the group involved in the last EU project, with additional academic partners from Warsaw and Swansea Universities who contribute their expertise on power electronics. To complete the concrete engineering Design, Dr. Tech. Olaf Olssen, a Norwegian engineering firm with extensive experience of construction and deployment in the North Sea, have joined. Finally Wave Dragon Wales Ltd, the UK subsidiary of Wave Dragon is the partner who will build, run and own the device.

European cooperation

Wave Dragon is currently actively involved in the following Europe Wide projects:

1. The EU funded WD-MW Research Project (contract number 019983). This three year project will provide a baseline design for a multi-*MW* sized Wave Dragon and develop the remaining subsystems. It will continue to monitor the deployment of the first multi-*MW* sized Wave Dragon. There are 10 partners involved from seven European countries.
2. The Co-ordinated Action on Ocean Energy (contract number 502701 (SES6) CA-OE), in which Wave Dragon ApS participates as developer. Wave Dragon partners SPOK ApS, the Technical University of Munich and Aalborg University participate as experts in their fields. This project enables a direct link to suppliers, utilities and other developers.

3. The Marie Curie Research Training Networks "WAVETRAIN" (Contract nr. MRTN-CT-2004-505166) in which SPOK is participating. This EC financed mobility network programme aims at knowledge transfer and training of researchers within the field of wave energy. The Wave Dragon Team already had a wealth of experience with this programme and have hosted a research fellow (James Tedd) for a period of three years. Further on Wave Dragon Team has recruited another fellow (Iain Russell) from the network to work permanently in the newly established UK company Wave Dragon Ltd, mainly dealing with the environmental development.

4. EU-OEA – Dr Hans Christian Sørensen is a founding member of the board of the European Ocean Energy Association. The association will act as the central network for its members on information exchange and EU financial resources, as well as promoting the ocean energy sector by acting as a single EU voice.

5. NEEDS – Dr. Hans Christian Sørensen is representing the ocean energy developers in the New Energy Externalities Developments for Sustainability project establishing the thorough picture of the full costs including externalities of future energy systems. A Life Cycle Analysis for Wave Dragon will be included.

Wave Dragon was previously involved in the following European projects:

1. EU Joule Craft project (phase 3a). This early project involved feasibility testing of the design, both by computer simulations in Denmark and Germany, as well as laboratory tests of a $1 : 50$ scaled device in Denmark and Ireland, and tests of the low-head turbines in Germany.

2. WaveDragon $1 : 4.5$ research project (contract number: ENK5-CT-2002-00603). This project supported the sea-testing and deployment of the $1 : 4.5$ scale demonstrator of Wave Dragon. It is described in great detail in section 7.2. and in the final report to the Commission (Sørensen et al., 2006).

In addition to these formal projects that Wave Dragon has been involved with the company has been very active in promoting itself and wave energy in general to the academic and business world within Europe. This has involved giving many papers and presentations at conferences, and other workshops. Many students from France, the UK and Denmark have benefited from internships of several months with Wave Dragon, learning how a new technology can be brought forward.

Summary

Wave Dragon has been instrumental in developing the Danish Wave Energy test centre at Nissum Bredning. The company is now developing a pilot zone in Pembrokeshire to deploy a multi-MW device. This has been achieved with good support from national governments and European co-operation projects.

7.7 Discussion

Chapter 7 has provided a comprehensive review on the concepts that have reached the full-scale stage, including details regarding their development programmes. All concepts presented in sections 7.1 to 7.4, some *nearshore* some *offshore*, some *submerged* some *floating*, have generated and delivered electricity to national grids. There are hundreds of patented wave energy conversion machines, but at a time when wave energy is reaching a pre-commercial stage these concepts are the most likely to become immediately available and readily installed. However it is impossible to say with absolute certainty if any of these concepts will be selected for future large scale wave farms, or if alternative devices will emerge.

A section was also dedicated to the operation experience gathered with these technologies (7.5). It is essential that all existent or new technology developers incorporate these lessons in their own development programmes, ensuring that the same mistakes are not repeated. Although the industry is at its infancy, this considerable amount of experience is encouraging when predicting the future of wave energy conversion.

An account of the available (and planned) test centres was also given (see also Chapter 8). Finally it is encouraging to witness that there has been a great deal of cooperation between research centres and technology developers, which is particularly clear in the number of European collaboration networks that have been established. These networks are likely to be extended to other geographical areas in the near future.

References

Oscillating Water Column

Wave Energy Research and Development at JAMSTEC: Offshore Floating Wave Energy Device Mighty Whale (http://www.jamstec.go.jp/jamstec/MTD/Whale/)

Falcão de O AF (2000) The shoreline OWC wave power plant at the Azores. Proc 4th Eur Wave Power Conf, paper B1. University of Aalborg, Denmark

Neumann F, Brito e Melo A, Sarmento A (2006) Grid connected OWC wave power plant at the Azores, Portugal. Proc Int Conf Ocean Energy. From innovation to industry, OTTI, ISBN 3-934681-49-2, pp 53–60

Gato LM, Falcão AF (1990) Performance of Wells turbine with double row of guide vanes. Int J Japan Soc Mech Eng II 33:265–271

Archimedes Wave Swing

Cruz J, Sarmento A (2007) Sea state characterisation of the test site of an offshore wave energy plant. Ocean Eng 34(5–6):763–775

Leijon M, Bernhoff H, Agren O, Isberg J, Sundberg J, Berg M, Karlsson KE, Wolfbrandt A (2005) Multiphysics simulation of wave energy to electric energy conversion by permament magnet linear generator. IEEE Transactions on Energy conversion 20: 219–224

Prado M, Neumann F, Damen M, Gardner F (2005a) AWS results of pilot plant testing 2004. In: Proc 6th Eur Wave Tidal Energy Conf. Glasgow, United Kingdom

Prado M, Gardner F (2005b) Theoretical Analysis of the AWS Dynamics during Submersion Operation. In: Proc 6th Eur Wave Tidal Energy Conf. Glasgow, United Kingdom

Prado M, Gardner F, Damen M, Polinder H (2006) Modelling and test results of the Archimedes Wave Swing. Proc IMechE Part A. J Power Energy 220:855–868

Rademakers LWMM, Van Schie RG, Schitema R, Vriesema B, Gardner F (1998) Physical Model Testing for Characterising the AWS. In: Proc 3rd Eur Wave Energy Conf, Volume 1. Patras, Greece, pp 192–199

Sarmento AJNA, Luís AM, Lopes DBS (1998) Frequency-Domain Analysis of the AWS. In: Proc 3rd Eur Wave Energy Conf, Volume 1. Patras, Greece, pp 15–22

Pelamis

Cook S (1998) Investigation into Wave Loads on Catamarans. Dep Appl Phys. Curtin Univ Techonology, Perth, Australia

Falnes J (2002) Ocean Waves and Oscillating Systems. Cambridge University Press

Henderson R (2006) Design, simulation, and testing of a novel hydraulic power take-off system for the Pelamis wave energy converter. Renew Energy 31(2).271–283

Pizer D, Retzler C, Henderson R, Cowieson F, Shaw F, Dickens B, Hart R (2005) PELAMIS WEC – Recent Advances in the Numerical and Experimental Modelling Programme. Proc 6th Eur Wave Energy Conf. Glasgow, UK, pp 373–378

Retzler C, Pizer D, Henderson R, Ahlqvist J, Cowieson F, Shaw M (2003) PELAMIS: Advances in the Numerical and Experimental Modelling Programme. Proc 5th Eur Wave Energy Conf. Cork, Ireland, pp 59–66

Yemm R, Pizer D, Retzler C (1998) The WPT-375 – a near-shore wave energy converter submitted to Scottish Renewables Obligation 3. Proc 3rd Eur Wave Energy Conf, Vol. 2. Patras, Greece, pp 243–249

Yemm R, Henderson R, Taylor C (2000) The PWP Pelamis WEC: Current Status and Onward Programme. Proc 4th Eur Wave Energy Conf. Aalborg, Denmark

Wave Dragon

Corona L, Kofoed JP (2006) Wave induced stresses measured at the Wave Dragon Nissum Bredning Prototype. Co-ordinated Action on Ocean Energy. 3rd Workshop proceedings. Amsterdam

EMU et al. (2000) Publishable Final Report – Low Pressure Hydro Turbines and Control Equipment for Wave Energy Converters (Wave Dragon). Contract JOR3-CT98-7027. Copenhagen, Denmark

Franco L, de Gerloni M, Van der Meer JW (1995) Wave Overtopping on Vertical and Composite Breakwaters 1. Proc 24th Int Conf Coastal Eng. Kobe, Japan, pp 1030–1044

Hansen LK, Christensen L, Sørensen HC (2003) Experiences from the Approval Process of the Wave Dragon Project. Proc 5th Eur Wave Energy Conf. University College Cork, Ireland

Knapp W, Holmen E, Schilling R (2000) Considerations for Water Turbines to be used in Wave Energy Converters. Proc 4th Eur Wave Energy Conf. Aalborg

Knapp W (2005) Water Turbines for Overtopping Wave Energy Converters. Co-ordinated Action on Ocean Energy 2nd Workshop proceedings. Uppsala

Kofoed JP, Frigaard P, Soerensen HC, Friis-Madsen E (2000) Development of the Wave Energy Converter – Wave Dragon. Proc 10th ISOPE Conf, vol. 1, No 10, pp 405–412

Kofoed JP (2002) Wave Overtopping of Marine Structures – Utilization of Wave Energy. PhD thesis, Aalborg University

Kofoed JP, Frigaard P, Friis-Madsen E, Sørensen HC (2006) Prototype testing of the wave energy converter wave dragon. Renew Energy 31

Kramer M, Frigaard P (2002) Efficient Wave Energy Amplification with Wave Reflectors. Proc 9th ISOPE conference

May J, Bean D (2005) Wave Dragon Pre-Commercial Demonstrator – Environmental Impact Assessment Scoping Report, Prepared by PMSS, available at www.wavedragon.co.uk

Neumann F, Tedd J, Prado M, Russell I, Patrício S, La Regina V (2006) Licensing and Environmental Issues of Wave Energy Projects. Proc World Renew Energy Congr IX. Florence, Italy

Sørensen HC et. al. (2003) Development of the Wave Dragon from scale 1:50 to prototype. Proc 5th Eur Wave Energy Conf. Cork, Ireland

Sørensen HC et al. (2005) The Results of Two Years Testing in Real Sea of Wave Dragon. Proc 6th Eur Wave Tidal Energy Conf. Glasgow, September

Sørensen HC et al. (2006) Sea Testing and Optimisation of Power Production on a Scale 1:4.5. Test Rig of the Offshore Wave Energy Converter Wave Dragon. Final technical report, project NNE5-2001-00444

Svendsen R, Frigaard P (2001) Calculation of the Wave Conditions in Nissum Bredning, Report. Aalborg University Department of Civil Engineering

Tedd J, Knapp W, Frigaard P, Kofoed JP (2005) Turbine Control Strategy Including Wave Prediction for Overtopping Wave Energy Converters. Co-ordinated Action on Ocean Energy. 2nd Workshop proceedings. Uppsala

Tedd J, Kofoed JP Knapp W, Friis-Madsen E, Sørensen HC (2006). Wave Dragon, prototype wave power production. World Renew Energy Congr IX. Florence

Van der Meer JW, Janssen JPFM (1994) Wave Run-up and Wave Overtopping at Dikes. In: Kobayashi N, Demirbilek Z (eds) Wave Forces on inclined and vertical wall structures. ASCE, pp 1–27 [Also Delft hydraulics, Publ. No 487]

Additional References

Sarmento A, Whitaker T, Brito e Melo A, Clement A, Salter S, Pontes T, Neumann F (2006) The European Research Training Network For Competitive Wave Energy. Proc WREC IX [ISBN 008 44671 X]

8 Environmental Impact Assessment

Cristina Huertas-Olivares[1] and Jennifer Norris[2]

[1]*Wave Energy Centre, Lisbon, Portugal*
[2]*European Marine Energy Centre, Stromness, Orkney, Scotland, UK*

Attempts to convert wave energy are not new: wave energy conversion began with the purpose of signalling the presence of navigation buoys. Wave energy was used to force air up through the central channel of the buoys which contained integral whistles to give an acoustic warning. This application was developed in the 1940s, with generators used to light the buoys.

Later work in the 1970s in Norway and the UK (Salter's Duck), although not progressed to commercial levels, saw the first real precursors of today's electricity generating converters based on wave power.

In contrast to today's more stringent regulatory requirements, these early wave energy devices were not subject to licensing constraints or assessments of their impacts. So, although the capture and exploitation of wave power for human use is not new, the associated studies into the possible environmental impacts of this industry are now key elements of its future success.

It has been stated that wave energy schemes produce no emissions in normal operation (Thorpe, 2001), which is true with regard to emissions typical of fossil fuel exploitation. However, there are other possible emissions covering both the operational phase and the associated installation, maintenance and decommissioning phases. During the operational phase there is likely to be noise emitted by devices under operation, a risk of emissions associated with any incidents involving pollutants (e.g., hydraulic oils), emissions from antifouling coatings, and so on. Like all the other renewable energies, wave energy will inevitably have some effects on the environment, positive as well as negative.

In the 21st century's developing wave power industry, efforts are being initiated at an early stage (e.g., WaveNet, 2003) to avoid possible criticism that the industry is simply transferring the problems associated with electricity production from land to sea. Studies of potential impacts of the industry are being carried out not only in response to legislative requirements but also to demonstrate sustainability which, if proven, will enhance the prospects for future commercialisation. Proven sustainability, with minimal environmental damage and maximal environmental benefit, will inevitably attract investors and reassure governments and decision makers who may otherwise view the environmental variable as a barrier. It will also encourage increased social acceptance of waves as a benign source of energy, a key issue in these types of development, where lack of knowledge plays a negative role (Firestone and Kempton, 2007). There needs to be increased public awareness of the enormous potential for the sustainable exploitation of wave power as a reliable source of energy (Ram et al., 2004). The journey to successful exploitation of this energy source will only run smoothly if commercial developments take full account of environmental and sustainability considerations from the outset.

A good example of the path to follow has been set by the Danish Offshore Wind Farm developments at Horns Rev and Nysted. After performing an extensive environmental monitoring programme they have found that this type of renewable energy industry is actively encouraged by the population; indeed, it has become an integral part of the identity of the country itself (Dong et al., 2006). There are other lessons to be learnt from the experience of the offshore wind energy industry, which are referred to during the course of this chapter.

The status of the legislation regarding the study of the receiving environment of proposed deployment areas for wave energy exploitation is outlined, giving some country-specific examples. Some possible environmental impacts are then identified, with discussion about the possible range of issues that should be covered in monitoring programmes, and mitigation measures that can be taken to reduce likely negative, and maximise any positive, impacts. Finally, there is a presentation of case-studies of both existing and planned pilot deployments of wave energy conversion installations. The chapter is concluded with a reflection on the future of this industry, based on the experiences of existing installations.

8.1 Legislation and Administrative Issues

8.1.1 The Need to Establish a Common Legislation

There are three levels at which wave energy schemes can be deployed: single units, small arrays of single devices, and larger wave farms. There are also non-commercial deployments installed in real open-sea conditions specifically for testing purposes, as at the EMEC test facility in the UK. The environmental requirements for these different scenarios need to be clarified, with recognition that testing a single unit in the open sea will not have the same scale of impact on the receiving environment as a wave energy farm.

Environmental legislation requirements for wave energy farms are uncertain in many countries at this moment in time. According to the European Union (EU) Directive on Environmental Impact Assessment (EIA) (Directive 85/337/CE), all EU member states must commission an assessment of the environmental consequences of certain types of projects, including power stations, before building permission is granted. Other European legislation amending or complementing this directive has appeared (Directive 97/11/EC, Directive 2003/35/EC) and has been adopted in the national legislation of the member states. This latter directive does not specifically include wave energy farms in its annexes, due to the relative timescales involved in the development of this technology.

It is, nonetheless, reasonable to presume that this and other Directives will be updated as the wave energy industry develops, in analogy with what occurred with the offshore wind energy industry. Therefore, it can be expected that EIA based on European Directives will become an essential element for allowing large-scale ocean wave energy schemes.

In the meantime, developers are looking towards national legislation, where it exists, which does leave the situation regarding EIA requirements for wave energy deployments unclear in many countries. To give an example: in early 2007 Portugal set itself the obligation to make an Incidence Environmental Study (Despacho 51/2004; Despacho 66/2005). This study is a less demanding administrative instrument than the so-called Environmental Statutory, used to evaluate projects like nuclear power stations. In Incidence Environmental Studies it is necessary to analyse *a gross mode* gas emissions, seascape, geology and geomorphology, natural values (flora, fauna and natural habitats), heritage, noise, land, territory planning and surrounding population. Yet even this list remains insufficient as, for example, underwater noise studies are not required. Since this issue has been identified as one of the main environmental uncertainties associated with wave energy schemes, its omission from the list of potential impacts is a good example of the disadvantage of a non-unified, country-specific approach.

The situation in the UK is again different, with consenting authorities requiring a complete EIA for commercial deployments. To obtain consent for a wave energy project in the UK developers require a number of licences under the following legislation: the 1985 Food and Environmental Protection Act (FEPA) licence relates specifically to proposed deposits on the seabed; Section 36 of the Electricity Act (1989) governs marine energy converters generating over 1 *MW*; Section 34 of the Coast Protection Act (1949) (CPA) concerns safety and navigation issues, as well as environmental issues; a European Protected Species (EPS) license may be required, if deemed necessary by the environmental regulator; a lease from The Crown Estate; and Planning Authority permissions to connect to the electrical grid. This legislation (and their Regulatory Authorities) requires assessment of a proposed project within the marine environment with regard to its potential for environmental impact (Russell et al., 2006) and navigational safety. In 2007 the UK began a consultation exercise on regulations for the application of the EIA specifically in regard to marine energy devices, which reviewed the existing range of applicable legislation, with a view to streamlining the licensing process for the marine energy industry.

At EMEC, and given that it is a test facility in the UK, the scenario is again slightly different. Although developers are not required to do full EIAs in relation to their proposed deployments, they do need to apply for the above-mentioned licences. Since many of the issues of concern that relate to the wave and tidal energy industries, respectively, are generic to the wave industry, EMEC has developed an advisory role with regard to the developers' consents. The experience at EMEC confirms that early communication with regulatory authorities can bring great benefits in terms of reduced effort and consistency of approaches to possible risks.

At this stage of development the European Water Framework Directive (Directive 2000/60/EC) seems likely to be extended to include wave (and tidal) energy devices. Its influence extends into coastal waters, therefore including onshore devices, and offshore devices at the early stages of development (since early deployments are likely to be installed in 50–80 m water depth). It is likely that the future Marine Strategy Directive (Borja, 2006; Salomon, 2006) will also consider wave devices.

Encouraging the right sort of development in appropriate locations should be an aim of future legislation (WWF-UK et al., 2001). Appropriate placing of wave farms is an essential precondition for maximising the positive environmental effects of devices (e.g., possible reef effects for fisheries) and limiting negative impacts, including full consideration of possible conflict of interest with other sea users (see 8.2.1). In this context, spatial planning is an important tool. For such, Strategic Environmental Assessment (SEA) (Directive 2001/42/EC) seems to be the most sustainable approach (Oñate, 2002). SEAs do have the potential to include full consideration of licensing issues, which may be of benefit particularly in countries where legislation remains piecemeal and widespread across numerous public departments (Sarmento et al., 2004). The time and effort saved by improved coordination of licensing issues could significantly increase the amount of interest shown – and confidence felt – by investors in this emerging technology.

The SEA approach has being used, for example, in the UK for offshore wind energy farms (SNH, 2004). In 2006 the Scottish Executive commissioned a marine SEA to look specifically at the potential for wave and tidal energy to make a substantial contribution to the energy generation in Scotland and the UK. SEAs can be used to help inform developers about which sites are likely to be economically viable and environmentally and socially acceptable, and can greatly assist in the short-listing of potentially viable deployment sites, particularly if integrated into national (and perhaps international) Marine Strategies. However, SEAs are unlikely to include sufficient detail to be able to specify precise device deployment areas, and individual EIAs are still likely to be required to assess the appropriateness of specific locations for individual proposed projects.

Such decisions on locations of wave energy deployments will also need to take full account not only of possible conflict of use, but also of any particular site designations, such as Nature 2000 sites: Special Areas of Conservation (SACs) (Directive 92/43/EEC) and Special Protection Areas (SPAs) (Directive 79/409/CEE), and any other national and locally designated sites.

It is clear that, until now, a legal framework remains to be established in a standardised way, and in some countries the relevant legislation is still being developed or refined. More agreement on environmental legislation is needed in order to establish a coherent approach across this field. There are clear benefits of international normalisation, which would ensure that no country could benefit from a more environmentally permissive legislative approach to potential deployments (Huertas-Olivares et al., 2006). Development of legislation should ideally go hand-in-hand with wave energy projects, in order to ensure that experience from actual deployments is reflected in legislative requirements. Whilst recognising that legislative systems cannot be fully flexible and immediately responsive to change, there are clear benefits to be gained by all parties if experience can be reflected in legislation as early as possible. This could produce a significant enhancement to the development of the wave industry (Cruz and Sarmento, 2004). This does not in any way imply that legislators should wait for mishaps to occur to guide the progress of the normative approach, as has happened on occasion in the offshore industry (Stacey and Sharp, 2007).

From a wider perspective, Brooke (2003) also pointed out the necessity of establishing international regulations due to the potential for extended problems in the future which may arise due to conflict between countries regarding 'ownership' of (or rights to exploit) the waves.

8.1.2 Structure of an EIA: Methodologies

Environmental Impact Assessment (in Europe) is the name given to the whole[1] administrative process which should be followed to establish if a project is acceptable from the environmental point of view (Directive 85/337/CE). In other words, it is an evaluation process which enables the team responsible for the project (the developer), those with interests in the project (stakeholders), and the statutory authorities to: identify and understand the significant environmental impacts and risks of the project, develop plans or procedures to mitigate and/or reduce significant risks, and appreciate the benefits that would be derived as a result of the implementation of the project (Talisman, 2005). Whilst the administrative process may be country-specific, the overall aims should be widely adopted, resulting in a consistent international approach to EIAs.

The Environmental Statement (ES) presents the findings of an EIA. Apart from the description of the project, the legal and administrative framework, analysis of alternatives etc., an ES should include the following stages:

1. Description of the existing environment
2. Scoping – identification and evaluation of the impacts
3. Public Consultation
4. Mitigation measures
5. Monitoring Programme
6. Environmental Management System

---

[1] Environmental Impact Assessment is sometimes referred to as 'Environmental Statutory', which can be a source of confusion. This is the case in the Spanish legislation (Ley 6/2001).

One of the objectives of the ES is to describe the environment that may be affected during construction, installation, operation and decommissioning of any proposed wave energy deployment (stage 1). It provides information that should enable the scoping process to identify the potential environmental impacts (stage 2). For example, if stage 1 identifies a migratory path near the proposed location of a project, it could reasonably be predicted that the development may disturb it.

A detailed appraisal of the minimum range of environmental factors that need to be addressed (in stage 2) by wave energy developers has been produced by EMEC in its guidelines for developers seeking to deploy their concepts at the EMEC test facility (EMEC, 2005; see 8.3.2). Although these guidelines have been developed specifically in relation to the EMEC test site, this document is freely available, and may constitute a useful starting point for developers looking to define the scope of possible impacts (in section 2 of their ES). It aims to guide developers from identification of possible impacts to mitigation and the formulation of project-specific commitments with the objective of minimising any negative effects.

A more general summary of the potential environmental factors that need to be studied in these types of projects is given in Table 8.1. The first column in this table corresponds to the environmental system: 'abiotic' system refers to the inert elements of nature; 'biotic' factors relate to live elements and understanding of the habitat ecosystem; and the socio-economic system is the interaction between environmental and human issues. The second column refers to the environmental factors that are included in each system, and that are considered 'significant' for the specific type of project. A review of the parameters to be study for each environmental factors was presented in Huertas-Olivares et al. (2007).

To identify environmental impacts, a number of 'EIA methodologies' are used. One of the goals of the methodologies is to ensure that all relevant environmental

Table 8.1 Summary of environmental factors to be addressed in baseline environmental studies of wave energy projects

System	Factors
Abiotic	Geology & factors effecting coastal processes
	Water quality
	Air quality
Biotic	Benthos
	Fish
	Marine mammals
	Other aquatic fauna
	Aquatic flora
	Terrestrial ecology
	Birds
	Land/ Sea uses
Socio-eco	Archaeology & cultural resources
	Socio-eco
	Landscape/ Seascape

factors have been included (Gomez, 2003). Currently there is no universal methodology that could be used for all types of projects and their sites (Conesa et al., 2003), and no EIA methodologies established for commercial wave energy projects. Several methods that have become widely used in the production of EIAs: examples include Checklist, Leopold matrix, Batelle-Colombus, etc., with advantages and disadvantages associated to each one (Canter, 1997).

Proposed projects can be assessed at three levels of detail: the identification and the qualitative and quantitative evaluations, with the latter two being different methods of the evaluation process. Quantification is important because it facilitates comparison between the environmental impacts of different projects (Cruz and Sarmento, 2004). The most commonly used methodology confronts the actions of the project (installation, operation and maintenance, and decommission) with the environmental factors susceptible to being effected (i.e., receptors) in a matrix, to identify when an impact may occur.

This is the approach taken in the EMEC EIA guidance document to developers. The guideline document specifies aspects of devices that must be considered in relation to key sensitivities in the receiving environment. It gives information about which ecological and socio-economic issues are important, and explains why they are important. Potential impacts need to be ranked, in accordance with normal methodology (outlined above), and mitigation measures specified. A recording format is provided, which requires developers to complete a matrix table to compile the likely effects of pre-specified activities on specific receptors. Whilst these guidelines are intended as guidance for developers deploying devices at the EMEC test site, all attempts have been made to make the coverage as comprehensive as possible in terms of possible impacts of the industry, and to adopt tested methodologies and reporting formats.

There are other aspects that have to be considered, such as the relative importance of different potential impacts. For instance, some authors see socio-economic impact assessment as an independent part of EIA, giving the essential 'human element' a more protagonist role (La Regina, 2006). On the other hand, the proposed scale of the proposed deployment is highly relevant here: for a test site, with a small numbers of single units (or small arrays), the socio-economic impact is not likely to be high. Again, there is a need for standardise the methodology used to ensure the consistency of the approach as the industry evolves.

In developing a coordinated strategy towards EIAs for new types of renewable energy, such as wave energy projects, the industry would benefit from a combination and refinement of the different methodologies and lists of factors to be considered, to ensure the inclusion of all relevant issues. The method of transparencies is essential for this type of project, where a number of conflicts of use have to be considered. Geographic Information Systems (GIS) are useful here, but are not necessarily easily available in all countries. This could be followed by a checklist, made through consulting a multidisciplinary panel and following some guidelines such as those of Europe Union (European Commission, 2001). The next step could be a simple Leopold matrix to identify all the potential impacts in order. Potential impacts would be characterised and assigned a magnitude and importance rating.

Table 8.2 Basic EIA matrix for a floating offshore wave energy device

Environmental factors	Actions	Installation				Operation				Decommissioning			
		S	C	M	D	S	C	M	D	S	C	M	D
Abiotic	Geology (coastal processes)		x	x		x	x	x			x	x	
	Water quality	x	x	x	x	x			x	x	x	x	x
	Air quality	x				x				x	x		
Biotic	Benthos		x	x			x				x	x	
	Fish		x	x	x	x	x	x			x	x	x
	Marine mammals	x		x	x		x	x	x		x	x	
	Other aquatic fauna		x	x			x				x	x	
	Marine birds							x					x
	Flora		x	x			x				x	x	
	Terrestrial ecology												
Socio-eco	Conflict of uses	x	x	x	x	x	x	x	x	x	x	x	x
	Archaeology (cultural resources)		x	x									
	Socio-eco	x				x	x			x	x		x
	Visual impact									x			

Note: S – ships; C – cables; M – moorings; D – device

The next step, which would need to be developed over a longer timescale, would be to establish a quantification method such us Batelle-Colombus. The adoption of this type of consistent approach would allow comparison to be made between different wave energy devices according to their impact on the environment. It would also help to establish limits of acceptability of impacts in relation to generic issues, although site-specific issues would still need to be assessed independently.

Where full EIAs are deemed necessary, they should include a requirement for public consultation (stage 3 of the EIA process). There must be an open dialogue between all concerned parties. A number of basic issues relating to the consultation process need to be specified at the outset, such as: exactly who should be consulted; when and how consultations should occur; full consideration of any relevant social context to ensure full understanding between all parties involved (Hansen et al, 2003). Many authors have highlighted the importance of early, effective and iterative consultation with relevant stakeholders (van Erp, 1996; Wolsink, 1990). The BWEA (2002) published guidelines to help offshore wind energy developers with this step, and the EMEC equivalent document points to initiate this process by a round of early scoping meetings with all relevant parties of the regulatory process. Early consultation is essential in highlighting any potential problem areas, which can then be most easily, effectively and economically addressed.

For all remaining negative repercussions, mitigating measures have to be proposed (stage 4). These must be undertaken as soon as the project starts, and properly established in advance of installation. These measures must be both technically and economically realistic, and the efficiency of each measure in reducing significant negative effects to an acceptable level must be assessed. An estimation of the required investment is necessary at this stage to verify the feasibility of the proposed measures from an economic point of view.

Stage 5 of the EIA process specifies the establishment of a monitoring programme. To monitor the environment in respect of possible changes is to follow the change of one or several environmental factors. To be able to measure a change, a monitoring programme should consist of three stages of data acquisition: before installation, during operation, and after the decommissioning of the project. This is known as 'Before and After Control Impact' design (BACI). Monitoring carried out in advance of device deployment provides the 'Environment Baseline Study', which is essential in establishing a reference data set for later comparison after devices have been deployed. Reference sites outside the area of influence of the project would normally be recommended as control sites (Bundesamt für Seeschifffahrt und Hydrographie, 2003).

There is a clear need for agreement on what parameters or indicators should be measured in an environmental monitoring programme. EMEC has made significant progress into this area, and is involved in ongoing consultations with regulators and their environmental consultees (especially Scottish Natural Heritage) about monitoring requirements in relation to device deployments at the specific locations of its test sites (see 8.3.2).

Agreement is also needed on recommendations for data collection in monitoring programmes, such as sampling frequency, date or sample point, and – just as important – on the analytical approach to be adopted. Significant effort should be made into this area to ensure that consistent 'best practice' methods are adopted across the industry. Guidelines with recommendations of an agreed environmental monitoring programme, which developers are encouraged to follow at all sites (bearing in mind particular sensitivities of each respective site), need to be developed (as did Ospar, 2004, with regard to the oil and gas offshore industry). The monitoring methods being development in relation to test sites could form a useful basis from which to begin wider discussions on the range of issues to be monitored, and the best methods to be used. A monitoring guideline is expected to be an output of task 5 of the European Thematic Network Wavetrain (www.wavetrain.org) in 2007.

Most monitoring programmes can require three/four months to become established and this must be taken into account in the developers timelines and budget. There is a potential conflict here, as developers need to deploy devices in the water at the earliest opportunity in order to demonstrate performance and build investor confidence.

The first monitoring programmes should run for extended periods of time, to allow for full investigation into any possible seasonal effects. Moreover, studies in the marine environment, especially in deeper waters, require the establishment of new techniques for field researchers to use (Sundberg and Langhamer, 2005) and these could be costly to develop. Therefore, continuous monitoring is likely to be very expensive, time consuming and potentially difficult, and there is a need for funding to be available to ensure that this essential information gathering does occur.

There are clear benefits in terms of cost and effort, as well as quality and comparability of data to be gained from a collaborative approach to developing monitoring for any new industry, and the wave energy industry is no exception. International collaboration between the key parties involved in early deployments is a desirable way forward, and could potentially involve EMEC, the Portuguese Pilot Zone, and Wavehub, depending on the state of development of, and devices deployed at, each of these facilities (see 8.3). All monitoring activity could be coordinated by an International Advisory Panel of Experts, with members having unique competence and expertise in the specialist areas. This is the approach which has been successfully followed by the Danish Monitoring Programme (Dong et al., 2006) in relation to the offshore wind industry. The Danish Energy Authority appointed five international experts to the International Advisory Panel of Experts on Marine Ecology (IAPEME). The panel's task was to comment on the environmental monitoring programme before, during, and after the establishment of the wind farms, and to assess the methods used in the programme. The panel also commented on the observed impacts. Similar proposals have been made in relation to other projects, for example, for the offshore petroleum industry in Canada by Curran et al. (2006).

In the absence of an existing international advisory panel for marine energy devices deployment, EMEC has adopted the same approach, but at a national level. Thus there are plans in place, and regulatory support, for the establishment of a monitoring advisory group of specialists to oversee the monitoring coordinated by EMEC at its test sites. If an international advisory group is created, then the experience and findings from EMEC and other facilities would be able to feed into it in a straightforward way.

Environmental management systems (stage 6 of the EIA process) should also be established to ensure that the regulatory environmental requirements and terms are efficiently anchored in all aspects of projects. These management systems include procedures and instructions for all parties involved in a project in relation to the handling of environmental issues (such as waste, noise measurement, contingency plans in case of environmental accidents such as oil spill, etc). The environmental requirements should also be incorporated into the requirement specifications to suppliers. Again presenting EMEC as a pioneer example, an integrated management system was developed, covering health, safety, quality and environmental issues under a single system. This ensures that environmental considerations are fully integrated into all works associated with the test facilities.

8.2 Scientific Matters

8.2.1 Potential Environmental Impacts and the Need for Monitoring

As mentioned at the start of the chapter, the greatest advantage of harvesting wave energy is that the devices themselves do not produce any greenhouse gases (during operation), harmful wastes or pollutants when converting wave into electrical energy (Thorpe, 1999). Hence, in normal operation, wave energy, like other renewable sources, is virtually non-polluting, with significant benefits in the form of mitigating climate change, securing energy supply, decoupling economic growth from resource use, and creating jobs.

However, regulators need to consider these substantial benefits in the light of the possible impacts that could arise as a result of deploying wave devices in the marine environment. Consideration must be given to any positive impacts as well as negative (Block, 2000), and must consider the mitigation that has been proposed by developers to reduce any negative impacts as far as possible.

Limited knowledge about the unknown yet possible negative impacts of harnessing wave energy remains a problem at this early stage of the industry. Substantially less is known about the ecological marine biota than terrestrial ecosystems (Sundberg and Langhamer, 2005). This limited knowledge contributes to the uncertainty over the potential impacts and their likely magnitude, which are also expected to be site specific, needing to be determined for each specific proposed wave energy scheme.

So far, test rounds have consisted of small scale deployments, usually single units which are either scale prototypes or full-scale versions (see 8.3). These small-scale wave energy installations are likely to have minimal environmental impacts. However, little is certain on how an installation of multiple wave energy converter units would impact on the environment and marine life. Although wave energy devices are likely to be installed in arrays, since they have a relatively low rated output (Cruz et al., 2005), single unit applications may be installed to serve small, isolated communities. There is certainly much to be gained from studying the effects of deployments of individual devices, as well as small arrays (in the scope of larger scale farms). Installed prototype devices give scope for both the investigation of generic impacts, and the establishment of draft methodologies for monitoring programmes, but it is important to remember that some potential impacts will be site-specific and device-specific.

It is of key importance to the development of the industry that all findings emerging from monitoring programmes are made available to regulators and other decision makers, and ideally to the wider public. It must be remembered that public support is needed for any new industry to be successful, and developers can play a key part in this process by ensuring the timely release of monitoring findings. These findings do not need to include commercially sensitive device-specific information, and such open sharing of relevant findings from monitoring is likely to be essential in guiding the further development of the industry.

The need for a consensus to be reached concerning the range of potential environmental impacts that may result from the industry has been raised above. Some lessons can be learnt from other marine development projects, such as oil platforms, offshore wind and aquaculture farms, etc., but it is important to ensure that any comparisons made are truly relevant to the wave energy industry. There has been more than one list drawn up of potential impacts of marine energy devices on the environment; see e.g. SNH (2004); Cooper and Kazer (2006). There is general agreement of the range of potential impacts that need to be investigated. Included in these lists are impacts that may arise due to the presence of cables (e.g., electromagnetic effects on fish and other marine species of underwater cables), as well as the effects of device installation, operation and decommissioning. Key issues, some of which pointed out in Cruz et al., 2005, are the disturbance to seabed habitats, the effects of energy extraction on coastal processes, the effects of underwater noise generated by devices on marine life, risk to marine mammals and diving birds (physical damage caused by collision with devices), disturbance to and consequent displacement of wildlife from its normal habitats, effects on fisheries, and a range of possible effects on human activities and livelihoods. Whilst the remainder of this section addresses the key issues relating to abiotic, biotic, and socioeconomic impacts of devices, detailed discussion of the potential for such impacts is beyond the scope of this chapter and the reader is referred to the growing number of documents on the subject.

The use of an 'actions' and 'receptors' matrix for the specification of the main impacts associated with proposed deployments of wave project has been described above. In any predictions of the range of potential impacts it is important that actions and receptors are fully considered in relation to one another, thereby enabling a full evaluation of their relative importance to be made (Peidro and Fernández 2006). Specification of actions should be as wide as possible in coverage, looking at all aspects of the project, from energy capture to transfer into the grid.

Whilst this discussion does not seek to cover the impacts related to a project in terms of a complete Life Cycle analysis, developers should be encouraged to give this full considering at an early stage. An approximation to a Life Cycle Analysis has been already made for some devices (Soerensen et al,. 2006). There are clearly some negative impacts associated with a full Life Cycle Analysis (such as emissions from lorries and ships used for transport, emissions associated with device fabrication, etc.) which should ideally be taken into account in an EIA.

However, with regard to specific device deployments, significant differences have been observed between impacts expected from the nearshore or shoreline deployments, and those associated with offshore deployments. It is therefore useful to consider the environmental impacts of these groups separately. Special attention is given to the offshore deployment case, as it is envisaged that these will be the larger arrays.

Shoreline / nearshore WECs

The short term impact of installation of shoreline devices is expected to have a greater impact than those deployed offshore, as shoreline devices may require excavation in the coastline/shoreline (Heath et al., 2000).

Wells and other turbines in onshore or nearshore oscillating water column (OWC) devices can emit uncomfortable levels of noise (Brooke, 2003). They can also generate noise and vibrations in the water, and are a potential source of disturbance to wildlife. Although noise suppression system can be installed, this is likely to be one of the most significant impacts related to OWC plants.

It can sometimes be difficult to ascribe a positive or negative 'label' to a potential impact. As an example, OWCs are also likely to attract seabirds, which may have a positive impact on local bird populations, providing nesting sites. However, if deployment times are short, this would loss of the nesting habitat after removal of the device.

A key potential socio-economic impact of shoreline or nearshore devices is that of visual impact (SNH, 2004). Whilst it is tempting to label this as a 'negative' impact, there is potential for it to become a positive impact if located in areas where there may be significant local or tourist interest.

Particular positive impacts can also include reduced dependence on energy from outside a remote area, if installed in small, remote island communities. In such cases associated benefits will include reduced dependence on oil or its derivatives. Some (or most) of the negative impacts related to offshore devices (particularly those identified as of less significance) are likely to be applicable to nearshore and shoreline concepts also.

Offshore WECs

Impacts on the abiotic system

Geology & factors affecting coastal processes

Hagerman (2004) predicts a wave height reduction of the order of 10–15 % immediately behind a wave energy plant, with diffraction substantially re-establishing uniformity of wave height within 3–4 *km* behind the plant. There is concern that this energy extraction may alter currents (Thorpe, 1999). The reduction of the wave energy levels reaching the coast may reduce longshore sediment transport, possibly reducing erosion in the vicinity of the deployment whilst increasing erosion. However this impact is likely to be significant only for devices located within 1 or 2 *km* of the shoreline, which is not the typical scenario (Hagerman, 2004). If the wave energy installation occupies a considerable area, it could alter the wave characteristics and produce small modifications in the coastline. It has been pointed out that in some cases there can be positive effects (Cruz and Sarmento, 2004). There is a detailed discussion of this issue in Cooper and Kazer (2006).

Sediment fraction could also be affected in the vicinity of cables and moorings/anchors. However, some authors have concluded that the impact on the hydrography and sediment transport from the installation of wave energy devices will be negligible (Edelvang et al., 1999).

Air quality

Although there is the potential for airborne noise emissions from offshore wave devices, this noise is expected to be masked by the surrounding noise generated by wind and waves (Halcrow, 2006), providing sound insulation. Careful assessment needs to be done for onshore devices using air turbines, although there are proved ways to mitigate such impact. Potentially problematic for the industry is the issue of noise generated and transmitted underwater. Transmission of sound through water has the potential to interfere with marine mammal communication, although the specific frequencies emitted may not be problematic. This issue has been tackled in offshore wind energy projects. Air quality issues are likely to be more significant for nearshore and shoreline concepts.

Water quality

The installation phase has the potential to generate, as its main impact, an increasing turbidity of the water. Emissions from wave-energy installations might occur as a result of bad practice or accidents involving hydraulic or other oils. This could possibly lead to oil leakage if hydraulic circuits are breached during operation and/or maintenance. However, selection of the most environmentally benign (biodegradable) oils is a significant mitigating measure that can lead to a reduction of the pollution risk. Furthermore, some concepts require the use of antifouling[2] and other agents, many of which can be toxic to aquatic species (Chambers et al., 2006), hence careful selection of the paints should be considered.

Impacts on the Biotic System

Bethos, fish, flora and other aquatic fauna

Wave energy technologies can also have impacts on the nearshore biological communities with the potential to disturb the biological balance when impacting on sedimentation processes (Hagerman, 2004) and the mixture of microscopic species (e.g., Thorpe, 1999). The turbidity generated in the installation phase (discussed above) could also temporarily affect primary producers. This predicted direct impact on the subtidal benthic resource represents the loss of a proportion of the

---

[2] Antifouling agents are used to protect the viability of a plant by preventing the growth of fouling organism that could reduce its efficiency.

potential food resource for fish (Erftemeijer and Lewis, 2006). The effect is not expected to influence feeding efficiency or fish populations and consequently no significant impacts on fish are predicted (Engell-Sorensen and Skyt, 2003). Moreover, temporal habitat loss of intertidal and subtidal benthic communities could be produced as a result of the installation of the submerged cables, which would form a corridor disturbance impact, although it should be rapidly colonised once the installation in the area is completed (Andrulewicz et al., 2003).

Device installation[3] may also have impact on local benthic flora and fauna, and disturbance to these communities may also affect local ecosystems. This depends on the type of device (submerged would have more impacts) or the types of moorings and anchors used, but these effects are expected to be of short duration, provided ecologically sensitive areas are avoided. The presence of wave energy farms will reduce the fishing in deployment areas, which is a positive impact from the conservationists' point of view (Hentrich and Salomon, 2006; Pickering and Whitmarsh. 1997), but may be seen as a negative impact by the fishing community. The creation of fish sanctuaries is likely to increase the sea's productivity.

The artificial habitats introduced, if adequately designed, will be suitable for colonisation by a variety of marine animals and algae, and the hard bottom substrate may act, both individually and collectively, as an artificial reef and as sanctuary areas for threatened or vulnerable species (Sherman et al.. 2002; Lan et al.. 2004). Furthermore, it is expected that the introduction of hard bottom based communities will increase the availability of food for fish, which again will lead to an increase in the available food to marine mammals and birds. There are studies of the use of drilling platforms as artificial reefs (e.g., Ponti et al., 2002) which have confirmed the utility of the drilling platform Paguro that was sunk in $24\,m$ water depth. As prevesouly mentioned it is desirable that once in operation a device stays in place, but the prediction of service life of structural materials is complicated by the complexity of the interactions of materials with the various marine environments (Shifler, 2005). The success depends on a variety of factors: site orientation may be important, because it determines exposure to prevailing currents; and some sites could have more inputs of particulate food, larger oxygen exchange, and greater supply of larvae (Ponti et al.. 2002). Decommissioning of installations should be studied carefully, due to the potential for impacts on both local populations, and on populations that are more distant, yet ecologically interconnected (Schroeder and Love. 2004).

Electromagnetism from electric cables could be also an issue; the experience from offshore wind farms greatly enhances the ability to tackle such problem. Wilson and Downie (2003) state that the artificial magnetic and electric fields associated with submarine electric cables can cause interference and disturbance to orientation in migrating animals and with the feeding mechanisms of elasmobranches[4]. Other investigations have suggested that human-made electromagnetic fields from undersea power cables could interfere with prey sensing or naviga-

---

[3] Offshore devices would be prefabricated modular devices that may require modifications to the seabed.

[4] Elasmobranches are a group of fish which includes the sharks, rays and skates.

tional abilities (of electro sensitive fish such as sharks and rays) in the immediate vicinity of the sea cables. In this development phase, transmission cables used are those of alternating current (AC) system. USACE (2004) stated that they would not result in measurable deflection of compasses or disruption of radio, GPS, or radio-beacon navigational equipment on ships passing over the cables. The number and strength of the transmission cables, the type of cable used and the type of cable sheathing, as well as the depth at which the cable is buried, all represent factors that will influence the degree to which sensitive species are affected by electro-magnetic fields (EMF) (ABPmer, 2005). At the time of writing there is no clear evidence of the significance or scale of these impacts (Dang et al., 2006).

Marine mammals

One of the main sources of concern about the potential impacts of wave energy devices concerns underwater noise and vibrations, which may disturb certain marine mammals (Thorpe, 1999) and fish (Parvin et al., 2006) depending on their frequency range. Koschinski (2003) showed that marine mammals are able to detect the low-frequency sound generated by offshore wind turbines, and there may be a similar impact from wave energy devices.

Effects on marine mammals may include attraction, displacement (Richardson, 1995) or short-term behavioural changes (Gentry et al., 1998; IACMST, 2006). In addition they could alter migratory routes of species with commercial value.

There is also the potential for acoustic output to mask important natural sounds, produce stress or cause hearing loss. There is a need to consider separately the impacts of noise from the installation, maintenance and operational activities, and more work is needed to inform on the possible effects of noise on individuals and on populations. General expectations are that such emissions will be low and that mammals will adapt to them.

Another point of concern is that there may be a risk of collision of marine mammals with the installations. There may be similarities with mammal strikes by high speed ferries, which is a bigger problem than is generally appreciated, and about which there is much information available. Some investigations have been made into techniques that could be used to alert mammals to the presence of man-made structures (André et al., 2006).

Hagerman (2004) has identified that devices that pierce the sea surface are likely to attract pinnipeds during calm wave periods. If devices have suitable platforms, this may result in the growth of populations of these species above levels that would occur in the absence of the project. Sundberg and Lanhamer (2005) have pointed out that climate change could have an unpredictable influence on the environmental impacts of wave energy devices. Routes used by migrating fish and marine mammals can be irregular and unpredictable, and are likely to be dependent on, for instance, season, regional and local climate and access to food resources.

Birds

Devices may also provide artificial nesting space for seabirds, and result in the growth of populations of these species greater than would otherwise occur in the absence of the project.

On the other hand, many species of seabird are attracted to artificial light sources. Wiese et al. (2001) reported that large attractions and mortalities of bird have been documented by lighthouses, navigational lights, offshore oil platforms, etc., mostly during overcast nights with drizzle and fog. Moreover, most birds that migrate at night climb to their migrating altitude almost immediately after takeoff and begin a gradual descent shortly after midnight (Weir, 1976). That was also found in the Horns Rev offshore wave farm (Dong et al., 2006). Therefore, there is a potential, although low, possibility of bird collision with the devices protruding above the surface. Fox (2006) and his colleagues are some of the authors who are actevly investigating this impact in relation to offshore wind farms.

Impacts on the socio-economic system

Conflict of uses

Most considerations about this topic, outside those of nature conservation aspects, are well covered in the results from the WaveNet network (WaveNet, 2003), a predecessor of the Coordinated Action on Ocean Energy (see section 7.6). It follows the classification for the competing uses as areas with restricted or prohibited access and areas with conflicting uses. The former include major shipping routes and military exercise grounds. Existing pipelines and cables of various sorts may further restrict the establishment of wave power schemes. Already existing offshore activities (offshore wind installations, oil and gas fields) can also be obstacles to wave power developments. There may also be conflicts over the use of the seabed for mining and dredging of sand and gravel.

There are potential conflicts with recreational use (particularly surfing, as experienced by Wavehub in its consenting issues). More difficult conflicts are expected to arise from shipping occurring outside designated shipping lanes, and with regard to fishery and diving activities. The scale of conflict will clearly relate to the scale of the proposed deployment areas, with possible calls for compensation for lost livelihood from fishermen, as has occurred in several offshore wind projects. The presence of submarine cables could also create conflict with the fishing industry (Coffen-Smout and Herben, 2000).

A final note to archaeological sites, which need to be fully considered as part of the EIA process, in order to avoid the possible destruction of historically valuable sites.

Socio-economic

Apart from conflict of use issues, the biggest impact that wave energy deployments are expected to have in this area is in employment generation. The industry is still too young to accurately predict the number of jobs that will be directly created by wave energy exploitation. Associated with these jobs will be those created by the demand for components and sub-components. Each job created is then serviced by a range of other jobs from other sectors. Each employee at each stage creates earnings that are passed on to other sectors of industry by the so-called multiplying effect (WaveNet, 2003).

Visual impact

Submerged devices and slack moored floating devices with low freeboard will not be visible from shore, or will be visible only in exceptionally calm and clear weather. Taut-moored or fixed devices with high freeboard will be visible more often. This will result in some visual impact of the devices during daylight hours, and of the associated navigational lights during darkness.

Comments

This section has discussed some of the potential impacts that may arise in relation to the wave energy industry. However, the high degree of uncertainty about these potential issues has been emphasised, with the consequent need to establish monitoring programmes that address the different unknowns using best-practice methods in a consistent way. The need for agreement over the range of issues that need to be monitored, and the need for consistent approaches to data collection and analysis has also been emphasised, in parallel with the importance of the dissemination of the main findings of monitoring programmes.

8.2.2 Monitoring

When defining the requirements for monitoring, it is important to focus on those areas where significant potential impacts may arise, or where there is a relatively high degree of uncertainty as to the nature of the potential impact.

Mention has been made of the Danish large monitoring programme established for Horns Rev and Nysted, the results of which were made available to the public in November 2006. They identified the key environmental factors for offshore wind as being fauna, epifauna and vegetation, fish, marine mammals, birds and socio-economic issues (Dong et al., 2006). Some of the findings will be directly relevant to the wave industry – most notably those relating to the effects of electromagnetic fields induced by the power cables transporting the electric power to the shore, and any reef effects surrounding the installations.

The hydroacoustics surveys[5] did not prove the expected reef effect but tendencies of local and regional effects were observed (Hvidt et al., 2005). This is because full development of the reef community typically takes several years and can therefore only be observed after several years.

The investigations carried out to detect the effects of the electromagnetic fields on fish were characterised by a high complexity and many difficulties. The investigations show some impact from the cable route on fish behaviour, but the data did not prove any correlation between the observed phenomena across and along the cable route and the strength of the electromagnetic field (Engell-Sorensen et al., 2002).

Looking at benthic communities, the Danish monitoring studies observed that the introduction of the turbine foundations and scour protections onto seabed that previously consisted of relatively uniform sand have increased habitat heterogeneity. Local changes to benthic communities have occurred, affecting typical fauna communities, with most aquatic animals living in the seabed, to hard bottom communities with increased abundance and biomass. There was a massive colonisation by common mussels, and a cover of algae was found shifting from an initial colonisation of filamentous green algae to a more diverse and permanent vegetation of green, brown and red algae (Leonard and Pedersen, 2005).

The findings in relation to marine mammals are not so directly transferable to the wave industry. Concerns about possible collision (with consequent physical damage), and possible displacement or reduction of populations (due to disruption of normal habitats) have already been mentioned. The effect of acoustic output of devices on communication frequencies is also of concern. Techniques like GPS tagging, or the use of sonar imaging can be used. A joint project is underway between the Sea Mammal Research Unit at St Andrews and EMEC, which aims to improve the current technological shortcomings of sonar devices, to ensure that they can be of use in the marine energy industries. Although this project is related to tidal devices, the technology is equally applicable to wave devices.

To gather information on the socio-economic effects, a sociological and environmental economic study was carried out. The sociological study consisted of in-depth interviews to expose the attitudes towards the two wind farms. It was supplemented with analyses of local media coverage of the wind farm projects. The environmental economic study used a quantitative questionnaire based on Choice Experiment approach to reveal the preferences for different location strategies. The questionnaire, based on a series of attitudinal questions on wind farms in general, could form a useful basis from which to develop a similar study in relation to wave farms.

---

[5] The hydroacustic technique used combined the use of scientific sonar acoustics with GPS to determine density, diversity and location of fish.

8.2.3 Mitigation Measures

Stage 4 of the EIA process requires developers to describe how negative impacts can be reduced or compensated. With good planning and using the principle of the 'best available techniques', several of these impacts can be reduced. Examples of mitigation measures include placing devices in less sensitive areas, introducing constraints on timing of noisy or otherwise disruptive activities, ensuring the use of the most environmentally benign materials, and putting in place monitoring of key sensitivities.

Regarding abiotic factors, measures can be adopted to confine the turbidity generated in the water due to installation works. The potential for collisions can be reduced by good navigational marking of devices. For the maintenance of water quality it is also essential to use environmentally safe options for antifouling coatings (Stupak et al., 2003) and ecologically 'friendly' anti-crustacean paints. Noise generated by Wells turbines can be reduced to acceptable levels (or possibly eliminated altogether) by careful design and/or through acoustic muffling techniques (Heath et al., 2000).

In relation to biotic factors, electromagnetic effects may be mitigated by burying cables into the seabed (Sunderberg and Langhamer, 2005). Devices can be designed to discourage or prevent marine mammal haulout, or seabird roosting, on equipment above the sea surface. Regarding the impact of underwater noise on marine mammals, Cox (2003) suggested the use of alarms in the installation and decommissioning phases (the noisiest periods). The approach adopted by EMEC (on advice from the UK regulators) is to have all potentially disruptive works overseen by marine mammal observers with the power to delay the start of disruptive (typically noisy) works if sensitive species are observed in the vicinity. An advantage of such an approach is that data is recorded on the presence of any key species during such works, and this helps to build up knowledge on the relative disturbance caused by different marine procedures.

A beneficial measure is to ensure that the correct specification of piles and pile driver is used for the construction work, with rotary methods being used wherever possible in preference to percussive techniques. This avoids the use of excessive energy and noise generation. A 'soft' start-up procedure, which entails commencing drilling works at low energy levels and gradually building up to full impact force, can be overseen by a marine mammal observer. Such an approach reduces the risk of injury to marine species and allows them to move away from the source of disturbance.

Also, other mitigation measures include minimising the rotation velocity of device motors and the utilisation of appropriate materials to muffle any noise. Socio-economic issues will be specific to the location of the deployment, the proximity of human population, and any conflicts over sea areas. Mitigation measures should be honed to the specific issues, and should be developed in discussion with all affected parties.

8.3 Case Studies

8.3.1 Tests of Single Devices

In Portugal, two full scale pilot plants of different technologies have already been tested, one of which is still being run. For the AWS pilot plant at Póvoa de Varzim, an EIA was not required due to the size and character of the undertaking. Observations on environmental issues were only sporadic due to the short time of the test period. In fact, the structure was extremely well accepted by shellfish and small fish. Dolphins were also observed in the direct vicinity of the plant during the test (Neumann et al., 2004). There is a small report done for this project with a broad description of physical and ecological aspects of the area, but it has not been published. In Pico (Azores), to date, mainly due to the short operational periods of the plant, there are no systematic observations regarding environmental issues. This is being changed under the present refurbishment project, where both acoustic (air) and hydrophone measures are scheduled for the next test phase (Sarmento et al., 2006).

In other parts of Europe, with Wave dragon (Denmark), for example, a full Environmental Impact Assessment was undertaken for the pilot plant in Nissum Bredning. For the ongoing Wales prototype project, a full Scoping Report has been carried out (PMSS, 2005). Baseline geological, benthic, ecological, visual and navigational surveys have all been completed and the results used to design complete surveys in the near future. Work is also ongoing in noise and wave modelling (Russell et al., 2006). It is expected that the complete EIA will be published at the end of 2007.

8.3.2 European Marine Energy Centre – EMEC

EMEC was established as an open-sea grid-connected test facility for developers of wave and tidal devices, with the wave facility becoming available in 2004.

The site has seen the repeated deployment of a single Pelamis unit by its developer, Pelamis Wave Power Ltd, with four additional devices (from different developers) due to be installed over the 2008–9 period. Being in repeated contact with developers and regulators over how best to deal with unknown impacts of developers' devices, EMEC has developed in-depth experience and knowledge of the issues regarding how best to mitigate against such unknowns. It is widely recognised across UK regulators and their consultees, and across academia, that the test facility provides an unrepeatable opportunity for the monitoring of potential yet unknown impacts.

Although EMEC has been established as an operational facility, it is well-placed to be involved in developing and testing monitoring methods, in conjunction with the appropriate experts. One of the essential lessons learnt by developers deploying their devices at EMEC concerns how to deal with unknown issues, and how best to mitigate any risks. Although its budgets did not include a provision for monitoring, EMEC has been successful in attracting funds for the joint development of a number of methods, including wildlife observations from land (in conjunction with the Sea Mammal Research Unit), methods for obtaining an acoustic characterisation of its sites (with the Scottish Association for Marine Science), and sonar visualisation methods (again in conjunction with the Sea Mammal Research Unit). EMEC has also attracted funding for a high resolution camera to be sited overlooking the wave test site, which will enable observations of wildlife interactions with above-surface parts of wave devices to be made. EMEC is also involved in the further development of an acoustic monitoring system for diving birds, based on a novel method developed for characterising the acoustic signatures of such birds (Norris, 2003).

As part of its service to developers, EMEC has produced a document giving guidelines on the documentation that needs to be produced in relation to deployments at its test sites. This document has been referred to during the course of this chapter, since its coverage directly relates to the range of potential impacts associated with wave (and tidal) energy devices. Although developers at EMEC do not need to undertake an EIA, they do need to provide documentation that addresses the potential impacts of their device on the receiving environment at EMEC.

Part of the service that EMEC can offer the developing wave (and tidal) industry is the sharing of methods developed for monitoring, and findings of monitoring undertaken in relation to generic impacts.

8.3.3 Tests of the First Arrays

The three units of Pelamis that are being installed in Aguçadoura (northern Portugal) needed an Incidence Environmental Study. This study was adequate for an extension up to five units. The negative impacts were classified as having very little significance and little probability of ever occurring. In most of the cases they were minimal and temporary. Monitoring has only been established for socioeconomic aspects due to the potential conflict of uses with local navigation, fisheries and others (de Jesus, 2005).

Plans to install an array of four Pelamis units at EMEC are well underway, with licenses about to be issued at the time of writing. The array should be in place during 2008. In line with EMEC requirements, the developer has produced a commitments table, in which mitigation and monitoring measures are summarised. Monitoring will initially include the use of ROV scans and observations on bird and marine mammal behaviour, in line with monitoring being undertaken by EMEC. The issue of underwater noise emissions will be addressed as part of EMEC's ongoing monitoring plans which relate to the whole wave test site.

Also in the UK, for the WaveHub project, a complete EIA has been undertaken. Plans include monitoring of seabirds, yielding information for the assessment of any future wave energy development. Underwater noise and cetaceans will be also monitored, with the aim of defining the level of noise generated by devices and verifying the findings of the ES with respect to the noise generated during construction work. Predictions made with respect to the potential impacts on fish and cetaceans will also be verified. For that, baseline noise surveys previous to construction, and monitoring of cetacean activity in parallel, are needed. The clarification of the level of noise generated by wave energy devices during the operational phase is considered particularly important, given that there is lack of information due to the fact that this is an emerging technology (Halcrow, 2006). In the USA, for the arrays of floating devices in Kaneohe bay (Hawaii) and Makah Bay (Washington), an EIA was conducted but results are kept confidential.

8.4 Discussion

This chapter has provided a broad overview of what a full Environmental Impact Assessment (EIA) study should focus in the scope of wave energy developments. There are some keys areas where experience from related fields, like the offshore and wind energies industries, can be emulated. An immediate need can be identified for guidelines and short documents outlining procedures and methodologies, particularly for the consulting and project monitoring stages. This is particularly relevant for the test centres that in a first phase will provide connection to the national electricity grids. Although it is still too early to fully understand and analyse the impacts to a full extent, the expected negative effects are in much smaller number and significance than the benefits, if appropriate mitigation measures are set in place.

References

ABPmer (2005) Potential nature conservation and landscape impacts of marine renewable energy developments in Welsh Territorial Waters. CCW policy research report n 04/08

André M, Coatanhay A, Gervaise C, Gracia J, Delory E, van der Schaar M (2006) Acoustic release of gas bubbles to prevent cetacean entanglement in fishing nets. 20th Conf Eur Cetacean Soc. Gdynia, Poland

Andrulewicz E, Napierska D, Otremba Z (2003) The environmental effects of the installation and functioning of the submarine SwePol Link HVDC transmission line: a case study of the Polish Marine Area of the Baltic Sea. J Sea Res 49(4):337–345

Arnold A, Spangler B, Peyser J (2006) Proceedings of the Hydrokinetic and Wave energy technologies. In: Savitt Schwartz S (ed) Technical and environmental issues workshop (Prepared by RESOLVE, Inc.). Washington, DC

BWEA (2002) Best practice Guidelines: Consultation for offshore wind energy developments. British Wind Energy Association, UK

Block MR (2000) Identificación de aspectos e impactos medioambientales. AENOR, Madrid

Borja A (2006) The new European Marine Strategy Directive: Dificultéis, opportunities, and challenges. Editorial. Marine Pollut Bull 52:239–242

Brooke J (2003) Wave Energy Conversion. Elsevier Ocean Energy Book Ser 6. ECOR, UK

Bundesamt für Seeschifffahrt und Hydrographie (2003) Standards for Environmental Impact Assessments of offshore wind turbines on the marine environment. Bundesamt für Seeschifffahrt und Hydrographie, Hamburg and Rostock, Germany

Cabral H, Costa JL, Chaves ML, Chainho P, de Almedida PR, Domingos I, Silva G, Pereira T, Costa MJ (2004). Avaliação de impactos ambientais sobre fauna aquática: selecção dos indicadores e abordagens metodológicas. Actas da 1 Conferência Nacional de Avaliação de Impactes, CNAI 2004: Que futuro para a avaliação de impactes? 3–5 Noviembre. Centro de Congressos Aveiro, Portugal

Canter LW (1997) Manual de Evaluación de Impacto Ambiental. McGrawHill, Madrid, Spain

Chambers LD, Stokes KR, Wood RJK (2006) Modern approaches to marine antifouling coatings. Surf Coat Technol 201(6)4:3642–3652

Conesa V, Conesa LA, Ros V (2003) Guia metodológica para la evaluación del impactos ambiental, 3rd edn. Ediciones Mundi-Prensa, Madrid

Cooper B, Kazer S (2006) The potential nature conservation impacts of wave and tidal energy extractions by marine renewable developments. Report from ABP Marine Environmental Research Ltd for CCW Policy Research

Coffen-Smout S, Herbert GJ (2000) Submarine cables: a challenge for ocean management. Marine Policy 24:441–448

COWRIE (2004) The Potential Impact of Electromagnetic Fields generated by offshore windfarm cables. Phase 1.5 report, available from the Crown Estate website www.thecrownestate.co.uk/35_cowrie_electromagnetic_fields

Cox TM, Read AJ, Swanner D, Urian K, Waples D (2003) Behavioural responses of bottle-noise dolphins, Tursiops truncates, to guillnets and acoustic alarms. Biol Conservat 115

Cruz JMBP, Sarmento AJNA (2004) Wave Energy (in Portuguese). Instituto do Ambiente, Alfragide, Lisboa, Portugal

Cruz J, Alves M, Sarmento A, Brito-Melo A, Neumann F (2005) Comparative study between wind and wave farms. Proc ENER 05 (Portuguese Renew Energy Conf) (in Portuguese)

Curran KJ, Wells PG, Potter AJ (2006) Proposing a coordinated environmental effects monitoring (EEM) programme structure for the offshore petroleum industry, Nova Scotia, Canada. Marine Policy 30(4):400–411

De Jesús J (2005) Estudo de Incidencias Ambientais. Parque de ondas da Aguçadoura. Report by Ecossitema for Grupo Enersis, Portugal

Dong Energy (2006) http://www.hornsrev.dk/Engelsk/default_ie.htm

Dong Energy, Vattenfall, Danish Energy Authority, Danish Forest and Nature Agency (2006) Danish Offshore Wind. Key Environmental Issues. Operate A/S

Dorian JP, Franssen HT, Simbeck DR (2005) Global Challenges in energy. Viewpoint. Energy Policy 34:1984–1991

Edelvang K, Moller A L, Steenberg CM, Zorn R, Hansen EA, Mangor K (1999) Environmental Impact Assessment of hydrography. Horns Rev Wind Power Plant

EMEC (2005) Environmental Impact Assessment (EIA) Guidance for Developers at the European Marine Energy Centre. Draft version 0. Available from EMEC website www.emec.org.uk

Erftemeijer PLA, Lewis RRR (2006) Environmental impacts of dredging on seagrasses: A review. Marine Pollut Bull 52(12):1553–1572

Engell-Sorensen K (2002) Possible effects of the Offshore wind farm at Vindeby on the outcome of fishing. The possible effects of electromagnetic fields. Report prepared by Bio/consult as to SEA

Engell-Sorensen K, Skyt PH (2003) Evaluation of the Effects of sediment spill from off-shore wind farm construction on marine fish. Report by Bio/consult as for SEAS

European Commission (2001) Guidance on EIA. EIS Review. Environmental themes: Assessment. European Communities. Luxemburg

Firestone J, Kempton W (2007) Public opinion about large offshore wind power: Underlying factors. Energy Policy 35(3):1584–1598

Fox AD, Desholm M, Kahlert J, Christensen KT (2006) Information needs to support environmental impact assessment of the effects of Europe marine offshore wind farms on birds. Ibis 148:129–144

Gentryn R, Boness D, Bowles AE, Insley S, Payne R, Schusterman R, Tyack P, Thomas J (1998) Behavioural Effects of Antropogenic noise in the marine environment. Proceedings of workshop on the effect of anthropogenic noise in the marine environment, 10–18 February

Gomez Orea D (2003) Evaluación del Impacto Ambiental. Un instrumento preventivo para la gestión ambiental, 2nd edn. Ediciones Mundi-Prensa, Madrid

Hagerman G (2004) Offshore wave power in the US: Environmental Issues. E21 Global EPRI report

Halcrow Group Limited (2006) Wave Hub Environmental Statement. Halcrow Group Limited for the South West of England Regional Development Agency

Hansen LK, Hammarlund K, Sorensen HC, Christensen L (2003) Public acceptance of wave energy. Proc 5th Eur Wave Energy Conf. Univeristy College Cork, Ireland

Heath T, Boake C, Whittaker T (2000) The design, construction and operation of the LIMPET wave energy converter (Islay, Scotland). Proc 4th Wave Power Conf. Denmark

Hentrich S, Salomon M (2006) Flexible management of fishing rights and a sustainable fisheries industry in Europe. Marine Policy 30(6):712–720

Huertas-Olivares C, Patricio S, Russell I, Gadner F, van t'Hoff J, Neumann F (2006) The EIA approach to wave energy within the European Research Training Network WAVETRAIN. Int Conf Ocean Energy. Bremerhaven, Germany

Huertas-Olivares C, Russel I, Patricio S, Neumann F, Sarmento A (2007) Comparative study of baseline environmental studies in offshore renewable energies. ISOPE 2007. Lisbon, Portugal

Hvidt CH, Brünner L, Knudsen FR (2005) Hydroacustic registration of fish abundance at offshore wind farms. Horns Rev Offshore Wind Farm. Annual Report 2004. Report prepared by Bio/consult as and Carl Bro as to Elsam Engineering

Koschinski S, Culin B, Henriksen O, Tregenza N, Ellis G, Jansen C, Kathe G (2003) Behavioural reactions of free-ranging porpoises and seals to the noise of a simulated 2 MW windpower generator. Marine Ecol Prog Ser 265:263–273

IACMST (2006) Inter agency committee on marine science and technology, report of IACMST Working Group on Underwater Sound and Marine Life. National Oceanography Centre, Southampton, UK

Neumman F, Tedd J, Prado M, Russell I, Patricio S, La Regina V (2006) Licensing and environmental issues of wave energy projects. World Renew Energy Cong IX. Florence, Italy

NRC (2005) National Research Council of the National Academies, Marine Mammals Populations and Ocean Noise. Determining when noise causes biological significant effects. Washington DC

Lan C-H, Chen C-C, Hsui C-Y (2004) An approach to design spatial configuration of artificial reef ecosystem. Ecol Eng 22(4–5):217–226

La Regina V, Patricio S, Neumann F, Sarmento A (2006) The Role of Socio-Economic Impact Assessment (SIA) and Environmental Impacts Assessment (EIA) for understanding benefits from wave energy deployment. World Renew Energy Congr IX. Florence, Italy

Leonard SB, Pedersen J (2005) Hard Bottom Substrate Monitoring. Horns Rev Offshore Wind Farm. Annual Status Report 2004. Report prepared by Bio/consult as for Elsam Engineering

Norris JV (2003) Acoustic Signatures of Diving birds. MSc dissertation. ICIT, Heriot Watt University, UK

Oñate JJ (2002) Evaluación Ambiental Estratégica. La evaluación ambiental de políticas, planes y programmeas. Ediciones Mundi-Prensa, Madrid

Ospar (2004) Guidelines for Monitoring the Environmental Impact of Offshore oil and gas activities. Ospar Commission

Parvin S, Nedwell JR, Workman R (2006) Underwater noise impacts modelling in support of the London Array, Greater Gabbard and Thanet offshore wind farm developments. Report prepared by Subacoustech Ltd. for CORE Ltd. UK

Peidro C, Fernández R (2006) Análisis ambiental de las energías renovables, verdes más verdes. CIERTA 2006. Comunicaciones a la Conferencia Internacional sobre Energías Renovables y Tecnologías del Agua. Almería, Spain

Perez F (1993) La economia social: concepto y entidades que comprende. Cuadernos de Trabajo 17, CIRIEC, Spain

Pickering H, Whitmarsh D (1997) Artificial reefs and fisheries exploitation: a review of the attraction versus production debate, the influence of design and its significance for policy. Fisheries Res 31(1–2):39–59

PMSS Ltd (2005) Wave Dragon Pre-Commercial Demonstrator. Environmental Impact Assessment Scoping Report. Southampton, UK

Ponti M, Abbiati M, Ceccherelli VU (2002) Drilling platforms as artificial reefs: distribution of macrobenthic assemblages of the Paguro wreck (northern Adriatic Sea). ICES J Marine Sci 59:S316–S323

Popper AN, Ketten D, Dooling R, Price JR, Brill R, Erbe C, Schusterman R, Ridway S (1998) Effects of antropogenic sounds on the hearing of marine mammals. Proceedings of workshop on the effect of anthropogenic noise in the marine environment, 10–18 february, 1998

Ram B, Thresher R, Fall NK, Bedard R (2004) Wave Power in the USA: permitting and Jurisdictional Issues (E21 Global EPRI DOE NREL report)

Richardson WJ, Greene JCR, Malme CI, Thomson DH (1995) Marine Mammals and Noise. Academic Press, San Diego

Russell I, Sorensen HC, Bean D (2006) Environmental Impact Assessment of a Wave Energy Converter: Wave Dragon. Int Conf Ocean Energy. Bremerhaven, Germany

Salomon M (2006) The European Commission proposal for a Marine Strategy: Lacking substance. Viewpoint. Marine Pollut Bull 52:1328–1329

Sarmento AJNA, Neumann F, Brito-Melo A (2004) Non technical barriers to large-scale wave energy utilisation. Renewables 2004. Int Conf New Eenew Energy Technol Sust Dev. Evora, Portugal

Sarmento A, Brito-Melo A, Neumann F (2006) Results from sea trials in the OWC European Wave Energy Plant at Pico, Azores. Invited paper for WREC-IX. Florence, Italy

Schroeder DN, Love MS (2004) Ecological and political issues surrounding decommissioning of offshore oil facilities in the Southern California Bight. Ocean Coastal Manage 47:21–48

SNH (2004) Marine Renewable Energy and the Natural Heritage: An overview and policy statement. Policy Statement n 04/01. Perth

Shifler DA (2005) Understanding material interactions in marine environments to promote extended structural life. Corrosion Sci 47.2335–2352

Stacey A, Sharp JV (2007) Safety factor requirements for the offshore industry. Eng Failure Anal 14(3):442–458

Stupak ME, García MT, Pérez MC (2003) Non-toxic alternative compounds fro marine antifouling paints. Int Biodeter Biodegrad 52:49–52

Sherman RL, Guilliam DS, Spieler RE (2002) Artificial reef design: void space, complexity, and attractants. ICES J Marine Sci 59(Suppl 1):S196–S200

Soerensen HC, Naef S, Anderberg S, Hauschild MZ (2006) Life Cycle Assessment of the Wave Energy Converter: Wave Dragon. Int Conf Ocean Energy. Fron Innovation to Industry. OTTI, Bremerhaven, Germany

Sundberg J, Langhamer O (2005) Environmental questions related to point-absorbing linear wave-generators: impact, effects and fouling. Presented at the 6th EWTEC conference in Glasgow, 28 August–3 September

Talisman Energy (2005) Beatrice Wind farm Demostrator Project. Environmental Statement. Aberdeen, UK

Thorpe TW (1999) A brief review of wave energy. A report produced for the UK Department of Trade and Industry

Thorpe TW (2001) Wave Energy-Current Status and Developments. www.nesea.org

USACE (2004) Draft Environmental Impacts Statement for the Cape Wind Project, Nantucket Sound, MA. US Army Corps of Engineers. New Englans District, Concord, MA

Van Erp F (1997) Sitting process for wind energy, project in Germany: Public participation and the response of local population. Arbeiten Risiko Kommunicat Forsch. Zentrum Julich KFA

WaveNet (2003) E5: Environmental Impacts- WaveNet Report. European Comission

Wiese FK, Montevecchi WA, Davoren GK, Huettmann F, Diamond AW, Linke J (2001) Seabirds at Risk around Offshore Oil Platforms in the North-west Atlantic. Short Communication. Marine Pollut Bull 42(12):1285–1290

Weir RD (1976) Annotated Bibliography of Birds Kills at Man-made obstacles: A Review of the state of the art and solutions. Canadian Wildlife Service, Ottawa

Wilson S, Downie AJ (2003) A review of possible marine renewable energy development projects and their natural heritage impacts from a Scottish perspective. Scottish Natural Heritage Commissioned Report F02AA414

Wolsink M (1990) The sitting problem. Windpower as a social dilemma. University of Amsterdam

Author Index

Stephen Barstow
Fugro OCEANOR
Trondheim
Norway

Ana Brito e Melo
Wave Energy Centre
Lisbon
Portugal

Lars Christensen
Wave Dragon ApS
Copenhagen
Denmark

João Cruz
Garrad Hassan and Partners Limited
Bristol
England, UK

Richard Curran
Queens University Belfast
Belfast
Northern Ireland, UK

Oskar Danielsson
Uppsala University
Uppsala
Sweden

Matthew Folley
Queens University Belfast
Belfast
Northern Ireland, UK

Erik Friis-Madsen
Wave Dragon ApS
Copenhagen
Denmark

Tom Heath
Wavegen
Inverness
Scotland, UK

Ross Henderson
Pelamis Wave Power Ltd
Edinburgh
Scotland, UK

Jens Peter Kofoed
Wave Dragon ApS
Copenhagen
Denmark

Wilfried Knapp
Technical University of Munich
Munich
Germany

Mats Leijon
Uppsala University
Uppsala
Sweden

Denis Mollison
Heriot-Watt University
Edinburgh
Scotland, UK

Gunnar Mørk
Fugro OCEANOR
Trondheim
Norway

Frank Neumann
Wave Energy Centre
Lisbon
Portugal

Miguel Prado
Teamwork Technology BV
Zijdewind
The Netherlands

Matthew Rea
Edinburgh Designs Ltd
Edinburgh
Scotland, UK

Ian Russell
Wave Dragon Ltd
Wales, UK

Stephen Salter
School of Engineering and Electronics
University of Edinburgh
Edinburgh
Scotland, UK

António Sarmento
Wave Energy Centre
Lisbon
Portugal

Hans Christian Sørensen
Wave Dragon ApS
Copenhagen
Denmark

Jamie Taylor
University of Edinburgh & Artemis Intelligent Power Ltd
Edinburgh
Scotland, UK

James Tedd
Wave Dragon ApS
Copenhagen
Denmark

Gareth Thomas
Dept. of Applied Mathematics
University College Cork
Cork, Ireland

Karin Thorburn
Uppsala University
Uppsala
Sweden

Richard Yemm
Pelamis Wave Power Delivery Ltd
Edinburgh
Scotland, UK

Index

Printing: Krips bv, Meppel, The Netherlands
Binding: Stürtz, Würzburg, Germany